Turning Dust to Gold

Building a Future on the Moon and Mars

Haym Benaroya

Turning Dust to Gold

Building a Future on the Moon and Mars

Professor Haym Benaroya
Rutgers University
Department of Mechanical and Aerospace Engineering
Piscataway
New Jersey
USA

SPRINGER–PRAXIS BOOKS IN SPACE EXPLORATION
SUBJECT *ADVISORY EDITOR*: John Mason, M.B.E., B.Sc., M.Sc., Ph.D.

ISBN 978-1-4419-0870-4 Springer-Verlag Berlin Heidelberg New York

Springer is part of Springer-Science + Business Media (springer.com)

Library of Congress Control Number: 2009939419

Cover design: Jim Wilkie
Project copy-editor: Rachael Wilkie
Author-generated LaTex, processed by EDV-Beratung, Germany

Printed in Germany on acid-free paper

Contents

Foreword

The goal of this book is to demonstrate that expanding our civilization to the Moon and beyond is not beyond our reach, intellectually or financially. Apollo was not our last foray into the Solar System. Science fiction, as good as some of it is, will have a difficult time staying ahead of science and engineering fact. However, this book is intended to be more than purely a technical manual. Space is not, nor should it be, only of interest to aerospace engineers and cosmologists. Space needs to be of interest and of concern to all of us. It is the world's West, as in Go West for new opportunities for freedom and limitless growth. What I hope to demonstrate to you is that not only *can* we go back to the Moon to stay, but that we *must* do so. And not only for esoteric reasons, but for many reasons, some mundane, but most related to our and our children's continued prosperity, and to our continued survival as a species.

Space, whether we or our children live in it or on the Moon or Mars, will be important to all of us because of the resources it will open for our industrial societies and for the markets it will create. It would be difficult to imagine the United States as a world power if its western border was the Mississippi River. Would a North American continent of several dozen independent states modeled after Western Europe have been able to resist the totalitarian attacks that occurred in the 20th century? Alone, neither Europe nor Asia was capable.

The expansion of the best Earth has to offer in science and culture to the Moon, then Mars, and eventually the Solar System, can only strengthen humanity's core positive achievements: democracy, individual rights, equal opportunities for an individual's achievements, and all that is inherent and is based on these being in place.

This book is written from the perspective of a future observer, more than 150 years into the future, and some of the narratives based on a fictitious group of documents discovered in an old repository that chronicled the settlement of the Moon during the last half of the 21st century and first half of the 22nd century. In order to make the documents more connected and interesting to the casual reader, they have been weaved into a chronological storyline meant to trace the early days of the human return to the Moon with the aim of permanent settlement.

The interviews in this book are real and original.

Preface

This book is an historical work based on a group of documents discovered in an old repository that chronicled the settlement of the Moon during the last half of the 21st century and into the 22nd century. There are gaps in the historical record for reasons that I won't get into here, so we have had to guess how we went from "point *a*" to "point *c*" without any detailed information about "point *b*." In order to make the documents more connected and interesting to the casual reader, I, Yerah Timoshenko, have tried as best as I can to weave the issues and concerns of an emerging spacefaring civilization into a coherent perspective. I have focused on the early days of the human return to the Moon with all its trials and tribulations. I am very pleased to have been able to include a good number of historical interviews with a cross-section of people who played a role in humanity's struggle into space. More about me and what I do later in the book.

The return to the Moon occurred at a time of great turmoil on Earth, especially the decades-long war on terrorism. Also, even though the economies of the larger countries were growing and successful, there were constant philosophical and real-world battles about the spending of scarce resources. Debt, both national and personal, was high and growing higher with no end in sight. As is common, most people were focused on the day-to-day activities and events that directly impacted their lives.

The early 21st century was heavily politicized, with all problems and all politics reduced to black and white. One was either on one side of the fence or the other. Those who recognized the gray aspects of reality usually ended up impaled on the fence that separated the camps. It was a time when democracy was spreading, but terrorism was also global. Mankind was making tremendous progress in the sciences and engineering, and the biological sciences were on the verge of breaking through to a deep understanding of the human body and the underlying causes of deadly diseases.

Many nations were once again interested in manned space travel. Rockets were the only way to bring large payloads and people into low Earth orbit and beyond. It was tremendously expensive but also the only option at the time. Space elevators, while studied and believed to be feasible, had not yet been prototyped and tested in the Earth–space environment. As we know now, the massive development of the space elevators led to a thousandfold increase in space activities by the early 22nd century.

Many of us are reminded of the California Gold Rush that began on 24 January 1848 when gold was discovered at Sutter's Mill in Coloma, California, now in the

Sacramento area. That news spread worldwide by the end of 1848, resulting in about 300,000 people coming to California over the next several years. The Rush converted San Francisco from a tent town to a small city, pushing the admission of California as a state in 1850. In the same way, once space elevator prototypes proved the technology viable, full-scale versions were built in orbit around Earth and here in orbit around the Moon.

After that, there was no stopping the flow of goods and people. The floodgates had been opened.

The California Clipper takes you to the California Gold Rush, February 1849.

Our transformation into a spacefaring species occurred in spurts. In the 1960s and 1970s we sent people to the Moon for short visits, and then we gave up on that. Then in the 1980s through the 2010s we sent people into low Earth orbit to build a space station. Somewhere around that time people in several of the more technologically advanced countries decided that it was time to rebuild the space effort with an eye to the permanent settlement of the Solar System, first the Moon, then Mars, and then – well then depended on how successful we were at the first two steps. But so far, in 2169, we have thousands of people on dozens of bodies in the Solar System.

We worked steadily at getting humans and machines into space and to the Moon. We always knew we could do it – what we did not know was whether the political support could be borne over such long time periods. So we started out as turtles with rockets, but ended up as hares with elevators. The rest is history.

This book is a commemorative as well as a selective accounting of some of the key events that punctuated our short history in space. It is published in 2169, two hundred years after the first men landed on the Moon. There are discussions of some of the technical and other issues that were of great concern – some of which are still baffling our pioneers today.

So this story is a snapshot of how all that we see today, on the Moon, Mars, and beyond, evolved.

Acknowledgements

I gratefully acknowledge all the people who have deliberated about space and its challenges. I could not have written about such a broad subject without the insights of the community of scholars and practitioners. My interpretation of their work is subject to my perspectives and may have missed the mark – for which I take full responsibility.

I am thankful for the support and encouragement of Praxis Publishing, in particular, Clive Horwood, who accepted this project and graciously and patiently waited for me to complete the book over many missed deadlines. I am grateful to Dr John Mason who provided insightful suggestions for improving the style, content and organization of the manuscript, as I am to Rachael Wilkie who edited this work. And I appreciate the fine work with the LaTeX formatting of the manuscript by Frank Herweg, without whose efforts the book would not look as good as it does.

The book is more interesting due to the numerous images, many of which come from the NASA archives. I tip my hat to them for the development of such a collection, one that is indicative of all the great work of the NASA teams over the past six decades of trailblazing the space frontier. I sincerely thank all the creative people who generously granted permission to use their images. And I can never repay those who granted me interviews and trusted me to put their words in a book that they will hopefully find to be a good home.

I am grateful to and wish to highlight the two images created by my daughter Ana. And, last but not least, my sincere thanks go to my son, Adam, and my closest friend, Mark Nagurka of Marquette University, for their major editing of this book – resulting in a much improved style.

Dedication

I humbly dedicate this book
to all those who commit their lives
to enable the exploration of space
by humans and machines.
To those who make it possible by their engineering,
scientific and medical research,
by their creative achievements and imaginative practice.
To those who are now still in elementary school,
are able to look forward to the joy of discovery,
and will be the ones who return to the Moon,
take the first steps on Mars, the outer planets and their moons.
And last but certainly not least,
this book is dedicated to the men and women who spend their lives
– sometimes sacrificing their lives –
protecting the rest of us and our democracies
so that we can spend our lives
creating new worlds.

1 Go west ... settle space

"Beautiful, beautiful. Magnificent desolation."

Buzz Aldrin

In the year 2169, as we commemorate the 200th anniversary of humans landing on the Moon in 1969, we try to imagine how difficult it was during the early part of the last century as proponents of manned space travel and space settlement found themselves regularly discouraged by the progress they were witnessing. After all, only a few viewed space as a critical avenue for human creativity. It was viewed by many as a "special interest" for those who would find financial benefits as a result of manned space activity.[1] At that time, many problems existed on Earth (when was this not true!). Therefore, many preferred to minimize public expenditures for NASA and space in the U.S. Many teenagers had no interest in a return to the Moon, and roughly a quarter thought that the Apollo landings were faked!

Some of that thinking changed as a result of 43rd U.S. President George W. Bush's 14 January 2004 speech challenging and charging the U.S. to return to the Moon, to stay, as a first step in mankind's expansion to Mars and the remainder of the Solar System. His speech: *New Vision for Space Exploration Program*, given at NASA Headquarters, Washington, D.C. is presented below. This speech can be viewed as a turning point in the world's view on space exploration and settlement. Subsequently, the European Union, China, Japan and Russia, all committed their nations and organizations to a manned return to the Moon. Unfortunately, funding levels did not match the boldness of the vision.

1.0.1 Bush 2004 speech

President Bush: Thanks for the warm welcome. I'm honored to be with the men and women of NASA. I thank those of you who have come in person. I welcome those who are listening by video. This agency, and the dedicated professionals who serve it, have always reflected the finest values of our country – daring, discipline, ingenuity, and unity in the pursuit of great goals.

[1] No one who worked in space and aerospace was viewed as a neutral party in supporting a manned return to the Moon. They were viewed as a special interest rather than as an educated participant and supporter of one of the greatest adventures of humanity!

America is proud of our space program. The risk takers and visionaries of this agency have expanded human knowledge, have revolutionized our understanding of the universe, and produced technological advances that have benefited all of humanity.

Fig. 1.1 President George Bush presenting his *New Vision for Space Exploration Program.* 14 January 2004. (Courtesy NASA)

Inspired by all that has come before, and guided by clear objectives, today we set a new course for America's space program. We will give NASA a new focus and vision for future exploration. We will build new ships to carry man forward into the universe, to gain a new foothold on the Moon, and to prepare for new journeys to worlds beyond our own.

I am comfortable in delegating these new goals to NASA, under the leadership of Sean O'Keefe. He's doing an excellent job. I appreciate Commander Mike Foale's introduction – I'm sorry I couldn't shake his hand. Perhaps, Commissioner, you'll bring him by – Administrator, you'll bring him by the Oval Office when he returns, so I can thank him in person.

I also know he is in space with his colleague, Alexander Kaleri, who happens to be a Russian cosmonaut. I appreciate the joint efforts of the Russians with our country to explore. I want to thank the astronauts who are with us, the courageous spacial entrepreneurs who set such a wonderful example for the young of our country.

And we've got some veterans with us today. I appreciate the astronauts of yesterday who are with us, as well, who inspired the astronauts of today to serve our country. I appreciate so very much the members of Congress being here. Tom DeLay is here, leading a House delegation. Senator Nelson is here from the Senate. I am honored that you all have come. I appreciate you're interested in the subject – it is a subject that's important to this administration, it's a subject that's mighty important to the country and to the world.

Two centuries ago, Meriwether Lewis and William Clark left St. Louis to explore the new lands acquired in the Louisiana Purchase. They made that journey in the spirit of discovery, to learn the potential of vast new territory, and to chart a way for others to follow.

America has ventured forth into space for the same reasons. We have undertaken space travel because the desire to explore and understand is part of our character. And that quest has brought tangible benefits that improve our lives in countless ways. The exploration of space has led to advances in weather forecasting, in communications, in computing, search and rescue technology, robotics, and electronics. Our investment in space exploration helped to create our satellite telecommunications network and the Global Positioning System. Medical technologies that help prolong life – such as the imaging processing used in CAT scanners and MRI machines – trace their origins to technology engineered for the use in space.

Our current programs and vehicles for exploring space have brought us far and they have served us well. The Space Shuttle has flown more than a hundred missions. It has been used to conduct important research and to increase the sum of human knowledge. Shuttle crews, and the scientists and engineers who support them, have helped to build the International Space Station.

Telescopes – including those in space – have revealed more than 100 planets in the last decade alone. Probes have shown us stunning images of the rings of Saturn and the outer planets of our Solar System. Robotic explorers have found evidence of water – a key ingredient for life – on Mars and on the moons of Jupiter. At this very hour, the Mars Exploration Rover Spirit is searching for evidence of life beyond the Earth.

Yet for all these successes, much remains for us to explore and to learn. In the past 30 years, no human being has set foot on another world, or ventured farther upward into space than 386 miles – roughly the distance from Washington, D.C. to Boston, Massachusetts. America has not developed a new vehicle to advance human exploration in space in nearly a quarter century. It is time for America to take the next steps.

Today I announce a new plan to explore space and extend a human presence across our Solar System. We will begin the effort quickly, using existing programs and personnel. We'll make steady progress – one mission, one voyage, one landing at a time.

Our first goal is to complete the International Space Station by 2010. We will finish what we have started, we will meet our obligations to our 15 international partners on this project. We will focus our future research aboard the station on the long-term effects of space travel on human biology. The environment of space is hostile to human beings. Radiation and weightlessness pose dangers to human health, and we have much to learn about their long-term effects before human crews can venture through the vast voids of space for months at a time. Research on board the station and here on Earth will help us better understand and overcome the obstacles that limit exploration. Through these efforts we will develop the skills and techniques necessary to sustain further space exploration.

To meet this goal, we will return the Space Shuttle to flight as soon as possible, consistent with safety concerns and the recommendations of the Columbia Accident Investigation Board. The Shuttle's chief purpose over the next several years will be to help finish assembly of the International Space Station. In 2010, the Space Shuttle – after nearly 30 years of duty – will be retired from service.

Our second goal is to develop and test a new spacecraft, the Crew Exploration Vehicle, by 2008, and to conduct the first manned mission no later than 2014. The Crew Exploration Vehicle will be capable of ferrying astronauts and scientists to the Space Station after the shuttle is retired. But the main purpose of this spacecraft will be to carry astronauts beyond our orbit to other worlds. This will be the first spacecraft of its kind since the Apollo Command Module.

Our third goal is to return to the Moon by 2020, as the launching point for missions beyond. Beginning no later than 2008, we will send a series of robotic missions to the lunar surface to research and prepare for future human exploration. Using the Crew Exploration Vehicle, we will undertake extended human missions to the Moon as early as 2015, with the goal of living and working there for increasingly extended periods. Eugene Cernan, who is with us today – the last man to set foot on the lunar surface – said this as he left: "We leave as we came, and God willing as we shall return, with peace and hope for all mankind." America will make those words come true.

Returning to the Moon is an important step for our space program. Establishing an extended human presence on the Moon could vastly reduce the costs of further space exploration, making possible ever more ambitious missions. Lifting heavy spacecraft and fuel out of the Earth's gravity is expensive. Spacecraft assembled and provisioned on the Moon could escape its far lower gravity using far less energy, and thus, far less cost. Also, the Moon is home to abundant resources. Its soil contains raw materials that might be harvested and processed into rocket fuel or breathable air. We can use our time on the Moon to develop and test new approaches and technologies and systems that will allow us to function in other, more challenging environments. The Moon is a logical step toward further progress and achievement.

With the experience and knowledge gained on the Moon, we will then be ready to take the next steps of space exploration: human missions to Mars and to worlds beyond. Robotic missions will serve as trailblazers – the advanced guard to the unknown. Probes, landers and other vehicles of this kind continue to prove their worth, sending spectacular images and vast amounts of data back to Earth. Yet the human thirst for knowledge ultimately cannot be satisfied by even the most

Fig. 1.2 This is an artist's concept depicting a possible scene of an observatory on the far side of the Moon. The artwork was part of NASA's new initiatives study which surveyed possible future manned planetary and lunar expeditionary activity. The objective of the lunar observatory case study is to understand the effort required to build and operate a long-duration human-tended astronomical observatory on the moon's far side. Some scientists feel that the lunar far side – quiet, seismically stable and shielded from Earth's electronic noise – may be the Solar System's best location for such an observatory. The facility would consist of optical telescope arrays, stellar monitoring telescopes and radio telescopes, allowing nearly complete coverage of the radio and optical spectra. The observatory would also serve as a base for geologic exploration and for a modest life sciences laboratory. In the left foreground, a large fixed radio telescope is mounted on a crater. The telescope focuses signals into a centrally located collector, which is shown suspended above the crater. The lander in which the crew would live can be seen in the distance on the left. Two steerable radio telescopes are placed on the right; the instrument in the foreground is being serviced by scientists. The other astronaut is about to replace a small optical telescope that has been damaged by a micrometeorite. A very large baseline optical interferometer system can be seen in the right far background. (The painting was done by Mark Dowman and Doug McLeod. S89-25054, January 1989. Courtesy NASA)

vivid pictures, or the most detailed measurements. We need to see and examine and touch for ourselves. And only human beings are capable of adapting to the inevitable uncertainties posed by space travel.

As our knowledge improves, we'll develop new power generation propulsion, life support, and other systems that can support more distant travels. We do not know where this journey will end, yet we know this: human beings are headed into the cosmos.

And along this journey we'll make many technological breakthroughs. We don't know yet what those breakthroughs will be, but we can be certain they'll come, and that our efforts will be repaid many times over. We may discover resources on the Moon or Mars that will boggle the imagination, that will test our limits to dream. And the fascination generated by further exploration will inspire our young people to study math, and science, and engineering and create a new generation of innovators and pioneers.

This will be a great and unifying mission for NASA, and we know that you'll achieve it. I have directed Administrator O'Keefe to review all of NASA's current space flight and exploration activities and direct them toward the goals I have outlined. I will also form a commission of private and public sector experts to advise on implementing the vision that I've outlined today. This commission will report to me within four months of its first meeting. I'm today naming former Secretary of the Air Force, Pete Aldridge, to be the Chair of the Commission. Thank you for being here today, Pete. He has tremendous experience in the Department of Defense and the aerospace industry. He is going to begin this important work right away.

We'll invite other nations to share the challenges and opportunities of this new era of discovery. The vision I outline today is a journey, not a race, and I call on other nations to join us on this journey, in a spirit of cooperation and friendship.

Achieving these goals requires a long-term commitment. NASA's current five-year budget is $86 billion. Most of the funding we need for the new endeavors will come from reallocating $11 billion within that budget. We need some new resources, however. I will call upon Congress to increase NASA's budget by roughly a billion dollars, spread out over the next five years. This increase, along with refocusing of our space agency, is a solid beginning to meet the challenges and the goals we set today. It's only a beginning. Future funding decisions will be guided by the progress we make in achieving our goals.

We begin this venture knowing that space travel brings great risks. The loss of the Space Shuttle Columbia was less than one year ago. Since the beginning of our space program, America has lost 23 astronauts, and one astronaut from an allied nation – men and women who believed in their mission and accepted the dangers. As one family member said, "The legacy of Columbia must carry on – for the benefit of our children and yours." The Columbia's crew did not turn away from the challenge, and neither will we.

Mankind is drawn to the heavens for the same reason we were once drawn into unknown lands and across the open sea. We choose to explore space because doing so improves our lives, and lifts our national spirit. So let us continue the journey.

May God bless.

* * *

Even with this shot in the arm, many were skeptical of the desirability for such an "adventure." Few had the vision to think beyond the predictable and consider the possible and the desirable. Such linear thinking almost always leads to lower expectations and less action. We generally think linearly, meaning that we extrapolate the present into the future without accounting for the unexpected. Of course, the unexpected is, well, not expected! For example, in the year 1807, very few people would have extrapolated to the Wright Brothers' human flight one hundred years later. In 1869, only science fiction writers envisioned landing people on the Moon in 1969. Similarly, other great inventions in mechanics and electronics were not envisioned and therefore the technologies to which those inventions gave birth could not have been foreseen except by a tiny group of visionaries.

Therefore, in 2004, when President Bush gave his speech, those who supported his vision understood the potential benefits of charting a course back to the Moon. They supported this action not because of what could be predicted as the possible benefits, but for all the unpredictable outcomes and synergies. This kind of non-linear thinking is useful because it opens up our minds to the impossible as well as the possible. "The possible requires a lot of hard work; the impossible takes a little longer."

Fig. 1.3 Yuri Alekseyevich Gagarin. (Official Soviet photograph)

The first person in space and the first to orbit the Earth on 12 April 1961 was the Soviet cosmonaut Yuri Alekseyevich Gagarin (9 March 1934 – 27 March 1968). On 5 May 1961, Mercury Astronaut Alan B. Shepard, Jr. blasted off in his Freedom 7 capsule atop a Mercury-Redstone rocket. His 15-minute suborbital flight made him the first American in space.

The race to the Moon began at this time.[2] It was an outgrowth of the political and military rivalry between the United States and the Soviet Union. We know that the Americans landed with men on the Moon first, but once this was achieved, interest quickly waned and resources were pulled from NASA. The initiation of the American response to Gagarin's flight was with President Kennedy's speech before Congress.

[2] *The Decision to Go to the Moon – Project Apollo and the National Interest*, J.M. Logsdon, MIT Press, 1970.

Fig. 1.4 Mercury astronaut Alan B. Shepard, Jr. and his Freedom 7 capsule atop a Mercury-Redstone rocket, 5 May 1961. (Courtesy NASA)

1.0.2 Kennedy 1961 speech

The following is the speech delivered by the 35th U.S. President John F. Kennedy before a joint session of Congress on 25 May 1961 titled "Special Message to the Congress on Urgent National Needs." This is the famous "Moon" speech where he committed the nation to send men to the Moon before the end of the decade. As we see, the space component of the speech is a small part of the overall speech, supporting the historical claim that the Apollo program was in reality a part of the Cold War and political program. The complete speech is reproduced here to provide the reader with the larger context of the "Space Race."

One cannot underestimate the importance of context in trying to understand how space plays out politically. Space enthusiasts from the dawn of Apollo have tried to justify spacefaring in isolation, trying to claim that the return to the Moon and then to Mars would justify the expense and benefit all of humanity. While true – and from this vantage point in the future, obviously true – it was always necessary to put space in the larger context. Once that was done, resistance died away quickly and spacefaring was "obvious."

President Kennedy: Mr. Speaker, Mr. Vice President, my copartners in Government, gentlemen and ladies:

The Constitution imposes upon me the obligation to "from time to time give to the Congress information of the State of the Union." While this has traditionally

Fig. 1.5 Special Message to the Congress on Urgent National Needs, delivered by President John F. Kennedy, delivered to a joint session of Congress, 25 May 1961.

been interpreted as an annual affair, this tradition has been broken in extraordinary times.

These are extraordinary times. And we face an extraordinary challenge. Our strength as well as our convictions have imposed upon this nation the role of leader in freedom's cause.

No role in history could be more difficult or more important. We stand for freedom.

That is our conviction for ourselves – that is our only commitment to others. No friend, no neutral and no adversary should think otherwise. We are not against any man or any nation – or any system – except as it is hostile to freedom. Nor am I here to present a new military doctrine, bearing any one name or aimed at any one area. I am here to promote the freedom doctrine.

I.

The great battleground for the defense and expansion of freedom today is the whole southern half of the globe – Asia, Latin America, Africa and the Middle East – the lands of the rising peoples. Their revolution is the greatest in human history. They seek an end to injustice, tyranny, and exploitation. More than an end, they seek a beginning.

And theirs is a revolution which we would support regardless of the Cold War, and regardless of which political or economic route they should choose to freedom.

For the adversaries of freedom did not create the revolution; nor did they create the conditions which compel it. But they are seeking to ride the crest of its wave – to capture it for themselves.

Yet their aggression is more often concealed than open. They have fired no missiles; and their troops are seldom seen. They send arms, agitators, aid, technicians and propaganda to every troubled area. But where fighting is required, it is usually done by others – by guerrillas striking at night, by assassins striking alone – assassins who have taken the lives of four thousand civil officers in the last twelve months in Vietnam alone – by subversives and saboteurs and insurrectionists, who in some cases control whole areas inside of independent nations.

[At this point the following paragraph, which appears in the text as signed and transmitted to the Senate and House of Representatives, was omitted in the reading of the message:

They possess a powerful intercontinental striking force, large forces for conventional war, a well-trained underground in nearly every country, the power to conscript talent and manpower for any purpose, the capacity for quick decisions, a closed society without dissent or free information, and long experience in the techniques of violence and subversion. They make the most of their scientific successes, their economic progress and their pose as a foe of colonialism and friend of popular revolution. They prey on unstable or unpopular governments, unsealed, or unknown boundaries, unfilled hopes, convulsive change, massive poverty, illiteracy, unrest and frustration.]

With these formidable weapons, the adversaries of freedom plan to consolidate their territory – to exploit, to control, and finally to destroy the hopes of the world's newest nations; and they have ambition to do it before the end of this decade. It is a contest of will and purpose as well as force and violence – a battle for minds and souls as well as lives and territory. And in that contest, we cannot stand aside.

We stand, as we have always stood from our earliest beginnings, for the independence and equality of all nations. This nation was born of revolution and raised in freedom. And we do not intend to leave an open road for despotism.

There is no single simple policy which meets this challenge. Experience has taught us that no one nation has the power or the wisdom to solve all the problems of the world or manage its revolutionary tides – that extending our commitments does not always increase our security – that any initiative carries with it the risk of a temporary defeat – that nuclear weapons cannot prevent subversion – that no free people can be kept free without will and energy of their own – and that no two nations or situations are exactly alike.

Yet there is much we can do – and must do. The proposals I bring before you are numerous and varied. They arise from the host of special opportunities and dangers which have become increasingly clear in recent months. Taken together, I believe that they can mark another step forward in our effort as a people. I am here to ask the help of this Congress and the nation in approving these necessary measures.

Fig. 1.6 Earth's Moon, just 3 days away, is a good place to test hardware and operations for a human mission to Mars. A simulated mission, including the landing of an adapted Mars excursion vehicle, could test many relevant Mars systems and technologies. (Artwork done for NASA by Pat Rawlings, of SAIC. S95-01563, February 1995. Courtesy NASA) See Plate 1 in color section.

II. ECONOMIC AND SOCIAL PROGRESS AT HOME

The first and basic task confronting this nation this year was to turn recession into recovery. An affirmative anti-recession program, initiated with your cooperation, supported the natural forces in the private sector; and our economy is now enjoying renewed confidence and energy. The recession has been halted. Recovery is under way.

But the task of abating unemployment and achieving a full use of our resources does remain a serious challenge for us all. Large-scale unemployment during a recession is bad enough, but large-scale unemployment during a period of prosperity would be intolerable.

I am therefore transmitting to the Congress a new Manpower Development and Training program, to train or retrain several hundred thousand workers, particularly in those areas where we have seen chronic unemployment as a result of technological factors in new occupational skills over a four-year period, in order to replace those skills made obsolete by automation and industrial change with the new skills which the new processes demand.

It should be a satisfaction to us all that we have made great strides in restoring world confidence in the dollar, halting the outflow of gold and improving our balance of payments. During the last two months, our gold stocks actually increased by seventeen million dollars, compared to a loss of 635 million dollars during the last two months of 1960. We must maintain this progress – and this will require the cooperation and restraint of everyone. As recovery progresses, there will be temptations to seek unjustified price and wage increases. These we cannot afford. They will only handicap our efforts to compete abroad and to achieve full recovery here at home. Labor and management must – and I am confident that they will – pursue responsible wage and price policies in these critical times. I look to the President's Advisory Committee on Labor Management Policy to give a strong lead in this direction.

Moreover, if the budget deficit now increased by the needs of our security is to be held within manageable proportions, it will be necessary to hold tightly to prudent fiscal standards; and I request the cooperation of the Congress in this regard – to refrain from adding funds or programs, desirable as they may be, to the Budget – to end the postal deficit, as my predecessor also recommended, through increased rates – a deficit incidentally, this year, which exceeds the fiscal 1962 cost of all the space and defense measures that I am submitting today – to provide full pay-as-you-go highway financing – and to close those tax loopholes earlier specified. Our security and progress cannot be cheaply purchased; and their price must be found in what we all forego as well as what we all must pay.

III. ECONOMIC AND SOCIAL PROGRESS ABROAD

I stress the strength of our economy because it is essential to the strength of our nation. And what is true in our case is true in the case of other countries. Their strength in the struggle for freedom depends on the strength of their economic and their social progress.

We would be badly mistaken to consider their problems in military terms alone. For no amount of arms and armies can help stabilize those governments which are unable or unwilling to achieve social and economic reform and development. Military pacts cannot help nations whose social injustice and economic chaos invite insurgency and penetration and subversion. The most skillful counter-guerrilla efforts cannot succeed where the local population is too caught up in its own misery to be concerned about the advance of communism.

But for those who share this view, we stand ready now, as we have in the past, to provide generously of our skills, and our capital, and our food to assist the peoples of the less-developed nations to reach their goals in freedom – to help them before they are engulfed in crisis.

This is also our great opportunity in 1961. If we grasp it, then subversion to prevent its success is exposed as an unjustifiable attempt to keep these nations from either being free or equal. But if we do not pursue it, and if they do not pursue it, the bankruptcy of unstable governments, one by one, and of unfilled hopes will surely lead to a series of totalitarian receiverships.

Earlier in the year, I outlined to the Congress a new program for aiding emerging nations; and it is my intention to transmit shortly draft legislation to implement this program, to establish a new Act for International Development, and to add to the figures previously requested, in view of the swift pace of critical events, an additional 250 million dollars for a Presidential Contingency Fund, to be used only upon a Presidential determination in each case, with regular and complete reports to the Congress in each case, when there is a sudden and extraordinary drain upon our regular funds which we cannot foresee – as illustrated by recent events in Southeast Asia – and it makes necessary the use of this emergency reserve. The total amount requested – now raised to 2.65 billion dollars – is both minimal and crucial. I do not see how anyone who is concerned – as we all are – about the growing threats to freedom around the globe – and who is asking what more we can do as a people – can weaken or oppose the single most important program available for building the frontiers of freedom.

IV

All that I have said makes it clear that we are engaged in a world-wide struggle in which we bear a heavy burden to preserve and promote the ideals that we share with all mankind, or have alien ideals forced upon them. That struggle has highlighted the role of our Information Agency. It is essential that the funds previously requested for this effort be not only approved in full, but increased by 2 million, 400 thousand dollars, to a total of 121 million dollars.

This new request is for additional radio and television to Latin America and Southeast Asia. These tools are particularly effective and essential in the cities and villages of those great continents as a means of reaching millions of uncertain peoples to tell them of our interest in their fight for freedom. In Latin America, we are proposing to increase our Spanish and Portuguese broadcasts to a total of 154 hours a week, compared to 42 hours today, none of which is in Portuguese, the language of about one-third of the people of South America. The Soviets, Red Chinese and satellites already broadcast into Latin America more than 134 hours a week in Spanish and Portuguese. Communist China alone does more public information broadcasting in our own hemisphere than we do. Moreover, powerful propaganda broadcasts from Havana now are heard throughout Latin America, encouraging new revolutions in several countries.

Similarly, in Laos, Vietnam, Cambodia, and Thailand, we must communicate our determination and support to those upon whom our hopes for resisting the communist tide in that continent ultimately depend. Our interest is in the truth.

V. OUR PARTNERSHIP FOR SELF-DEFENSE

But while we talk of sharing and building and the competition of ideas, others talk of arms and threaten war. So we have learned to keep our defenses strong – and to cooperate with others in a partnership of self-defense. The events of recent weeks have caused us to look anew at these efforts.

The center of freedom's defense is our network of world alliances, extending from NATO, recommended by a Democratic President and approved by a Republican Congress, to SEATO, recommended by a Republican President and approved by a

Fig. 1.7 Apollo 11 Commander Neil Armstrong working at the modularized equipment stowage assembly of Eagle, the lunar module. (Courtesy NASA)

Democratic Congress. These alliances were constructed in the 1940's and 1950's – it is our task and responsibility in the 1960's to strengthen them.

To meet the changing conditions of power – and power relationships have changed – we have endorsed an increased emphasis on NATO's conventional strength. At the same time we are affirming our conviction that the NATO nuclear deterrent must also be kept strong. I have made clear our intention to commit to the NATO command, for this purpose, the 5 Polaris submarines originally suggested by President Eisenhower, with the possibility, if needed, of more to come.

Second, a major part of our partnership for self-defense is the Military Assistance Program. The main burden of local defense against local attack, subversion, insurrection or guerrilla warfare must of necessity rest with local forces. Where these forces have the necessary will and capacity to cope with such threats, our intervention is rarely necessary or helpful. Where the will is present and only capacity is lacking, our Military Assistance Program can be of help.

But this program, like economic assistance, needs a new emphasis. It cannot be extended without regard to the social, political and military reforms essential to internal respect and stability. The equipment and training provided must be tailored to legitimate local needs and to our own foreign and military policies, not to our supply of military stocks or a local leader's desire for military display. And military assistance can, in addition to its military purposes, make a contribution to economic progress, as do our own Army Engineers.

In an earlier message, I requested 1.6 billion dollars for Military Assistance, stating that this would maintain existing force levels, but that I could not foresee how much more might be required. It is now clear that this is not enough. The present crisis in Southeast Asia, on which the Vice President has made a valuable report – the rising threat of communism in Latin America – the increased arms traffic in Africa – and all the new pressures on every nation found on the map by tracing your fingers along the borders of the Communist bloc in Asia and the Middle East – all make clear the dimension of our needs.

I therefore request the Congress to provide a total of 1.885 billion dollars for Military Assistance in the coming fiscal year – an amount less than that requested a year ago – but a minimum which must be assured if we are to help those nations make secure their independence. This must be prudently and wisely spent – and that will be our common endeavor. Military and economic assistance has been a heavy burden on our citizens for a long time, and I recognize the strong pressures against it; but this battle is far from over, it is reaching a crucial stage, and I believe we should participate in it. We cannot merely state our opposition to totalitarian advance without paying the price of helping those now under the greatest pressure.

Fig. 1.8 The official emblem of Apollo 11, the United States' first scheduled lunar landing mission. The Apollo 11 crew were the astronauts Neil A. Armstrong, commander, Michael Collins, command module pilot, and Edwin E. Aldrin, Jr., lunar module pilot. The NASA insignia design for Apollo flights is reserved for use by the astronauts and for the official use as the NASA Administrator may authorize. (S69-34875, June 1969. It is being reproduced here with the specific permission of NASA.)

VI. OUR OWN MILITARY AND INTELLIGENCE SHIELD

In line with these developments, I have directed a further reinforcement of our own capacity to deter or resist non-nuclear aggression. In the conventional field, with one exception, I find no present need for large new levies of men. What is needed is rather a change of position to give us still further increases in flexibility.

Therefore, I am directing the Secretary of Defense to undertake a reorganization and modernization of the Army's divisional structure, to increase its non-nuclear firepower, to improve its tactical mobility in any environment, to insure its flexibility to meet any direct or indirect threat, to facilitate its coordination with our major allies, and to provide more modern mechanized divisions in Europe and bring their equipment up to date, and new airborne brigades in both the Pacific and Europe.

And secondly, I am asking the Congress for an additional 100 million dollars to begin the procurement task necessary to re-equip this new Army structure with the most modern material. New helicopters, new armored personnel carriers, and new howitzers, for example, must be obtained now.

Third, I am directing the Secretary of Defense to expand rapidly and substantially, in cooperation with our Allies, the orientation of existing forces for the conduct of non-nuclear war, paramilitary operations and sub-limited or unconventional wars.

In addition our special forces and unconventional warfare units will be increased and reoriented. Throughout the services new emphasis must be placed on the special skills and languages which are required to work with local populations.

Fourth, the Army is developing plans to make possible a much more rapid deployment of a major portion of its highly trained reserve forces. When these plans are completed and the reserve is strengthened, two combat-equipped divisions, plus their supporting forces, a total of 89,000 men, could be ready in an emergency for operations with but 3 weeks' notice – 2 more divisions with but 5 weeks' notice – and six additional divisions and their supporting forces, making a total of 10 divisions, could be deployable with less than 8 weeks' notice. In short, these new plans will allow us to almost double the combat power of the Army in less than two months, compared to the nearly nine months heretofore required.

Fifth, to enhance the already formidable ability of the Marine Corps to respond to limited war emergencies, I am asking the Congress for 60 million dollars to increase the Marine Corps strength to 190,000 men. This will increase the initial impact and staying power of our three Marine divisions and three air wings, and provide a trained nucleus for further expansion, if necessary for self-defense.

Finally, to cite one other area of activities that are both legitimate and necessary as a means of self-defense in an age of hidden perils, our whole intelligence effort must be reviewed, and its coordination with other elements of policy assured. The Congress and the American people are entitled to know that we will institute whatever new organization, policies, and control are necessary.

VII. CIVIL DEFENSE

One major element of the national security program which this nation has never squarely faced up to is civil defense. This problem arises not from present trends but from national inaction in which most of us have participated. In the past decade we have intermittently considered a variety of programs, but we have never adopted a consistent policy. Public considerations have been largely characterized by apathy, indifference and skepticism; while, at the same time, many of the civil defense plans have been so far-reaching and unrealistic that they have not gained essential support.

This Administration has been looking hard at exactly what civil defense can and cannot do. It cannot be obtained cheaply. It cannot give an assurance of blast protection that will be proof against surprise attack or guaranteed against obsolescence or destruction. And it cannot deter a nuclear attack.

We will deter an enemy from making a nuclear attack only if our retaliatory power is so strong and so invulnerable that he knows he would be destroyed by our response. If we have that strength, civil defense is not needed to deter an attack. If we should ever lack it, civil defense would not be an adequate substitute.

But this deterrent concept assumes rational calculations by rational men. And the history of this planet, and particularly the history of the 20th century, is sufficient to remind us of the possibilities of an irrational attack, a miscalculation, an accidental war, [or a war of escalation in which the stakes by each side gradually increase to the point of maximum danger] which cannot be either foreseen or deterred. It is on this basis that civil defense can be readily justifiable – as insurance for the civilian population in case of an enemy miscalculation. It is insurance we trust will never be needed – but insurance which we could never forgive ourselves for foregoing in the event of catastrophe.

Once the validity of this concept is recognized, there is no point in delaying the initiation of a nation-wide long-range program of identifying present fallout shelter capacity and providing shelter in new and existing structures. Such a program would protect millions of people against the hazards of radioactive fallout in the event of large-scale nuclear attack. Effective performance of the entire program not only requires new legislative authority and more funds, but also sound organizational arrangements.

Therefore, under the authority vested in me by Reorganization Plan No. 1 of 1958, I am assigning responsibility for this program to the top civilian authority already responsible for continental defense, the Secretary of Defense. It is important that this function remain civilian, in nature and leadership; and this feature will not be changed.

The Office of Civil and Defense Mobilization will be reconstituted as a small staff agency to assist in the coordination of these functions. To more accurately describe its role, its title should be changed to the Office of Emergency Planning.

As soon as those newly charged with these responsibilities have prepared new authorization and appropriation requests, such requests will be transmitted to the Congress for a much strengthened Federal-State civil defense program. Such a program will provide Federal funds for identifying fallout shelter capacity in existing, structures, and it will include, where appropriate, incorporation of shelter in Federal buildings, new requirements for shelter in buildings constructed with Federal assistance, and matching grants and other incentives for constructing shelter in State and local and private buildings.

Federal appropriations for civil defense in fiscal 1962 under this program will in all likelihood be more than triple the pending budget requests; and they will increase sharply in subsequent years. Financial participation will also be required from State and local governments and from private citizens. But no insurance is cost-free; and every American citizen and his community must decide for themselves

Fig. 1.9 This artist's concept of a lunar base and extra-base activity was revealed during a 1986 Summer Study on possible future activities for NASA. A roving vehicle similar to the one used on three Apollo missions is depicted in the foreground. (Artwork was done by Dennis Davidson. S86-27256, June 1986. Courtesy NASA)

whether this form of survival insurance justifies the expenditure of effort, time and money. For myself, I am convinced that it does.

VIII. DISARMAMENT

I cannot end this discussion of defense and armaments without emphasizing our strongest hope: the creation of an orderly world where disarmament will be possible. Our aims do not prepare for war – they are efforts to discourage and resist the adventures of others that could end in war.

That is why it is consistent with these efforts that we continue to press for properly safeguarded disarmament measures. At Geneva, in cooperation with the United Kingdom, we have put forward concrete proposals to make clear our wish to meet the Soviets half way in an effective nuclear test ban treaty – the first significant but essential step on the road towards disarmament. Up to now, their response has not been what we hoped, but Mr. Dean returned last night to Geneva, and we intend to go the last mile in patience to secure this gain if we can.

Meanwhile, we are determined to keep disarmament high on our agenda – to make an intensified effort to develop acceptable political and technical alternatives to the present arms race. To this end I shall send to the Congress a measure to establish a strengthened and enlarged Disarmament Agency.

IX. SPACE

Finally, if we are to win the battle that is now going on around the world between freedom and tyranny, the dramatic achievements in space which occurred in recent weeks should have made clear to us all, as did the Sputnik in 1957, the impact of this adventure on the minds of men everywhere, who are attempting to make a determination of which road they should take. Since early in my term, our efforts in space have been under review. With the advice of the Vice President, who is Chairman of the National Space Council, we have examined where we are strong and where we are not, where we may succeed and where we may not. Now it is time to take longer strides – time for a great new American enterprise – time for this nation to take a clearly leading role in space achievement, which in many ways may hold the key to our future on Earth.

I believe we possess all the resources and talents necessary. But the facts of the matter are that we have never made the national decisions or marshalled the national resources required for such leadership. We have never specified long-range goals on an urgent time schedule, or managed our resources and our time so as to insure their fulfillment.

Recognizing the head start obtained by the Soviets with their large rocket engines, which gives them many months of lead time, and recognizing the likelihood that they will exploit this lead for some time to come in still more impressive successes, we nevertheless are required to make new efforts on our own. For while we cannot guarantee that we shall one day be first, we can guarantee that any failure to make this effort will make us last. We take an additional risk by making it in full view of the world, but as shown by the feat of astronaut Shepard, this very risk enhances our stature when we are successful. But this is not merely a race. Space is open to us now; and our eagerness to share its meaning is not governed by the efforts of others. We go into space because whatever mankind must undertake, free men must fully share.

I therefore ask the Congress, above and beyond the increases I have earlier requested for space activities, to provide the funds which are needed to meet the following national goals:

First, I believe that this nation should commit itself to achieving the goal, before this decade is out, of landing a man on the Moon and returning him safely to the Earth. No single space project in this period will be more impressive to mankind, or more important for the long range exploration of space; and none will be so difficult or expensive to accomplish. We propose to accelerate the development of the appropriate lunar space craft. We propose to develop alternate liquid and solid fuel boosters, much larger than any now being developed, until certain which is superior. We propose additional funds for other engine development and for unmanned explorations – explorations which are particularly important for one purpose which this nation will never overlook: the survival of the man who first makes this daring flight. But in a very real sense, it will not be one man going to the Moon – if we make this judgment affirmatively, it will be an entire nation. For all of us must work to put him there.

Secondly, an additional 23 million dollars, together with 7 million dollars already available, will accelerate development of the Rover nuclear rocket. This gives promise of some day providing a means for even more exciting and ambitious exploration of space, perhaps beyond the Moon, perhaps to the very end of the Solar System itself.

Third, an additional 50 million dollars will make the most of our present leadership, by accelerating the use of space satellites for world-wide communications.

Fig. 1.10 NASA artist rendering of the Altair lunar lander's ascent stage docking with the Orion crew exploration vehicle after a mission to the lunar surface. (JSC2007-E-113274 Courtesy NASA)

Fourth, an additional 75 million dollars – of which 53 million dollars is for the Weather Bureau – will help give us at the earliest possible time a satellite system for world-wide weather observation.

Let it be clear – and this is a judgment which the Members of the Congress must finally make – let it be clear that I am asking the Congress and the country to accept a firm commitment to a new course of action, a course which will last for many years and carry very heavy costs: 531 million dollars in fiscal '62 – an estimated seven to nine billion dollars additional over the next five years. If we are to go only half way, or reduce our sights in the face of difficulty, in my judgment it would be better not to go at all.

Now this is a choice which this country must make, and I am confident that under the leadership of the Space Committees of the Congress, and the Appropriating Committees, that you will consider the matter carefully.

It is a most important decision that we make as a nation. But all of you have lived through the last four years and have seen the significance of space and the adventures in space, and no one can predict with certainty what the ultimate meaning will be of mastery of space.

I believe we should go to the Moon. But I think every citizen of this country as well as the Members of the Congress should consider the matter carefully in making their judgment, to which we have given attention over many weeks and months, because it is a heavy burden, and there is no sense in agreeing or desiring that the United States take an affirmative position in outer space, unless we are prepared to do the work and bear the burdens to make it successful. If we are not, we should decide today and this year.

This decision demands a major national commitment of scientific and technical manpower, materiel and facilities, and the possibility of their diversion from other important activities where they are already thinly spread. It means a degree of dedication, organization and discipline which have not always characterized our research and development efforts. It means we cannot afford undue work stoppages, inflated costs of material or talent, wasteful interagency rivalries, or a high turnover of key personnel.

New objectives and new money cannot solve these problems. They could in fact, aggravate them further – unless every scientist, every engineer, every serviceman, every technician, contractor, and civil servant gives his personal pledge that this nation will move forward, with the full speed of freedom, in the exciting adventure of space.

X. CONCLUSION

In conclusion, let me emphasize one point. It is not a pleasure for any President of the United States, as I am sure it was not a pleasure for my predecessors, to come before the Congress and ask for new appropriations which place burdens on our people. I came to this conclusion with some reluctance. But in my judgment, this is a most serious time in the life of our country and in the life of freedom around the globe, and it is the obligation, I believe, of the President of the United States to at least make his recommendations to the Members of the Congress, so that they can reach their own conclusions with that judgment before them. You must decide yourselves, as I have decided, and I am confident that whether you finally decide in the way that I have decided or not, that your judgment – as my judgment – is reached on what is in the best interests of our country.

In conclusion, let me emphasize one point: that we are determined, as a nation in 1961 that freedom shall survive and succeed – and whatever the peril and set-backs, we have some very large advantages.

The first is the simple fact that we are on the side of liberty – and since the beginning of history, and particularly since the end of the Second World War, liberty has been winning out all over the globe.

A second real asset is that we are not alone. We have friends and allies all over the world who share our devotion to freedom. May I cite as a symbol of traditional and effective friendship the great ally I am about to visit – France. I look forward to my visit to France, and to my discussion with a great Captain of the Western World, President de Gaulle, as a meeting of particular significance, permitting the kind of close and ranging consultation that will strengthen both our countries and serve the common purposes of world-wide peace and liberty. Such serious conversations do not require a pale unanimity – they are rather the instruments of trust and understanding over a long road.

A third asset is our desire for peace. It is sincere, and I believe the world knows it. We are proving it in our patience at the test ban table, and we are proving it in the UN where our efforts have been directed to maintaining that organization's usefulness as a protector of the independence of small nations. In these and other instances, the response of our opponents has not been encouraging.

Yet it is important to know that our patience at the bargaining table is nearly inexhaustible, though our credulity is limited that our hopes for peace are unfailing, while our determination to protect our security is resolute. For these reasons I have long thought it wise to meet with the Soviet Premier for a personal exchange of views. A meeting in Vienna turned out to be convenient for us both; and the Austrian government has kindly made us welcome. No formal agenda is planned and no negotiations will be undertaken; but we will make clear America's enduring concern is for both peace and freedom – that we are anxious to live in harmony with the Russian people – that we seek no conquests, no satellites, no riches – that we seek only the day when "nation shall not lift up sword against nation, neither shall they learn war any more."

Finally, our greatest asset in this struggle is the American people – their willingness to pay the price for these programs – to understand and accept a long struggle – to share their resources with other less fortunate people – to meet the tax levels and close the tax loopholes I have requested – to exercise self-restraint instead of pushing up wages or prices, or over-producing certain crops, or spreading military secrets, or urging unessential expenditures or improper monopolies or harmful work stoppages – to serve in the Peace Corps or the Armed Services or the Federal Civil Service or the Congress – to strive for excellence in their schools, in their cities and in their physical fitness and that of their children – to take part in Civil Defense – to pay higher postal rates, and higher payroll taxes and higher teachers' salaries, in order to strengthen our society – to show friendship to students and visitors from other lands who visit us and go back in many cases to be the future leaders, with an image of America – and I want that image, and I know you do, to be affirmative and positive – and, finally, to practice democracy at home, in all States, with all races, to respect each other and to protect the Constitutional rights of all citizens.

I have not asked for a single program which did not cause one or all Americans some inconvenience, or some hardship, or some sacrifice. But they have responded and you in the Congress have responded to your duty – and I feel confident in asking today for a similar response to these new and larger demands. It is heartening to know, as I journey abroad, that our country is united in its commitment to freedom and is ready to do its duty.

* * *

President Kennedy was not really interested in space. The motivation for his proposal to go to the Moon was the Bay of Pigs fiasco which was shortly followed by Gagarin's flight. This has been the subject of much research and is well documented. "John Kennedy made the lunar landing decision because his definition of the national interest led him to conclude, under the stimulus of the Gagarin flight, that a prestige-oriented space program was an appropriate, even though very costly, instrument of American foreign policy."[3] As we read in Kennedy's speech, space

[3] "The Apollo Decision and Its Lessons for Policy-Makers," John M. Logsdon, from *The Moon Decision: Project Apollo and the National Interest*, MIT Press, 1970.

was viewed as one arena of the Cold War between the United States and the Soviet Union. There was no way to deliberate on what path the nation should follow in space without putting the decision in the larger context. "The notion that the United States should enter a contest with the Soviet Union for the prestige accruing from space success was based on the rationale that this prestige was an element in the Cold War competition for national power and international influence. Such a competition was, from the U.S. point of view, part of an effort to contain the expansion of Soviet power."[4]

1.0.3 A positive view of space

Thousands of essays have been written on the importance of space for humanity's development. In the United States, there was a perpetual battle between those who saw space exploration and the human settlement of the Moon and Mars as an obvious path to take, and those who believed that, while worthy, the money could be better used for more pressing needs.

Americans were shaken by the difficult economic times at the beginning of the 21st century and the worldwide battle with terrorism. The world went through a serious economic crisis between 2008 and 2012. There were numerous financial fluctuations locally and globally. The United States saw a major increase in government involvement in the running of businesses while Europe generally moved to a more free market system. National debts continued to rise. Countries from the former Soviet block had adopted flat-tax systems that free market economists supported.

The rapid economic development of China and India, as well as other countries, created a more competitive world economy. With that interest in space, activities emerged at various levels around the world. There were conflicting approaches to space. Some nations, such as China, became leaders – national pride drove them to Apollo-like efforts. Other nations saw themselves in supporting, and important, roles whereby they created niches at which they excelled, for example, in robotics and manufacturing.

The creation of space technologies, especially for human spaceflight, drew the greatest technological talents of a nation. The "best and the brightest" gravitated to the mental challenges and large time commitments required to be successful as engineers or scientists because of the space program. Nations knew that their whole economics would benefit as a result of a robust space program – it was viewed as a win-win effort.

The following discussions highlight some of the issues that were examined during those early days.

A post-Apollo review

A post-Apollo evaluation of the need for a lunar base had been made in the U.S. with the following reasons given for such a base: advance lunar science and astronomy, provide economic rewards for the benefit of Earth, support general scientific

[4]Loc. cit.

advancement, develop technologies that would support national security, stimulate other space technologies, provide a test bed for future human expansion into the Solar System, establish a U.S. presence, stimulate interest among American students in science and engineering, help in the long-range survival of the species,[5] support manifest destiny, and create an epic vision.

During the late 1990s and early 2000s, few Americans pursued an engineering or scientific education. This, coupled with major increases in the numbers of engineers that graduated in the developing world, in particular China and India, was of concern to Americans. A robust and significant space program was viewed as a magnet to quantitatively capable students. A case in point was that during the Apollo program there was a spike in enrollment in engineering and science programs in the United States. This benefit alone attracted the attention of all nations.

Back to the Moon for the sake of America and humanity

By virtue of its dominance of the night sky, the Moon has embedded itself in humanity's psyche, as part of our lore, as the moonlight that has turned battles, and as the familiar "face" upon which we have all gazed many times. As a destination for people, it has been the focus of science and science fiction. Science fiction dreamt the impossible that science and engineering later made possible, often much sooner than anyone could have anticipated.

On 21 July 1969, at 0256 GMT, Neil Armstrong and "Buzz" Aldrin of Apollo 11 walked for the first time on the lunar surface to the amazement and cheers of most people around the world. These first steps were repeated with two men from each of Apollos 12, 14, 15, 16 and 17. These first expeditions were intended to mark the initiation of a permanent manned presence on the Moon, and eventually Mars and beyond. That dream was abruptly put on hold even before the first men walked on the Moon, until it was reinvigorated in a serious way by President G.W. Bush in 2004.

With President Bush's visionary and very specific speech of 14 January 2004, we were finally placed on track for the return to the Moon, this time to stay and settle, and then after creating a lunar infrastructure move onward to Mars and beyond. The President resumed the journey abandoned over 30 years earlier. With this act, and assuming the goals are fulfilled, he initiated what can arguably be viewed as one of the most far-reaching efforts of humanity.

While most individuals supported the vision and the need for our return to space, the cry was often heard that problems still existed on Earth – let us solve them first, went the refrain. There always have been and always will be problems on Earth. Certainly if discretionary spending were to be zeroed out until all problems are solved, we should also not spend any money on museums and concert halls, and similar luxuries. It became clear that if we were to wait until all the problems were solved before going back to the Moon, then we, as a species, would end our lives on Earth. Those problems will never be fully solved.

Expenditures on space result in a significant return to the civilian economy in the form of advanced technologies across all sectors. Particular examples include

---

[5] *The Survival Imperative*, W.E. Burrows, Forge Books, 2006.

Fig. 1.11 Neil Armstrong in full suit, 1 July 1969. (Courtesy Neil Armstrong and NASA)

medicine, materials, and electronics. The economy grows because of the advanced technologies that are derived from space development. In other words, space development increases the size of the pie, but it has always been difficult to sustain political support for space exploration except during the Cold War.[6]

There were numerous reasons to rebuild the space program around a manned return to the Moon. These included the recovery of resources from the Moon, and the setting up of outposts to monitor meteorite activity far enough in advance to deflect those that might approach Earth. But it was also understood that the return to the Moon would be a vision to excite young people on how fantastic the future could be, and how they could play a role in creating that future. What follows is a quote from an article of the time:

> In the United States there are difficulties in attracting enough young people to the disciplines of engineering, science and mathematics. Society depends on there being enough technically versed people who are eager to address the problems we all face. Whether in medicine, environmental protection, agriculture, electronics, or the design of a multitude of products,

[6] D.A. Broniatowski, A.L. Weigel, "The political sustainability of space exploration," *Space Policy*, Vol. 24, 2008, pp. 148–157.

> we need enthusiastic and talented American engineers, scientists and mathematicians to spend the many years they must in college in order to begin to understand how to solve today's problems and to anticipate the potential problems of the next generation.
>
> Just as Apollo brought thousands of Americans to the technical arts, President Bush's vision invigorates young people to study subjects that are the foundation and backbone of our modern civilization. This they do whether or not they become rocket scientists. The goal is to be part of the larger community of people who can intellectually appreciate what it takes to do these enormous and profound tasks.
>
> The nation needs its brightest in the technical arts. We are at risk because fewer of us pursue such disciplines. Our competitiveness with nations who appreciate the importance of engineering and science continues to erode. The settlement of the Moon and then Mars, and the manned exploration of the Solar System are an ideal focus for a nation that is used to forging its massive energies for the betterment of humanity.
>
> Sometimes the argument is made that instead of spending all this money on space to develop technology, just invest directly in the technology. While this appears reasonable, it is important to note that science and engineering are not spiritless professions. The people who spend their days doing science and engineering need to be excited by the adventure and purpose of it all. The best and the brightest are attracted to visionary activities. They are willing to work endlessly and tirelessly for a goal they view as noble, for a goal that allows them to feel that they have made an impact on the path that humanity takes.
>
> So space exploration and settlement satisfies two crucial needs of a flourishing civilization, the needs of the spirit and the needs of sustenance – in its broadest sense – that engineering and science provide.

How can the excitement of engineering problem-solving be explained? It is not the excitement of action movies, although the movie *Apollo 13* did convey the exhilaration of figuring out how to get three astronauts in a crippled spacecraft back to Earth alive.

Engineering problem-solving is in a metaphorical sense similar to the work of a sculptor. A sculptor begins with raw material and has a vision of what the artwork will look like. Often the sculptor sketches concepts to help guide the process of cutting away pieces of stone or wood or clay.

An engineer goes through a similar process. The engineer's raw material is a knowledge of math and science prescribing the rules of what can and cannot be done. Science is a description of the physical world. It tells the engineer what can be ideally expected from the behavior of materials, or chemicals, for example. Mathematics is the language of science and engineering by which concepts are quantified, worked with, and used to derive new understanding based on earlier knowledge.

With math and science as the basis, the engineer's vision or aim is to build something, whether a computer, a car, a space station, or a city on Mars. The engineer analyzes the vision and figures out how to make it a reality. This process

Fig. 1.12 A large Arecibo-like radio telescope on the Moon uses a crater for structural support. In the background are 2 steerable radio telescopes. (Artwork done for NASA by Pat Rawlings of SAIC. S95-01561, February 1995. Courtesy NASA) See Plate 2 in color section.

is called *design.* The engineer has many options in a design; different designs can meet the same vision. Factors such as cost and construction difficulty are taken into account in the selection of a design.

So the excitement of engineering problem-solving is one of considering many design options, and solving many problems along the way leading to a viable design that can be built. This process is a very difficult one and thus very satisfying when achieved. Seemingly ironic, the more difficult the design and the more problems that must be tackled, the more enjoyable the process.

There were thousands of problems that needed to be solved for our permanent return to the Moon. Each problem that was solved had a positive impact on Earth. Each solution also solved a problem on Earth, resulting in advanced medical equipment, stronger and less expensive materials, and faster manufacturing robots. These advances created new industries, new jobs, and the benefits of these "dual-use" technologies were – and are – far and wide. Space settlement was and is a wonderful vision around which our brightest focused their talents and energies, with the satisfaction that they have truly made a difference.

We have experienced many profound natural tragedies over the centuries, and if there is but one lesson to be learned, it is that regardless of how technologically sophisticated we become, there is always something that can tip the delicate balance of our society. Greater technological sophistication goes hand-in-hand with greater vulnerability; we depend on layers of technology in our everyday lives. One of the goals of a space program that placed a significant number of people on the Moon and Mars was to safeguard the species; with people populating the Solar System, a devastating event on Earth will not wipe out the human race.

Lunar Exploration Analysis Group

A group known as the Lunar Exploration Analysis Group (LEAG) was founded in the early 21st century that brought together scientists, engineers and business people to help mold the lunar exploration and settlement program into a coherent, integrated, exciting, and productive venture. It was an opportunity for experts from diverse fields to share ideas and form collaborations. In conjunction with this group, two other groups also had similar aims, the Space Resources Roundtable, and the Lunar Commerce Executive Roundtable.

These groups focused – and continue to focus – on how to make the most of space for humanity. At its origin the group was working on the development of business opportunities for the lunar site. The focus was on creating business plans for lunar-based enterprises that had a chance of being funded by an investor group.

Hard-nosed questions were posed by business leaders from large aerospace companies such as Boeing to small start-ups that came from other sectors of the economy. Space tourism drew much attention then, as did the utilization of lunar resources – these resources are now valuable for the needs of the lunar settlement as well as for export to Earth. To this day, we aim to evolve these sectors of our lunar/Martian economies. We continue to make hard-nosed demands of our entrepreneurs. A number have moved permanently to the Moon so that they can begin to capitalize on Martian settlement efforts.

Energy creation on the Moon, from Helium-3 and solar energy, was also viewed as a business opportunity. Helium-3 was viewed as a limitless source of energy predicated on the engineering of fusion reactors – energy could be beamed back to Earth.

What was to be the role of NASA in all of this excitement in the early 21st century? On this question, opinions generally agreed that NASA's role was to prepare the infrastructure, that is, create transportation and facilities for the Moon and Mars. Then NASA bowed out and businesses and investors could do what they knew best: create wealth by meeting the needs of people, on Earth and on the Moon. NASA and nations became part of the customer base and not competitors to the lunar and Martian business enterprises. This economic development model parallels the way that the West was developed and the way that many technologically-based businesses evolved – an example being the communication satellite business.

Businesses appreciated this economic development model. They did not have the resources needed to build the transportation infrastructure – government via NASA would do that. Such an approach also opened up space for businesses that were not traditionally viewed as "space businesses." Many of the aerospace corporations recognized the need to broaden their markets, and to bring non-traditional companies into the space venture. With the development of a broad economic base on the Moon, companies had many potential clients with whom to do business. The high risk of depending solely on government contracts – as had been the case – could be avoided.

The number of small space companies grew every year, as did their profitability. This was because many of the technologies that needed to be developed for space initiatives were "dual-use" for Earth-based markets.

We, our children and their children benefit today in 2169 from the visionary decisions made in the early 21st century.

Science and technology evolve in unpredictable ways – actually in revolutionary ways. In 2009, our ancestors only had to look back 20 years (no iPods, hybrid cars, flat panel TVs, cell phones, the WWW or disposable contact lenses), 30 years (no personal computers, artificial hearts or bar codes), 40 years (no cable TV, electronic devices, video games, Valium or internal heart pacemakers) or 50 years (no color TV, radial tires, integrated circuits, solar cells, microwave ovens or credit cards) to understand how much the world had changed since the first manned rockets were attempted. Many of those technologies had their roots in the Apollo era.

And today in 2169, we look back 20 years (variable gravity field generation, hyper-dimensional communications that provide us with effectively faster-than-light communications, and cybernetic beings), 40 years (radiation resistant materials and carbon-based machines) 60 years (perpetual battery power, the ability to loft giant turbines to dissipate hurricanes in the Caribbean, and terascale technology) and 80 years (the construction of massive water tunnels under the Western U.S. to move flood waters from the Midwest to the Southwest, and genetic engineering to help humans survive in space) and are in awe of the developments we have witnessed in the past century – all of which were not even on the drawing table in 2069.

Fig. 1.13 Astronaut Edwin E. Aldrin, Jr., lunar module pilot, descends the steps of the Lunar Module (LM) ladder as he prepares to walk on the Moon. He had just egressed the LM. This photograph was taken by astronaut Neil A. Armstrong, commander, with a 70mm lunar surface camera during the Apollo 11 extravehicular activity (EVA). While Armstrong and Aldrin descended in the LM "Eagle" to explore the Moon, astronaut Michael Collins, command module pilot, remained with the Command and Service Modules (CSM) in lunar orbit. (AS11-40-5868, 20 July 1969. Courtesy NASA)

Funds for engineering research: Crucial to economy – key to the future

There was much concern in the 21st century in the United States about its technological competitiveness. The following essay was passed on to me by a great-aunt of my father.

* * *

> "Between my two visits to Europe in 1992 and 2002, I was alarmed to see the relative decline of the United States in wealth, reputation and hard science. In my visit to China this summer [2006] I was promised full support in the costly cyclic tests if I can run my research in China every summer." – A report by one researcher.

Another researcher, who was read this statement, said: "I don't think he adds much more to what I have already told you about my own area of research. For me, I replace China by Belgium, Germany, and France."
These two quotes are by colleagues who expressed serious discouragement at the sight of the decline of American scientific and engineering research. This correlated with two reports on engineering research released by the preeminent organizations that study and fund such activity.
A National Academy of Engineering (NAE) report,[7] evaluated the past and potential impact of the U.S. engineering research enterprise on the nation's economy, quality of life, security and global leadership, and whether public and private investment is adequate to sustain U.S. preeminence in basic engineering research. A 15-member National Academy of Engineering committee conducted fact-finding activities and prepared a brief draft report with findings and recommendations. Their basic finding was, no surprise, that funding for engineering research was dismal, especially compared with medical funding. The basic recommendation was — a lot more funding.
In the same time period, a National Science Foundation funded report, "Making a Case for Engineering," also makes much the same argument, that engineering research, which is the foundation upon which the nation's wealth is built, was woefully underfunded. This was at the same time that other nations, including some potential adversaries, were spending tremendous amounts of money to build up their economies and their military.
Adding insult to injury, the same NAE report also stated that more and more research work at corporations would be sent to the fast-growing economies of the time with strong educational systems, such as China and India. In a survey of more than 200 multinational corporations on their research center decisions, 38 percent said they planned to "change substantially" the worldwide distribution of their research and development work over the near term with the booming markets of China and India, and their world-class scientists attracting the greatest increase in projects. The outsourcing of blue-collar jobs led to the outsourcing of white-collar and then research jobs.
One of the greatest obstacles to recruiting more Americans to engineering has been the "disconnect" between engineering innovations and the people who use, rely on and prosper as a result of the innovations. Surveys demonstrate that the general public was not well aware of the nature of the engineering profession and its impact on quality of life, even though engineering has compelling success stories to tell. We need to remember that engineering research is every bit as creative and challenging, requiring talent as much as any art or science.

* * *

These trends, had they continued, would have posed a strategic threat to the United States. Our leaders recognized this problem and put it on the front burner for deliberations and solutions. But still the response at the national level was

[7] "Assessing the Capacity of the U.S. Engineering Research Enterprise," National Academy of Engineering of the National Academies, The National Academies Press, 2005.

neither focused nor fast. And even more ironic was that the taxpayers funded students from overseas to earn their Ph.D.s in the U.S., students who eventually went back to their home countries to compete with the U.S.

The confusion that resulted from the economic crises of 2008 and onwards prevented a focused and rational approach to the development of research funding policies that would have encouraged Americans to go into the sciences and especially engineering. It was not until almost the end of the second decade of the 21st century that an enrollment upsurge was observed in graduate programs in engineering. There were several reasons for this increase: the accelerating progress of the human return to the Moon, an increase in defense spending in response to that of the Chinese and Russians, a massive increase in public works spending – in particular, an upgrade of the electrical grid and its placement underground, the building of a subterranean network of very large (30 ft) diameter pipes that could carry water from flood-prone regions of the country to areas in perpetual drought and the fast-tracking of the design and construction of the latest and safest generation of nuclear power plants.

Technological literacy is crucial to public policy

The dichotomy between the "good" and "bad" uses of science and engineering has been a part of the public discourse for a very long time. The atom bomb especially invigorated such discussions, with the most heated ones occurring among those who actually performed the research and built the hardware. These same people agreed, however, that working toward the national defense was a worthy and honorable profession.

Scientists and engineers will often discuss the morality of certain inventions that have dual civilian benefits and military capabilities. The difficulties and the ironies of these discussions are twofold. The first is that most technology benefits civilians and the military. In fact, the military has been the largest single supporter of basic research in the United States — and the motivation for many of the most important civilian inventions. The second is that too few engineers and scientists are in high-level public policy positions and therefore have little control over how "their" technology is used.

Why are so few technical types in policy positions? Generally, because people who have scientific interests spend years to learn and therefore enjoy "doing science." They usually do not have an interest in public policy or in elected office. This is understandable but regrettable for society, since those in elected office tend to be lawyers with a very limited understanding of the technological workings of society, except perhaps as it involves the law and personal opinion. Similarly, economists are heavily involved in governing, but again their expertise is generally not in technology, even though economists tend to be more mathematically trained than are lawyers.

The problem with this situation is that many decisions at the highest levels depend on an understanding of the science and the technology at some meaningful level. Examples where a scientific and technological understanding is needed for a better public policy process are stem-cell research, pharmaceutical/medical advances, bioengineering, nanotechnology, advanced materials, energy generation and space and aerospace activities, to name a few. The government deliberates and creates rules and laws on all of these and many more issues that are fundamentally based on technology.

While it is possible to make reasonably good decisions without knowing the underlying science, better decisions require more understanding. For this reason, scientists and engineers need to get more involved in the larger issues beyond the technology or science of their work. The populace needs to be continually informed and educated on how engineering affects their lives. Such information does not have to be overly technical, but of sufficient depth so that the larger issues are understood.

Some scientists and engineers write books that help non-specialists understand scientific basics or to explain how various engineering marvels were created. There are also numerous books on technology and its impacts on society.

The scientific understanding of matter at the atomic level has led to both nuclear power (which in most parts of the world is very useful and supported) and nuclear weapons, which continue to be a great worry. Another major example is the environment. The same technologies that pollute have some positive benefits to society. They provide us with energy, manufacturing resources, or entertainment. The delicate balancing act of safeguarding the environment while reaping the benefits of a technology is important to all of us. The reality is that technology is going to help us solve the negative side effects.

The lines have generally been blurred in this debate. Often, information that is misleading or wrong is published in order to push the debate and, ultimately, the decisions along a particular path. The public (including too many reporters) has been easily misled. The government, with all of its expert witnesses, can also be misled, since the issues are complex. The public needs to spend some time to understand the essential technical aspects of a problem. Our kids need to study math and science, more so today because of the ever-increasing complexity of society. For if we become less technologically literate, we will be led by the nose to places that will be very unpleasant and dangerous, to say the least.

1.1 The epic vision of and for space

When we consider the lives of great people, we generally see a vision that they held on to from the time that they were young their goals were eventually achieved at great personal sacrifice and cost. A long-term view was crucial to guide the day-to-day efforts of such a person. Similarly, when we view the histories of nations, we can trace the successful efforts of a nation and a people to a single-minded focus – where the nation acts as one, with determination, usually under a single leader that has articulated that goal.

This is true in war and in peace. In war, outside forces focus the nation's attention effectively since the nation's survival is at stake. In peace, that focus must evolve through a national discussion. Peacetime, or a lack of *obvious* national threat, can be a difficult time for a nation because democracy allows multiple

Fig. 1.14 Astronaut Edwin E. Aldrin, Jr., lunar module pilot of the first lunar landing mission, poses for a photograph beside the deployed United States flag during an Apollo 11 Extravehicular Activity (EVA) on the lunar surface. The Lunar Module (LM) is on the left, and the footprints of the astronauts are clearly visible in the soil of the Moon. (Astronaut Neil A. Armstrong, commander, took this picture with a 70mm Hasselblad lunar surface camera. AS11-40-5875, 20 July 1969. Courtesy NASA) See Plate 3 in color section.

factions with different goals to evolve, making the achievement of any one of those goals very challenging.

Effective leaders are those with a vision for their nation that goes beyond the day-to-day. While visions that benefit those currently alive can be effective, a nation truly needs a vision that benefits the children and their children. In such an instance, politics may take a back seat to the shared and larger view of the human purpose. This is an epic vision. The goals of the nation, its leaders, and its people synchronize – friction to this effort becomes minimal – and achieving the vision becomes paramount to all.

The first trek to the Moon by the United States during the decade of 1960 had its focus enforced by the Soviet space program and the Cold War. Had the Soviets not succeeded in placing man and machine into orbit, it seems unlikely that the U.S. would have viewed a similar effort as urgent. There were many events that pointed the U.S. in the direction of the Moon. The marshalling of resources was unprecedented outside of a war effort. Kennedy's goal of sending men to the Moon before the end of the decade fell within the calculus of the Cold War between the U.S. and the Soviet Union. The U.S. space effort gained its determination from the competition of an effective and powerful Soviet space effort.

An *epic* (from Greek: $\varepsilon\pi o\varsigma$) is a lengthy narrative poem, normally concerning a serious subject containing details of heroic deeds and events significant to a culture or nation. An epic hero embodies the values of the civilization. The hero generally participates in a repeated journey or quest, faces adversaries that try to defeat him in his journey, and returns home significantly transformed by his journey. The epic hero illustrates traits, performs deeds, and exemplifies the values of the society from which the epic originates. Many epic heroes are recurring characters in the legends of their native culture.

An epic vision for a nation has similar traits. The vision is a reflection of how that society views itself and its place in the world and across time. Effective leaders of nations have created such epic visions for their nation. This has happened in democracies as well as dictatorships. In dictatorships, the people are coerced into accepting the vision and have little choice about their participation. In democracies, the leaders have the more difficult role since they must convince by dint of rational and emotional discourse that their view for the nation is a good one and it must resonate with the views of a substantial majority of the people. It is that resonance that focuses everyone's energies along the same vector that points to the long-term goals of the society.

Space has been an epic vision. Kennedy created that resonance of purpose with the citizenry who were equally worried that a Soviet advantage in space would create multiple dangers to the United States – physically, socially, and psychologically. Similarly, the Soviets viewed success in space as a measure of the worthiness of their society and its structure. Bush's 2004 speech was an attempt to recreate that epic vision for the American return to the Moon. Unfortunately, much in the same way that Apollo faded away in the early 1970s due to the Vietnam war and government deficits, the return to the Moon was always on society's back burner. Barely enough resources were allocated to NASA. Whenever the manned space program appeared in the national media it was for negative reasons, either budget or technical woes.

But there had been general public support of NASA, and Congress also believed in the program even though it was in a major struggle over enormous budget deficits and international crises. But with time NASA solved the technical problems and landed astronauts on the Moon in 2024.

Once that threshold was passed, inertia developed along with a broad public interest. Success breeds support. People were excited and many started to feel as though they were a part of the adventure. The space program was now owned by the general population and the media had numerous stories about the technologies and the people making it happen. Space became a very positive force in American society as well as with the peoples of nations that were part of the effort.

1.1.1 The nature of complex technical projects

It cannot be repeated often enough that technology is very challenging. The creation of a jet engine, an LED flat screen, a nano-chip, an automobile, a space shuttle, a space station, a rocket, a robotic asteroid miner, or an android surgeon is extremely complex. While mathematical and physical theories are derived to understand the fundamental elements, engineering designs must take into account that the theories are idealizations and therefore have uncertainties.

Before any airplane is put into production, test pilots must fly prototypes to discover and weed out surprises – surprises being behavioral characteristics that the mathematical model "missed." The recursive process of design is one of simulating, building and testing, then simulating again, rebuilding and retesting until the product performs as required and does so in a reliable way.

This process is true for all manufactured items. The more complex the item, the more effort that is required at all levels. The theories are more intricate, the tests are more elaborate, and a larger number of simulate–build–test cycles are required. So when NASA in the early 21st century found flaws in initial designs of the new Constellation rockets, that was to be expected. As was the need to perform repeated tests that allowed the engineers to better understand the systems – how the systems interact – and to improve the designs to the point where they met the specified performance levels.

We learned much from our initial forays into automated in-situ resource utilized (ISRU) construction on the Moon and Mars. We had numerous failures of machines, sometimes due to software glitches, sometimes due to hardware malfunctions, and sometimes – unavoidably – due to human error. But each failure – some of which cost lives! – took us one step closer to a viable and reliable system. When we first landed on Mars in 2034, we had a very steep learning curve to master. Our goal was to transfer the automated ISRU construction teams from the Moon and to get them working. But the dust storms on Mars along with the double lunar gravity led to unanticipated difficulties.

However, today on Mars we have dozens of automated ISRU construction teams building the future Martian cities. All of this was inconceivable even 70 years ago.

1.1.2 Chief Executive Timoshenko

As a commemorative celebration of humanity in space, this book draws on documents from the pioneering era between the late 1950s and the early 2040s. After that era, we accelerated our settlement activities with many projects, some of which are mentioned in this volume. This book also draws upon speeches and interviews from that pioneering era. And it incorporates personal anecdotes that are intertwined with our history in space of the past two centuries.

The following essay was one that I presented in my capacity as lunar Chief Executive at ceremonies that celebrated our relatively short history as a spacefaring civilization. I am no longer functioning in that capacity, but I continue to be interested in chronicling our settlement of space.

Chief Executive Yerah Timoshenko: It gives me great pleasure to mark the beginning of the third century of the human lunar civilization. I will leave it to the historians to chronicle all the details of the last two hundred years. My goal here is to look back at some of the visionaries that made all of this possible. These people who understood how important it was, and is, to have a long view of the human experience. The visionaries of the twentieth and twenty-first centuries were able to imagine the steps needed to get us to the Moon, not once, but twice after the half-century hiatus that followed Apollo. There are visionaries who today are mapping our steps to Mars, the outer Solar System, and even beyond. These are the steps that we need to follow. They are not optional for society, no more so than breathing is to a person.

While the decision to go the Moon, Mars and beyond seems to be an obvious one today – we can't even imagine humanity without all of our outposts in the Solar System – the period between the late 20th century and early 21st century was one of black and white views. It was a common belief that the world operated as a "zero-sum game." Whatever was spent on one thing could not be spent on something else. There were few who believed that the pie could grow if investments were chosen wisely.

In the space arena there was government spending through the national space agencies, and there were fledgling entrepreneurial forays into space. Certainly one great success in space business was in communications.

I now live on the Moon, not too far from the Apollo 11 landing site in *Mare Tranquillitatis*, which is a protected Historic Site. Actually, I live below the surface of the Moon. Most habitats are primarily underground – radiation and micrometeorites being the primary hazards – but we don't feel underground. Three-dimensional visualization screens with views of our choosing provide us with exact visions of the lunar surface, or any view we prefer. I enjoy looking out at the lunar surface some days, and other days I enjoy a view of the Victorian houses in Cape May, New Jersey – and perhaps a vision of MD at Gunnison.

I descend from a line that traces back on my father's side to a prominent engineer of the 1900s in the United States, of Russian origin, who developed many theories. His descendents, including my great-great-great grandmother, emigrated to a very early scientific station on the Moon when she was twenty-four years old in 2029. This scientific station became the first permanent colony on the Moon. From 2029 on, there were people here who had made the Moon their permanent residence.

My mother's side was more artistic, both with the written and the visual arts. My grandmother was commissioned to paint the large murals that are displayed in the first Hilton Hotel. Her father wrote the story of the first scouting teams that roamed the Moon for months at a time, long before communication and power networks were set up.

I was born on the Moon in the early part of the 22nd century – never mind the date! – and have never visited Earth. One day maybe I will, but the older I get the less likely it is that I will do so because Earth's gravity is six times that of the Moon – I don't think that I could handle the stress. I have been to Mars once, and that was very difficult given that its gravity is over twice that of the Moon. My children were born on the Moon, of course, but they have plans on moving out to

Mars, or maybe one of the outer planets. They continually train so that they can make the transition to a larger gravity.

I studied engineering and psychology at the Lunar Institute. So much of a successful life on the Moon depends on a broad knowledge base that it is common for our students to learn at least two fields that were once viewed as disparate. Some students on the Moon major in physics and art, others major in medicine and philosophy. We have grown to believe that all knowledge is connected.

Life on the Moon is very constrained when compared to life on Earth. Due to the lack of atmosphere, we are severely limited in our "travels." We always use a spacesuit when traveling beyond our cities. We need to be in a pressurized cabin at all times. But even though our cities are large – we are aware of their limits and boundaries – we are not claustrophobic. For those of us who are multi-generational inhabitants of the Moon, it is all we know and we are content with the many benefits accrued by living here – for example the fantastic mobility we have here, and the stark beauty of the place.

So enjoy the celebrations. We have special speeches by the grandchildren of the descendents of the pioneering settlers. We have amusement rides – roller coasters that bring you up to one g, and the gravity-free pool.

* * *

The celebrations were fantastic. I will provide you with more details of life on the Moon later in this book. Interspersed with these are interviews with some of the visionaries of two centuries ago, and a series of discussions that dominated humanity's return to the Moon.

The physicist Kai Multhaup had the following view of human spaceflight. "I see it as a driver for evolution. We are an exploratory species, and when we have the technology to go somewhere, we do. It's about culture and the human desire to evolve and expand, and to protect ourselves against catastrophes which can erase life on planets and end civilizations. ... The space shuttle and [space]station are often criticized. They have their flaws, but they are important. You can't take the shuttle to the Moon or attach an engine to the International Space Station and send it to Mars, because they aren't designed to explore. They are meant to operate in low-Earth orbit, and teach us how to live and work there."[8] They served their purposes well – they taught us how to take our first steps in space.

The American return to the Moon was attempted by both Presidents Bush, father and son. On July 20, 1989, President of the United States George H. W. Bush (the father) announced plans for the *Space Exploration Initiative*, calling for the construction of the Space Station *Freedom*, sending humans back to the Moon, and ultimately sending astronauts to Mars. He proposed not a 10-year Apollo-style plan, but a long-range continuing commitment ending with "a journey into tomorrow – a journey to another planet – a manned mission to Mars." The President noted it was humanity's destiny to explore, and America's destiny to lead. He ended by asking Vice President Dan Quayle to lead the National Space Council in determining what was needed to carry out these missions in terms of money, manpower and

[8]K. Multhaup, Westfälische Wilhelms-Universität Münster, Germany. Quote posted on 22 January 2009, in *Live Science*, courtesy Imaginova.

technology. Preliminary cost estimates were in the hundreds of billions of dollars, and even though funding would be over many years and decades, political support evaporated during the Bush term. The first Gulf War took place during his term as President and dominated his time in office. A weakening economy also doomed his Initiative.

However, there was the belief that a return to the Moon was something worth doing. "Exploration inevitably includes a human component ... although [it is unclear] how to define the long-term prospects, limitations and specific assignments of this component with certainty. Both robotic and human exploration ... are necessary and interdependent. Human exploration is obviously more demanding and expensive than robotic exploration, but also capable of more immediate and creative, less structured solutions to unforeseen discoveries and problems. Shielding from radiation, countermeasures against microgravity, relief from the tensions of long-term human interaction at very close quarters while maintaining high performance levels, nuclear propulsion to reduce transit duration and exposure – all of these are new and complex R&D arenas in which combinations of productive outcomes must be sought. But it is important to add that such outcomes not only open the way to successful near-term missions but to a permanently enhanced set of human potentialities for movement and discovery."[9]

Krafft Ehricke's philosophy, called the *Extraterrestrial Imperative,* identified scientific and evolutionary facts that formed the basis for the manned industrialization of space. The Moon was the focal point for such an expansion. His famous quote is:

> "It has been said, "If God wanted man to fly, He would have given man wings." Today we can say, "If God wanted man to become a spacefaring species, He would have given man a Moon."

He believed that we were fortunate to have a nearby sister planet. For without the Moon, the nearest body – Mars – was a year travel time in each direction. The Moon, no more than three days' flight from Earth, was an ideal body upon which we could test our wings. "No other celestial body and no orbiting space station can more effectively permit development of habitats, material extraction and processing methods, and in essence, all the science, technology, and sociology required for a responsible approach to extraterrestrial operations."[10] He called the evolving lunar civilization *Selenopolis*. He had hopes that such a civilization would evolve humanity to a higher level of wisdom.

> "Human growth is contingent not only on the absence of war, or overcoming hunger, poverty, and social injustice – but also on the presence of overarching, elevating goals, and their associated perspectives. Expanding into space needs to be understood and approached as world development, as a positive, peaceful, growth-oriented, macrosociological project whose goal is to ultimately release humanity from its present parasitic, embryonic bondage

[9] R. McC. Adams, "Why Explore the Universe," 30th Aerospace Sciences Meeting & Exhibit, 6–9 January 1992, Reno.

[10] *Krafft Ehricke's Extraterrestrial Imperative*, M. Freeman, Apogee Space Books, 2009.

Fig. 1.15 Krafft A. Ehricke (24 March 1917 – 11 December 1984).

> in the biospheric womb of one planet. That will demand immense human creativity, courage, and maturity."[11]

Krafft Arnold Ehricke was a space visionary, a German-born rocket-propulsion engineer who was the chief designer of the Centaur and who produced many other ideas for the development of space including a space plane design and a strategy for lunar colonization. As a child, he was influenced by Fritz Lang's film "Woman in the Moon" and formed a rocket society at age 12. He studied celestial mechanics and nuclear physics at Berlin Technical University. Injured during World War II, he was transferred to Peenemünde where he served as a propulsion engineer from 1942 to 1945. Upon moving to the United States, he became an American citizen in 1954 and during the 1950s with General Dynamics helped develop the Atlas missile and then the Centaur upper stage. Later, he carried out advanced studies at Rockwell International while also working independently on schemes for the commercialization and colonization of space.

The dreams of human exploration of space and human settlement of the Moon and Mars have captured the imagination of many. Some of those who were so inspired eventually rose to positions where they could work to create the reality that we know today. Much has been accomplished in 200 years – 1969 to 2169. It is really fantastic. And while many wished that it had happened faster and sooner, two hundred years is not really a very long time for humanity.

[11] K.A. Ehricke, "Lunar Industrialization and Settlement – Birth of a Polyglobal Civilization," in *Lunar Bases and Space Activities of the 21st Century*, edited by W. Mendell, Lunar and Planetary Institute, 1985.

1.2 Quotes

- "In time, [a Martian] colony would grow to the point of being self- sustaining. When this stage was reached, humanity would have a precious insurance policy against catastrophe at home. During the next millennium there is a significant chance that civilization on Earth will be destroyed by an asteroid, a killer plague or a global war. A Martian colony could keep the flame of civilization and culture alive until Earth could be reverse-colonized from Mars." Paul Davies
- "I have learned to use the word 'impossible' with the greatest caution." Wernher von Braun
- "The creation of something new is not accomplished by the intellect but by the play instinct acting from inner necessity. The creative mind plays with the objects it loves." Carl Jung
- "Scientists discover the world that exists; engineers create the world that never was." Theodore Von Kármán
- "I like to think that the Moon is there even if I am not looking at it." Albert Einstein
- "I came to realize that exaggerated concern about what others are doing can be foolish. It can paralyze effort, and stifle a good idea. One finds that in the history of science, almost every problem has been worked out by someone else. This should not discourage anyone from pursuing his own path." Theodore Von Kármán
- "Every child is an artist. The problem is how to remain an artist once he grows up." Pablo Picasso

2 A short retrospective of our recent history in space

"Here men from the planet Earth first set foot upon the Moon. July 1969 AD. We came in peace for all mankind."

Neil Armstrong

This essay[1] provides the reader with a brief historical overview on how we came to be a spacefaring people during the 22nd century. We will focus on the early post-Apollo era on Earth more than one hundred and fifty years ago, the difficulties faced by those who proposed an accelerated space program, how our grandparents and parents were finally able to colonize the Moon and Mars, and how we began our young lunar and Martian civilizations. While it is hard to imagine the present without these outposts of human civilization, the beginning of the 21st century saw our predecessor generations concerned primarily with the day-to-day, completely ignoring how future generations would benefit from a development into space.

Let us first set the backdrop so that the reader can better understand the human mindset of the late 20th century. Apollo epitomized to some a peaking of Western technology and perhaps civilization. Almost as soon as Man walked on the Moon, with that crowning achievement at hand, serious space efforts on the scale of Apollo were given up. Although advancements in science, technology, and society continued, the seventies were the beginning of a long economic and spiritual decline worldwide that even the fall of the Soviet Empire could not alter.

The Space Shuttle was proposed and approved for development as a reusable launch vehicle that ostensibly could fly once per week and become the cargo hauler to low Earth orbit. Unfortunately, it did not meet the vision of its creators, even though it was a remarkable feat of engineering. The original shuttle concept was downsized and altered, with its final manifestation a reflection of political rather than technical deliberations.

The 1990s saw strong economies in the West, but there lacked a "larger" vision that should have paralleled a strong economy. In the United States, which was still the preponderant space power in the 1980s and 1990s, public interest in space was primarily confined to the movies that showed off computer-generated visions of a spacefaring future. The American Space Shuttle and Russia's *Soyuz* fleets transported space station components into orbit. Figure 2.1 shows Shuttle *Discovery* in 2005.

[1]This retrospective was written in 2149 by Yerah Timoshenko as a commemoration of the 50th anniversary of the first human birth on the Moon.

Fig. 2.1 One of a series of photographs showing the Space Shuttle Discovery as taken from aboard the International Space Station during rendezvous and docking operations. The Italian-built Raffaello Multi-Purpose Logistics Module (MPLM) is visible in the Shuttle's cargo bay and a Soyuz spacecraft docked to the Station is at the right. A blue and white Earth and the blackness of space provide the backdrop for the image. (ISS011-E-11236, 28 July 2005, Courtesy NASA)

By the middle of the decade 2000–2010, the Space Shuttle was getting old and expensive to maintain. Two Shuttles had been lost in tragic accidents with all aboard lost in the line of duty. Seven astronauts were aboard shuttle *Challenger* when it exploded during liftoff on 28 January 1986. School teacher Christa McAuliffe and six other astronauts were killed when a solid-fuel booster rocket leak led to a massive liquid-fuel tank explosion during lift off from a Cape Canaveral launch pad. The *Endeavour*, a name chosen by American students, was constructed as a replacement, and flew in 1992.

The shuttle *Columbia* broke up 200,000 feet over Texas on 1 February 2003 as it descended from orbit into the atmosphere toward a landing at Kennedy Space Center in Florida. Seven astronauts aboard the shuttle were lost in the disaster. Aboard the shuttle during the 16-day flight had been commander Rick D. Husband, 45; pilot William C. McCool, 40; payload commander Michael P. Anderson, 42; mission specialists David M. Brown, 46; Kalpana Chawla, 41; and Laurel Clark, 41; and Israel's first astronaut, payload specialist Ilan Ramon, 47. Ramon[2] had been a national hero in Israel for taking part in the 1981 bombing of a nuclear reactor in Iraq.

Political turmoil between the United States and Russia led to a cooling-off period when the West was uncomfortable relying on Russian lift capability for the

[2]In one of those ironic twists of fate, his son, Lt. Assaf Ramon, died on 13 September 2009 when piloting an F-16 in a training mission near the Israeli community of P'nei Chever in the southern Hebron Hills.

Fig. 2.2 STS-51L crew photo. Seated are Pilot Michael J. Smith, Commander Francis R. Scobee, Mission Specialist Ronald E. McNair, and standing are Mission Specialists Ellison S. Onizuka on the left and Judith A. Resnik on the right, and Payload Specialists Sharon Christa McAuliffe and Gregory B. Jarvis. (Courtesy NASA)

completion of the Space Station. The Americans were just gearing up with their Apollo follow-on, the *Constellation* program. But the first rockets were not ready until 2014. In addition, under the new Obama Administration, a review of the program was undertaken, with some modifications possible. Figure 2.4 shows a Soyuz vehicle and Figure 2.5 shows the Ares cargo and crew launch vehicles.

While it was originally planned that the Shuttle would be retired in 2010, the potential gap in lift capability was worrisome to the Americans, who still had to continue building and maintaining the International Space Station (ISS). Concerns grew to a crescendo after the Russian invasion of Georgia, and a rekindling of the Cold War. Due to needs on the ISS, the Shuttle operated a few years longer, and some of the gap in launch capability was taken over by American military launch vehicles. There is some ambiguity about the dates.

The major concerns about placing humans in a lunar facility for very long periods of time were the human factors. Even though humans lived in the Russian space station *Mir* for as long as 14 months, and survived, there were (and still are) physiological risks to humans in micro- and low-gravity that need resolution. The

Fig. 2.3 The crew of Columbia shuttle flight STS-107. Seated from left are Rick D. Husband, mission commander; Kalpana Chawla, mission specialist; and William C. McCool, pilot. Standing from left are David M. Brown, Laurel B. Clark, and Michael P. Anderson, all mission specialists; and Ilan Ramon, payload specialist from the Israeli Space Agency. (Courtesy NASA)

human body evolved for life on Earth under an atmosphere that shields against micrometeorites and radiation, and under the gravitational pull of Earth. The Moon was viewed as a test bed not only for technology, but for Man and plant. We certainly understand much more about this today, but occasionally, even we are surprised both by genetic "hiccups" and by how robust the human body can be.

The *Mir* space station, shown in Figure 2.6, holds the record for the longest continuous human presence in space at eight days short of 10 years and was made internationally accessible to cosmonauts and astronauts of many countries. The most notable of these, the Shuttle-Mir Program, saw American Space Shuttles visiting the station eleven times, bringing supplies and providing crew rotation. Mir was assembled in orbit by successively connecting several modules, each launched separately from 1986 to 1996. The station existed until 23 March 2001, when it was deliberately de-orbited, breaking apart during atmospheric reentry over the South Pacific Ocean.

Fig. 2.4 Rollout of Soyuz TMA 2 aboard an R7 Rocket on 10 September 2005. It takes a big rocket to go into space. In 2003 April, this huge Russian rocket was launched towards the Earth-orbiting International Space Station, carrying two astronauts who made up the Expedition 7 crew. Seen here during rollout at the Baikonur Cosmodrome, the rocket's white top is actually the Soyuz TMA-2, at the time the most recent version of the longest-serving type of human spacecraft. The base is a Russian R7 rocket, originally developed as a prototype Intercontinental Ballistic Missile in 1957. The rocket spans the width of a football field and has a fueled mass of about half a million kilograms. Russian rockets like this were a primary transportation system to the International Space Station. (Photo by Scott Andrews. Courtesy NASA)

Skylab was America's first experimental space station and it pre-dated the Mir. Designed for long duration missions, the Skylab program objectives were twofold – to prove that humans could live and work in space for extended periods, and to expand our knowledge of solar astronomy well beyond Earth-based observations. Skylab was launched on 14 May 1973. Successful in all respects despite early mechanical difficulties, three 3-man crews occupied its workshop for a total of 171 days, 13 hours. It was the site of nearly 300 scientific and technical experiments – medical experiments on humans' adaptability to zero gravity, solar observations, and detailed Earth resources experiments. The empty Skylab spacecraft returned to Earth on 11 July 1979 scattering debris over the Indian Ocean and the sparsely settled region of Western Australia. Studies to save Skylab led to the conclusion that a lack of funds, spaceworthy vehicles, and a very active sun – warming up the atmosphere and increasing drag on the spacecraft – would make a rescue almost impossible.

Fig. 2.5 Engineering concept of NASA's launch vehicles designed to support the Bush Vision of 2004, the Ares V cargo vehicle, left, and Ares I crew launch vehicle. (Circa 2007. Courtesy NASA)

In addition to physiology, long duration space psychology was essentially a guess based on what we knew about humans in isolation in various Earth locations. Remote locations such as those in the Antarctic were domestic testbeds. McMurdo was a major research station there, and we can see how pioneering was the Crary Lab facility, shown in Figure 2.8.

Fig. 2.6 Russia's Mir space station on 12 June 1998 backdropped over the blue and white planet Earth in this medium-range photograph recorded during the final fly-around of the members of the fleet of NASA's shuttles. Seven crew members, including Andrew S.W. Thomas, were aboard the Discovery when the photo was taken; and two of his former cosmonaut crewmates remained aboard Mir. Thomas ended up spending 141 days in space on this journey, including time aboard the Space Shuttles Endeavour and Discovery, which delivered and retrieved him to, and from, the Mir. (Courtesy NASA)

Such isolated regions were also utilized to test concepts for habitation and survival. Figure 2.9 shows an inflatable structure placed in Antarctica and tested for eventual use on the Moon.

At the same time, worldwide interest in returning to the Moon grew. The Chinese were making major inroads in their technology developments, with their first spacewalk on 27 September 2008. The South Koreans were earnestly building up technical capabilities and recruiting lunar researchers worldwide. The Indians were becoming a space powerhouse. And many other countries were able to launch satel-

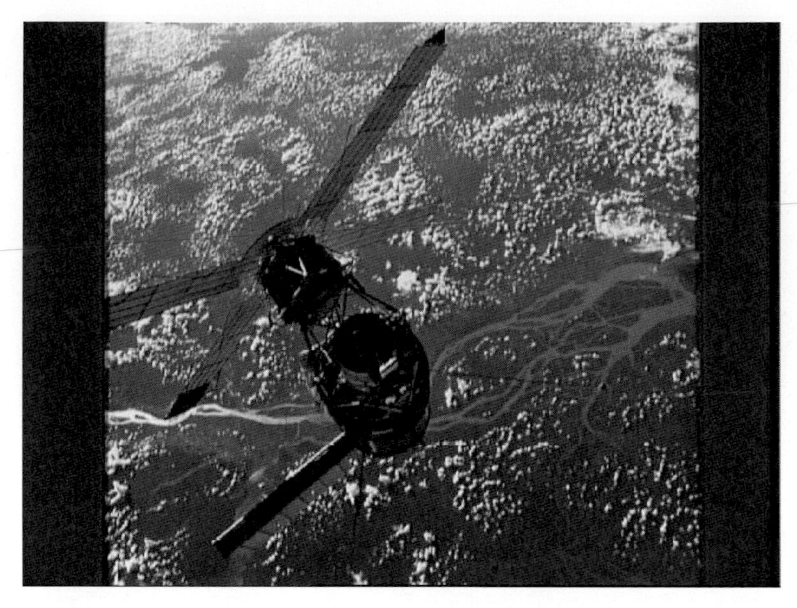

Fig. 2.7 Skylab in orbit with panels above gathering solar energy. (Courtesy NASA)

lites into orbit. Manned flights into space were still a major challenge, but several countries had that ability.

Economic problems towards the end of the first decade of the 21st century drained financial resources from the United States and other major economic powers. The wars in Afghanistan and Iraq were also a major financial burden for the U.S. as was the worldwide war on terror. In a general sense, there were similarities to the political and economic environment when Apollo was cut short and the U.S. focused its manned efforts to low Earth orbit. Yet, there was widespread interest in the early 21st century in manned space exploration as well as lunar and Martian habitation.

Few in positions of political or moral leadership were willing to be bold enough regarding their vision for the future of their nation or of humanity. As a result, the public's perception and the reality of the political hierarchy as being self-serving grew to the point where it became impossible to propose and promote initiatives with long term goals. The public had grown cynical of elected officials who in the end could do little that was right by the average voter. So the last quarter of the 20th century and the beginning of the 21st century were marked by stagnating standards of living if one uses quality of life measures, with many finding themselves in more harried lives with less security, not only for themselves; more importantly, most could only see a more dismal existence for their children. Some economists predicted a long slow decline lasting many decades. Of course, we now know that those predictions could only be extrapolations of the past and could not have really predicted the unique events that changed the whole course of history.

As NASA was returning to the Moon, a private space development group was formed in the United States that had as its mission the creation of a new self-

Fig. 2.8 Albert P. Crary Science and Engineering Center, McMurdo Station, dedicated in November 1991. The laboratory is named in honor of geophysicist and glaciologist Albert P. Crary (1911–1987), the first person to set foot on both the North and South Poles. The Crary Lab has five pods built in three phases to make 4,320 square meters of working area. Phase I has a two-story core pod and a biology pod. Phase II has earth sciences and atmospheric sciences pods. Phase III has an aquarium. (Courtesy National Science Foundation)

sufficient civilization on the Moon. There were other such groups in other countries, but we will focus on this group since their efforts are well documented. It should be clear, however, that while earlier space developments were wholly government led, the next generation of serious development – beyond the creation of the space transportation system and initial lunar infrastructure by NASA – was borne by the efforts of motivated investment groups coupled with visionary technologists.

Somehow, such groups formed almost simultaneously and independently around the world. Each reflected the culture that spawned it, but they all had one thing in common: their vision of the future was based on space exploration and settlement. The Japanese group saw hotels in orbit, such as the Shimizu Hotel shown in Figure 2.11, lunar cities, and Martian convention centers. Apollo 17 astronaut Harrison Schmitt promoted Lunar Helium-3 early on as the fuel of the future – fuel for nuclear fusion reactors.[3] Many such groups saw that they could transfer space technology to the Earth-based consumer markets. These efforts were a boon in three ways: first, the technology needed to settle the Moon was created, second, the money spent on such development was spent on the Earth boosting that economy, and third, that "space" technology found applications on Earth.

[3] *Return to the Moon*, H.H. Schmitt, Copernicus Books, 2006.

Fig. 2.9 This inflatable building was tested by NASA in Antarctica to determine if it could someday be used on the Moon. NASA's Innovative Partnership Program investigated using inflatable structures for future long-term lunar habitats. Weighing less than 1,000 pounds, the building inflates in fewer than 10 minutes. This photo was taken at McMurdo Station. (Courtesy NASA)

The following overview chronicles how the American group began as a small enterprise, but excited enough interest within the industrial and investment communities to build a development engine that was to drive American commercial and civilian space activities to the present day.

The primary difficulty in privately financing a venture such as a space station or a lunar base was that the time frame was too long, well beyond the several years that venture capital was willing to wait for its heavy return on investments. The key then was to *deconstruct* the process of going to the Moon into its smallest financeable units. For example, essential technologies required to settle small groups on the Moon included: rocket technology, space communications, electronics and computing equipment, self-sustaining life support systems, space medicine, materials for radiation/impact shielding, tools for construction in space and on the Moon, software for monitoring and control, materials for structures for space and the Moon, and rover technologies.

In addition, a minor space station was built to assist in the transfer of structures and materials through low-Earth orbit (LEO) to lunar orbit and finally the lunar surface.

Each of these technologies was (and is) an industry unto itself, with specific R&D, manufacturing, and implementation issues and, most importantly, potential profits. Each had (and has) potential terrestrial applications and avenues for investor returns. As I will explain, this was crucial for the success of the venture, which would utilize both tried and true technologies as well as develop new technologies as

Fig. 2.10 An early lunar habitat has been assembled out of components delivered by automated cargo flights. Pressurized rovers, logistics modules, and a spacesuit maintenance and storage module combine to provide the living and working quarters for the crew. This concept from the late 1990s became a reality circa 2030 after the creation of permanent lunar settlements in 2029. (The above artistic vision is NASA/JSC image #S93-45585. Courtesy NASA)

required. The appropriate financing mechanism would provide intermediate returns on milestone developments.

The American group initially found itself with little political support. It had to structure an initial fifteen-year development plan that would have as its goal a small facility on the Moon with the staff, equipment and ability to begin preliminary operations. It turned out that this fifteen-year plan was not ambitious enough, and had to be continually updated with new milestones as the original ones were met ahead of schedule. Progress was always underestimated.

The lunar operations included preparations for the expansion of the facility for larger teams, and to prepare for near-term mining and processing operations. Once the initial plan had been completed, venture financing was set in place for the first phase of such operations. It is important to note that the plan stipulated utilizing as much existing technology and facilities as possible. All operations were led by an in-house engineering and scientific team, coupled with business, marketing and legal teams that could help prepare for the transfer of the intermediate technologies and by-products of this effort. It is also important to note that in order to attract corporate users to the facilities being created, a guarantee was required that certain facilities would be in place at certain times.

The continual slippage of milestones along with the reduction of the scale of what was to be constructed was historically the problem with the American space

Fig. 2.11 Shimizu space hotel. This illustration shows a large facility that is accessed by a space plane, shown docked at the lower part of the station. The hotel is within the centrally-located inverse pyramid. (Courtesy Shimizu Corporation) See Plate 4 in color section.

station – originally called *Freedom* – and eventually becoming the *International Space Station* (ISS) when the U.S. invited other participants in order to help finance the station's construction. The ISS was eventually placed into orbit. Figure 2.12 shows how it looked in 2009, but it was a minimalist version of the original concept. It did not live up to its expectations because sights were lowered so much so that the final structure did very little to advance science. It did, however, teach us much about building structures in orbit, a skill whose value should not be underestimated. We drew on this knowledge and experience when we began to build space elevators in 2046.

However, once NASA and the other world space agencies lost interest in the ISS, private enterprise became interested. An investment group offered to purchase the station, but kept its plans for the ISS a secret for a number of years. After the ISS was transferred into private hands in 2014, it was redeveloped into an orbital service station for satellites by 2019, then configured by 2025 to help with the critical efforts to clear the destructive orbital debris field around Earth. Two

Fig. 2.12 Backdropped by the blackness of space and Earth's horizon, the International Space Station is seen from Space Shuttle Discovery as the two spacecraft begin their relative separation. Earlier the STS-119 and Expedition 18 crews concluded 9 days, 20 hours and 10 minutes of cooperative work onboard the shuttle and station. Undocking of the two spacecraft occurred at 2:53 p.m. (CDT) on 25 March 2009. (S119-E-009768. Courtesy NASA)

decades later, actually in 2046, it was the staging point for the robotic space elevator deployment teams that led to the erection of the first space elevator in orbit around Earth. For this purpose, the ISS was propelled to geosynchronous orbit.

Some uncertainties can be built into high-risk, new ventures, but no one can be expected to base business decisions on schedules that slip many years behind promised dates, as did the ISS. All the delays and changes prevented industry from making plans to utilize the microgravity and hard vacuum environment. The need to build confidence by utilizing existing, or near-existing, technologies was understood by the private venture group.

My great-grandparents – who were in their twenties – worked in space during the late 2040s and early 2050s in support of space elevator construction and deployment. They migrated to where the most exciting jobs were. Sometimes that was in space, sometimes it was back on Earth. In 2059 they were on the Moon for a while. That is where my mother was conceived. Many were conceived on the Moon, but born on Earth, as was my grandmother in 2060. The first birth on the Moon was in 2099.

The two key elements of the successful American venture were that (i) the business plan had a sufficient number of intermediate, independently-profitable milestones, assuring that products would be available at reasonable intervals for marketing on Earth with returns for the investors, as well as for the continuation of the larger scale space project, and (ii) to ensure that enough existing technology was utilized so that an acceptable rate of progress could be maintained while assuring the end user that all milestones would become available for contracted usage. A milestone is defined as the availability of a facility or product by a promised date.

While the following ventures were born of the American effort, some aspects grew out of partnerships with other national efforts.

An essential intermediate milestone was a space station (not the ISS) built not for science, but as an in-orbit construction and transfer point for lunar materials. While this was its nominal purpose, once it was in place, industry and others could rent space on it. After all, this space platform was powered and had easy access to microgravity and vacuum. When it was conceived, it was with some excess capacity and designed for general use.

In the first year of operations, the first customers set up small, automated R&D modules to test concepts. Of course, the venture group assisted in the design of the space canister containing the modules to be placed in orbit, arranged transportation to orbit, used its astronauts to place and connect the canister to the platform grid, and maintained communications and module monitoring. All of this was part of the platform milestone of the original venture plan.

This space station served as a way-station, a transfer point for cargo on its way to the Moon. In most instances, items bound for the Moon had to be repackaged for the microgravity environment and low lunar gravity destination. For the trip from launch to the station, goods required significant protective packaging in order to withstand the high gravitational acceleration of the launch. Eventually, the space elevators came online, and their acceleration was modest, providing a savings due to very minimal packaging needs.

During the second year of operations, some profits went to the expansion of capacity. This permitted the rental of larger and longer-term facilities for industry and others. In this second phase, long-term facilities were in use by pharmaceuticals, materials and semiconductor manufacturers, a global mapping company, an entertainment conglomerate, a news organization, a communications giant, a genetic engineering company, and several branches of the American Department of Defense.

(Much later it was revealed that both the Russians/Soviets and the Americans had small space stations in orbit since the late 1970s. These were primarily observational and training facilities. In addition, they were eventually used to track and eliminate the orbital debris problem that became catastrophic in 2019 when a manned launch vehicle collided with a relatively large piece of an old satellite that de-orbited into an erratic path.)

By the fifth year of platform operations, the revenue stream was wide enough to warrant plans for a second platform to be totally devoted to the servicing industry. This platform milestone was but one of the numerous venture milestones in the American plan. It was, however, only the second most profitable venture milestone. The first was the development of lunar mining facilities that dwarfed platform revenues by a factor of ten! We all know that what began as one orbital platform, grew to two, and today stands at fifteen Earth-orbiting platforms and seven lunar-orbiting platforms. The lunar platforms were built after lunar surface operations grew to a point where the exporting of local resources required an orbital facility for materials processing before shipment to Earth. As with the original low-Earth orbit (LEO) platforms, first there was one, then others could be financed. Some of the lunar-orbiting platforms currently service vehicles on their way to Mars.

Even the development of the space elevator did not minimize the usefulness of the space stations. They remained useful for commerce, tourism and defense. Goods were transferred between the elevators and the platforms.

Beyond the major venture milestones, there were numerous minor ones. An example is the development of the lunar facility used for rehabilitating accident victims. The low gravity lunar environment was found to be a benign starting point to recovery and to accustom people to advanced prosthetic devices. Patients were sent to the Moon in special canisters, under anesthesia. After three to six months of rehabilitation on the Moon, patients could be transferred back to Earth. Some chose to relocate to one of the lunar cities where the 1/6 g allowed them to have a very high quality of life.

There were more than a dozen international ventures that staked a claim to some aspect of the space enterprise. A British–Indian group, with American venture backing, entered into a partnership to send robotic miners to several asteroids and return valuable minerals. An Israeli–Korean group sent unmanned craft to numerous lunar sites to retrieve soil samples and to perform engineering tests for future facilities. They later developed teams of autonomous construction robots that formed the backbone of lunar development in the 22nd century.

Also of note in the early 21st century were three major ventures by business tycoons who earned their billions from very different businesses. These were Jeff Bezos, who founded Amazon.com, Robert Bigelow, who founded the hotel chain Budget Suites of America, and Richard Branson's Virgin Group in conjunction with Burt Rutan. Each in their unique way developed part of the space business enterprise. From Bezos' spaceport and related businesses in Texas, to Bigelow's aerospace structures venture in Las Vegas, to the Branson–Rutan spaceships also of Nevada, the businessmen of space were racing to stake claims to the greatest of Man's adventures. This era saw the development of spaceports, first the Mojave Spaceport in Southern California, and then Spaceport America for which construction began in New Mexico on 19 June 2009.

While our spacefaring humanity was born with rockets, it grew up and matured with space elevators. We all know from our history books how expensive rocket transportation systems were – thus the many decades of delay between Apollo and our lunar settlements. With the introduction of space elevators in lunar and Earth orbit, costs dropped dramatically and our ascent to the Moon, Mars, and the outer Solar System accelerated exponentially. Currently, space elevators number in the dozens and dot every major port of call for our explorers and settlers.

Of course, when families started to move to the Moon in 2084, schools were built for the young. If you thought it was impossible to keep kids in their seats in sixth grade on Earth, try to do it on the Moon under 1/6 g! Within ten years of the first families, colleges were created as were vocational schools that specialized in lunar needs. Specialties included low-gravity construction, HVAC[4] and the management of lunar sports facilities. Engineering was the most popular major in college, but eventually joint majors with various social sciences took hold.

The lunar civilization is now quite robust, but the Martian one is still at the frontier stage, with only four settlements, three mining camps, several elevators and two orbital platforms; total population: 25,000. By contrast, the Moon has 11 cities with populations of between approximately 2,000 to over 100,000. The total population is approximately 250,000 people. In addition, there are dozens of privately-held commercial bases with operations across the business spectrum. There are also a half-dozen or so scientific bases (some really cannot be called bases, since they are more like outposts) with astronomy and planetary geophysical research in full swing.

Martian civilization requires a different breed of person than the lunar settler. Even with all of our space flight capability, the trip to Mars still requires five days (from the Moon) in rather small quarters. Moreover, once on Mars, the facilities are sparse even by lunar standards. However, the opportunities abound! I hope to make the trip to Mars one day soon and experience it myself.

When the Martian venture milestones were proposed over eighty years ago, they seemed bold. In fact, the milestones were exceeded rapidly, and new milestones were quickly created. By far, the most exciting Martian milestone is the one for the year 2220: the beginning of the establishment of an Earth-like atmosphere, that is, the beginning of *terraforming*. The mix of gases will have to be somewhat different than Earth's due to the lower gravitational field, but it will be sufficiently oxygen-based to support a growing indigenous Martian civilization in the next century. Terraforming Mars is the venture milestone of the millennium.[5]

Many of the technological issues that relate to the settling of people on the Moon and Mars seemed tame by comparison to the others that needed to be addressed. The two major non-technical and non-biological issues were the definition of ownership and the development of governance. We should not be under the illusion that these are resolved. Every new foray to unsettled regions yet again brings up these concerns. There are settlements on the Moon and Mars from many countries, and many by private entities. A whole new body of laws has evolved to define the rules of engagement and governance. At the present time, the most serious such issue is the evolving desire by locals for self-determination. Some of us do not want to be ruled by Earth. These are contentious challenges that evolve with every turn of events.

To conclude this brief retrospective of the last almost two hundred years in space, it is important to remind ourselves of two important factors that led to mankind's successful evolution into space. The first is that the possibilities are bounded only by our imaginations. If we can conceive it, historically humans have shown that they can build it and do it. The speed of our progress is only a question of will. The second factor is the recognition that the evolving civilizations on the Moon and Mars are in reality separate entities from Earth. They are evolving into

[4] Heating, ventilating and air conditioning.

[5] *Terraforming: Engineering Planetary Environments*, M.J. Fogg, Society of Automotive Engineers, 1995.

unique groups of people who demand self-determination in much the same pattern witnessed over the last five hundred years on Earth.

Now in the present – 2169 – Earth has projected the species to the Moon and Mars. We are all cousins and those of us on the Moon and Mars welcome trade, communication and immigration under our free systems of government. But we anticipate and do not fear the days when the Moon and Mars are independent planets. Having anticipated such events – having learned how humans demand self determination – we have in place an economic and political system that encourages free will, enterprise and individual rights. We sincerely hope that the evolution to independence is a peaceful one.

* * *

There were numerous visionaries who thought of the manned exploration and settlement of space as crucial to humanity's survival. One was Sergei Pavlovich Korolev (1907–1966) who is widely regarded as the founder of the Soviet space program. Involved in pre-World War II studies of rocketry in the Soviet Union, Korolev, like many of his colleagues, went through Stalin's prisons and later participated in the search for rocket technology in occupied Germany. His belief in the prospects of rocket technology, and his skills in design integration, organization and strategic planning, led to his becoming the head of the first Soviet rocket development center, later known as RKK Energia. He is credited for turning rocket weapons into an instrument of space exploration and making the Soviet Union the world's first spacefaring nation.

Fig. 2.13 Sergei Korolev (12 January 1907 – 14 January 1966), the chief designer of the Soviet Space Program. (Courtesy of the New Mexico Museum of Space History)

Fig. 2.14 Wernher von Braun (23 March 1912 – 16 June 1977). (Courtesy NASA)

Another such visionary was Wernher von Braun, a German rocket physicist and astronautics engineer who became one of the leading figures in the development of rocket technology in Germany and later the United States. Wernher von Braun[6] is sometimes said to be the preeminent rocket engineer of the 20th century.

In his 20s and early 30s, von Braun was the central figure in Germany's pre-war rocket development program, responsible for the design and realization of the deadly V-2 combat rocket during World War II. After the war, he and some of his rocket team were taken to the United States as part of the then-secret Operation Paperclip. In 1955, ten years after entering the country, von Braun became a naturalized U.S. citizen.

Von Braun worked on the American intercontinental ballistic missile program before joining NASA, where he served as director of NASA's Marshall Space Flight Center and the chief architect of the *Saturn V* launch vehicle, the superbooster that propelled the *Apollo* spacecraft to the Moon. According to NASA, he was "without doubt, the greatest rocket scientist in history. His crowning achievement as head of NASA's Marshall Space Flight Center, was to lead the development of the Saturn V booster rocket that helped land the first men on the Moon in July 1969."[7] He received the 1975 National Medal of Science.

H.-H. Koelle was also a pioneer of the American space program. He was a colleague of von Braun. Here is an interview given at the beginning of the American return to the Moon.

[6]http://history.msfc.nasa.gov/vonbraun/index.html

[7]R. Williams, http://earthobservatory.nasa.gov/Features/vonBraun/

2.0.1 An historical interview with Heinz-Hermann Koelle (January 2008)

Can you give us a one or two paragraph bio?

I was born in 1925 in the former Freestate of Danzig, and was a pilot during World War II, and founder of the postwar German Society of Space Research – GfW (1948). I earned a Dipl.-Ing. (MS) Mechanical Engineering, Technical University Stuttgart (1954). I was a member of the Dr. W. von Braun team in Huntsville, Alabama, USA (1955–1965); Chief, Preliminary Design, U.S. Army Ballistic Missile Agency, member of the launch crew of Explorer I, U.S. Citizen (1961); Dr.-Ing. (Ph.D.) Technical University Berlin (1963); Director, Future Projects Office NASA/MSFC, (1961–1965), responsible for the preliminary design of the Saturn family of launch vehicles and planning of the MSFC share of the Apollo and follow-on programs; Editor-in-Chief: *Handbook of Astronautical Engineering*, McGraw-Hill (1961); Professor of Space Technology, Technical University Berlin (1965–1991); Vice-President International Astronautical Federation (1967–1969); Dean, Department of Transportation, Technical University Berlin (1989–1991).

I was elected a member of the International Academy of Astronautics (IAA) (1966), was Chairman, IAA Subcommittee on Lunar Development (1985–1997), and have written over 350 publications. I am the recipient of the Medal of the Aeroclub of France, Hermann Oberth Medal of the DGRR – the German Society for Rocket Technology and Astronautics, the Hermann Oberth Award of the American Institute of Aeronautics and Astronautics, the Eugen Saenger Medal of the DGLR – the German Society for Aeronautics and Astronautics, the Patrick Moore Medal of the British Interplanetary Society, and the Engineering Sciences Award of the International Academy of Astronautics. In 2003, I was elected Honorary Member of the International Academy of Astronautics and in 2007 given the Space Pioneer Award of the National Space Society, USA.

You were von Braun's right-hand man in the Future Projects Office at NASA's Marshall Space Flight Center (MSFC) when the Apollo project was proposed by President Kennedy. Before Kennedy's speech, what were you working on and was there any sense that the Soviets would move rapidly into space with men?

To those following the rocket development in detail it was clear for some time that the Russians were ahead of the United States because they had the bigger rockets, and had demonstrated that they could launch satellites large enough to transport people. Thus, the first flight of a man (Gagarin) did not come as a surprise to us and we were prepared to respond.

In these months, Dr. von Braun had a good working relationship with Vice President L.B. Johnson, Chairman of the Space Council, and he was his principal consultant on the options available on how to respond to the Russian space lead. Our team was ahead of the "pack" in space planning activities since the Nation discovered the "booster gap" emphasized by JFK in his campaign for the Presidency. We were the only government organization that had developed a "National Booster Plan" outlining plans for more powerful launch vehicles. We were also the group that was in charge of the Saturn I development, at that time the biggest booster coming online.

Also, we had developed a plan for a lunar base. All of this material was available in the Future Projects Office and was constantly updated and used as ammunition by Dr. von Braun in his contacts with Washington. We were in the middle of this national effort and helped to shape the decision of President Kennedy in the process to select an option challenging the Russians.

Who were the people within the Kennedy Administration that advised Kennedy to go forward with Apollo?

The decision process leading to the Apollo program was analyzed and documented by Professor John Logsdson of George Washington University. He can identify the people involved much better than I can.[8]

The President had assigned the task of analyzing the options available to counteract the Russian space accomplishment to his Vice-President, L.B. Johnson, who was a strong supporter of the Space Act of 1958, and was Chairman of the Space Council, which was comprised of representatives of all relevant departments of the Administration. The technical and scientific aspects were discussed particularly with NASA, The National Science Foundation and the Chief Science Advisor. At NASA, the Top Management (the Trinity), supported by the Office of Manned Spaceflight (OMSF) and the relevant Center Directors, were called upon to make feasible proposals. Also, the relevant committees of Congress participated in the discussions.

What were the thoughts and feelings of the group after the Kennedy speech?

Let's go with full speed ahead! Our dreams would come true if we did it right! We concentrated on the transportation system needed to get the Astronauts to the Moon. Its size and availability would determine crew size and schedule. We began immediately the preliminary design of the Saturn C-3 vehicle, which originally was our Saturn II concept proposed in the lunar base study already in 1959. We began with a booster that had 2 F-1 engines from Rocketdyne that were already under development. We soon realized that this concept was not satisfactory to transport a lunar crew to the Moon and back with the desired performance margin. We also looked at the NOVA concept of Milt Rosen[9] that would have 6 F-1 engines in the first stage. After detailed analysis of the entire mission concept and the resources required we dropped the initial two-step concept of a Saturn C-3 vehicle to be followed by a NOVA, and proposed instead the Saturn C-4 as a compromise. By adding a fifth engine in the process the performance reserve was increased.

Some say that Apollo was primarily a political program. Of course it was also an extremely challenging engineering and scientific journey. What was the greatest challenge of Apollo?

[8] *The Decision to Go to the Moon, Project Apollo and the National Interest*, J.M. Logsdon, MIT Press, 1970.

[9] http://www.astronautix.com/astros/rosilton.htm

Sports Highlights

The Philadelphia Phillies reinstate star Richie Allen, 1C.

Chicago Cubs increase their National League with two victories over the slumping Phillies, 1C. (Complete Sports, 1-5C)

Monday, July 21, 1969

TODAY

Florida's Space Age Newspaper

Published in Brevard County, Florida

Next Space Shot

Delta rocket will orbit Intelsat communications satellite Tuesday, exact time not announced.

TODAY's Weather

Considerable cloudiness with thundershowers. High 88, low 75. Easterly winds 8 to 15 mph. (Complete weather Page 2A)

10 Cents

MAN ON MOON

NEIL ARMSTRONG, ED ALDRIN ERECT AMERICAN FLAG ON SURFACE OF MOON ON 'LIVE' TELEVISION

'That's One Small Step for Man; One Giant Leap for Mankind'

By SANDERS LaMONT
TODAY Aerospace Writer

HOUSTON — Man landed and walked on the moon for the first time Sunday, July 20, 1969.

Astronaut Neil Armstrong stepped onto the surface at 10:56 p.m. As his left foot touched the surface, he said solemnly:

"That's one small step for man."

As his right foot touched the powdery surface, he added, "One giant leap for mankind."

He was joined on the surface by companion astronaut Buzz Aldrin slightly less than 20 minutes later.

"Beautiful, beautiful, beautiful. A magnificent desolation," Aldrin said.

The pair put on a spectacular live television show for millions for viewers on earth.

Even as Armstrong stepped onto the surface, he described it as "fine and powdery—I can kick it up with my toe. It is like fine charcoal."

Before he reached the bottom rung on the ladder from the spaceship he said the surface "looks like very, very fine grains as you get closer to it—almost like a powder."

The two astronauts flawlessly completed their scientific tasks and clambered back into their space ship after two hours and 17 minutes on the lunar surface.

Both had returned to the moonship "Eagle" at 1:11 a.m., more than an hour ahead of the time they were originally to have started it.

Continued Page 2A

FIRST STEP TO MOON

PRESIDENT SPEAKS TO ASTRONAUTS FROM WHITE HOUSE TELEPHONE . . . "This has to be the most historic telephone call ever made"

Apollo 11 Headlines

Astronauts aboard Lem were calm, but bosses in Houston were biting their nails. (2A)

Neil Armstrong's first step quote goes down in world's history books. (3A)

Last man to check Lem before Apollo flight gives "wow" at landing. (4A)

Astronaut Mike Collins' sister and mother watch from South Merritt Island home. (4A)

The nightclub scene: Confetti splashed and people cried as Armstrong touched the moon. (4A)

Armstrong's parents ask world to pray for astronauts' safe return. (5A)

Spokesman says Aldrin family "thrilled" by lunar landing. (5A)

Landing brought whoops of joy from homes of three astronaut families. (5A)

Veteran broadcaster Walter Cronkite is all "Oh Boy" for Lem moon landing. (6A)

Astronauts moon walk suits cost NASA $300,000. (6A)

Seven minutes of engine burn will mean a safe return Tuesday. (7A)

Apollo 12, second U.S. lunar landing mission, will almost double time for surface exploration. (7A)

Director of U.S. space program says successful moon landing opens "a new era." (7A)

Man wouldn't be on moon now if it weren't for NASA engineer who fought for Lem landing concept. (7A)

Dingy Connie Mack Stadium in Philadelphia gives Apollo 11 lusty roar of approval. (1C)

Manufacturers key ad campaigns to products astronauts use. (12C)

Everyday Features

Four Sections

Fig. 2.15 Man on Moon cover from *Florida Today*, Monday 21 July 1969. Top photo from live television. Bottom left photo shows President R.M. Nixon speaking to the astronauts: "This has to be the most historic telephone call ever made... ." Bottom right photo shows first step on the Moon. (Courtesy *Florida Today*)

The primary challenge was to find a feasible transportation concept that could be developed in the time available using efficiently the resources that could be expected, and a management structure that would work.

Who were some of the key people around you during Apollo?

The NASA Administrator Jim Webb was a true leader who did handle the political problems with the Administration and Congress. Bob Seamans, Associate NASA Administrator, was the executive manager to organize NASA and the support of the industrial contractors. Brainerd Holmes[10] was the Head of the human space program, with George Mueller being the Chief of the Apollo program. He was assisted by George Low and Joe Shea in Washington, and the Center Directors Wernher von Braun, Bob Gilruth and Kurt Debus, who comprised the Apollo Management Council.

In what ways were these people special? What were their special talents?

These people were all very experienced managers and were able to motivate all those required to accomplish the job. They were very cooperative and could get along with each other surprisingly well. They were team people and they did know that the whole enterprise could be successful only by achieving compromises acceptable to all key participants.

It must have been tremendously exciting to be in the middle of the Apollo project. Was it very dynamic?

It was not only dynamic, it was hectic! We had the DX priority nationwide and the order to beat the Russians to the Moon! The selection of the lunar orbital mission mode proved to be very useful. It allowed a clear cut interface between the launch vehicle (developed by MSFC) and the spacecraft (developed by JSC). This minimized the number of interdependencies and lots of friction. But nevertheless it was a tremendous job to coordinate the efforts of the NASA centers and thousands of contractors. For most key people the week had 60 work hours. Mistakes could not be avoided and required a lot of reprogramming and double work. Cost control was very difficult. But there was a general feeling that after the race would be over, the question would be: "Did we win?" and not "How much did it cost?" The detrimental effect was that industry was spoiled and space flight became very expensive.

How do we generate that dynamism among the population? Space activity is multi-generational and people bore easily today. How do we create that continued interest?

First of all, we must find ways and means to contain terrorism on Earth within acceptable proportions that would lead to a reduction of military expenditures on Earth. That would leave room for other initiatives. Space applications that benefit people on Earth have to be expanded and commercialized. Commercial

[10] http://www.astronautix.com/astros/holinard.htm

space tourism would help greatly to make people aware of the potentials of space. High quality public relations must make it clear to all peoples that we are jointly on a spaceship called EARTH that travels through the universe in a fairly dangerous environment. People will then realize more and more that we all will have the same fate on this journey, and that we may be able to influence it in a modest way. Thus, current and future generations will learn to recognize that space travel will enhance the survival chances of our species. This is a gradual process that will take several decades. However, there are three potential events that would accelerate greatly the development of space travel:

1. The impact of a major meteorite on Earth.
2. Discovery of biological life on other planets of our own Solar System.
3. Establishing contact with intelligent life in another Solar System.

These events are unlikely to happen in the near future. But who knows?

Why is settling the Moon so important for civilization?

If the human civilization wants to survive in the long term it has to expand in space. The resources on our home planet are limited; the secrets of life can be found on other celestial bodies. The Moon is the "Panama Canal" to open the Solar System to mankind. Without adequate manpower, equipment, facilities, supplies and energy on the lunar surface we will not be able to explore and utilize the lunar resources. This experience and development of know-how will enable man to send expeditions to other planets eventually. This is a project that will be more demanding than many people think today; it requires a careful step-by-step approach to be successful!

How do we answer critics who say space is too expensive and that there are numerous problems on Earth to take care of first?

Nearly 1,000 billion dollars are currently spent annually on defense matters. Space research and development is supported by national budgets on the order of about 50 billion dollars annually. This is certainly not enough to solve all the problems on Earth, i.e., that is not a realistic alternative. It must be recognized that development of space travel is an international endeavor, requiring less than one percent of defense expenditures. This enterprise will unite people on Earth and lead to a reduction of military strife on Earth and thus reduce defense expenditures accordingly.

What do you see as the major hurdles for our return to the Moon for permanent manned settlements?

The rate of progress in the development of lunar resources is driven by the performance and cost of the space transportation system providing the logistics support of the lunar operation. Expendable space transportation systems are too expensive and not safe enough. We have to realize that it will take fully reusable transportation systems as we have them on Earth to solve this problem on land, water and air.

Assuming that we develop the reusable transportation system, it appears that the physiological challenges on the Moon – from radiation, to regolith dust, to micrometeorites, to low gravity – are poorly understood. But what we do know is that humans and machines have severe challenges ahead of them before they can call the Moon a permanent new home. Would you agree with this assessment, and from your knowledge of these problems, are you confident that we can address these issues as we return to the Moon?

Cosmonauts and Astronauts have been on board manned satellites up to a year and have survived without suffering major injuries. The environment on the Moon is similar, but can be better controlled by human-made equipment. We simply have to learn to make the best use of lunar resources. The lava tubes available on the Moon should provide adequate protection against radiation and meteorites. Solar energy, supplemented by nuclear systems, can provide the energy required for life support and production. We will gradually increase the stay time of the lunar crew from a few months hopefully to more than a year. This is important because of the high transportation cost of people on round trips from the Earth to the Moon. Based on total cost per lunar labor-year there should be an optimum stay time for people on the Moon varying with the life of the lunar installation. This we have to find out by going there. As seen today, there seem to be no hurdles in the exploration and development of the Moon that could not be overcome.

How optimistic are you that President Bush's timeline for the return to the Moon will approximately be kept?

This depends on the future Administrations. At this time there is a 50:50 chance; it might get worse before it gets better. It depends also on the performance of the teams involved. But we must realize that there are great risks involved and must be prepared to accept them.

When you envision the future, where do you see us in 50 years? In 100 years?

If we stakeholders can agree on a feasible and affordable logistics concept, and can retain public support (that we currently have in some countries), I believe and hope that we will be able to establish a first lunar laboratory that will provide working and living space for about 30 people by the middle of this century.

It seems to me, that there is a fair chance to have a lunar population of about one hundred people at the end of this century, who are preparing the Moon for a lunar settlement that can provide services and products that cover more than fifty percent of its own resources required to sustain such an expansion into space of our civilization.

* * *

Another major contributor was Jack Crenshaw. His work was crucial to the mechanics of space flight and getting the spacecraft with the three men to the Moon and back, accurately and safely.

Fig. 2.16 Cover of NASA's Office of Manned Space Flight *Lunar Construction*, prepared by the Office of the Chief of Engineers, Department of the Army, April 1963. (Courtesy Al Smith and the U.S. Army Corps of Engineers)

2.0.2 An historical interview with Jack Crenshaw (July 2009)

Can you give us a one or two paragraph bio?

I have been working with physics, math, computers, and dynamic simulations for over 50 years. I wrote my first computer program in 1956, and my first dynamic simulation in 1959. I have developed microprocessor-based systems since 1975. My specialty is putting advanced methods of math and physics to work to solve practical, real-world problems. I am a Contributing Editor for *Embedded Systems Design* magazine, and author of the bimonthly column *Programmers Toolbox.* I wrote the book, *Math Toolkit for Real-Time Programming*, and the online tutorial series, "Let's Build a Compiler."

I hold a Ph.D. in Physics from Auburn University and currently live in Florida.

What kind of work did you do for Apollo?

Trajectory analysis and mission planning. I worked in Bill Michael's Lunar Trajectory Group at NASA's Langley Research Center. We were simulating lunar trajectories on digital computers, almost as soon as we had digital computers.

This was before I had even heard of Project Apollo. With Luna-3, the Russians had taken the first image – albeit of very poor quality – of the back side of the Moon. We hoped to top the feat by sending a Brownie-quality camera around the Moon, taking photos of the back side, and returning the film to Earth. NASA was developing Scout, a solid-fuel research rocket. Theoretically, it would have enough energy to send a small payload to the Moon. My job was to help design the trajectories.

About the same time, Lieske and Buchheim, two researchers at the RAND Corporation, wrote a paper describing free-return, circumlunar trajectories – the now-familiar figure-8 paths. It was immediately obvious that this class of trajectories was the only reasonable way to go to the Moon and back. I began studying the problem intensely, first using a two-dimensional simulation of the restricted three-body problem, and later a 3-D, exact simulation. I was determined to do more than just plot some graphs; I wanted to understand the underlying physics.

One of the things that concerned us about using the Scout was its very poor injection accuracy. I did a sensitivity study, evaluating the effects of injection errors on Earth reentry conditions. Today, we'd call that a state transition matrix, but we didn't have that term then. To my knowledge, my work on that topic was one of the earliest done, though I suspect the fellows from MIT and the West Coast were doing similar things.

The results of the study were sobering; swinging around the Moon by just the right distance to get back home is challenging, a little like a three-cushion bank shot in billiards. The sensitivity was extreme, and the accuracy of the Scout boost guidance was orders of magnitude too low. Midcourse guidance would be necessary, and we had precious little mass to work with.

However, I discovered an interesting family of trajectories that were much less sensitive to injection errors. You see, for orbits as highly elliptical as translunar orbits, the perigee is almost totally dependent on the angular momentum. For a perigee near the Earth's surface, this momentum translates to about 600 ft/s of tangential velocity at the distance of the Moon. That's how much velocity the spacecraft will have as it arrives at the Moon, and it had better be the velocity when it leaves again. It's this requirement that the angular momentum be unchanged that forces the figure-8 trajectory to be symmetrical.

But that's only if we want that grazing reentry back at the Earth. Suppose instead that we require the spacecraft to leave the vicinity of the Moon with zero angular momentum. It will fall straight down towards the Earth, and re-enter vertically – fatal for astronauts, but not for instruments.

Now the situation is much different. Even if the Moon imparts as much as 600 ft/s of unwanted velocity, we would still enter the Earth's atmosphere and land, somewhere. If you don't care where, this kind of accuracy was possible even with the crude guidance of the old Scout. If you aim to hit the Earth dead center, instead of the usual grazing incidence, you'd have to work hard to miss it entirely.

Of course, the mission never flew, so the trajectory is now nothing more than a footnote to space history. But the concept might still have some legs for low-cost sample return missions.

Generating a lunar trajectory is not easy. The problem is a two-point boundary-value problem, complicated by the fact that both end points (the launch and return sites) are fixed on a rotating Earth, and we have the "minor" midpoint constraint that the trajectory come somewhere near the Moon. The transition from the restricted three-body model to an exact N-body model was pretty traumatic. The restricted three-body model is a two-dimensional one, so all the bodies move in the same plane. When your spacecraft arrives at the Moon, it's either ahead of the Moon, behind it, or just right, and a small change in lead angle will make it just right.

In the real world, and therefore in an exact N-body simulation, things are not nearly so simple. In it, the Earth's equator, the Moon's orbit, and the transfer trajectory all lie in different planes. Meanwhile the launch site and splashdown site are describing their own little orbits as the Earth rotates. It takes a good understanding of the geometry to get anything useful.

To tame the problem, we had to invent quite a few approximations, rules of thumb, and design aids like the patched-conic method and the angular momentum rule. To help visualize the geometry, Bill Michael had a design aid built that included a small globe and a celestial sphere, with a couple of extra plastic disks that he could move to represent the Moon's orbit plane and trajectory plane. Without such aids, generating N-body trajectories was a slow and painful – and expensive – process. With them, you could almost design the orbit on the back of the proverbial envelope.

In 1961 Bill and I presented a paper ("Trajectory Considerations for Circumlunar Missions," January 1961, IAS Paper #61-35) at the IAS (later AIAA) annual meeting. It was the second paper published on circumlunar trajectories.

Also in 1961, I moved to General Electric in Philadelphia to work out the Apollo study contract and proposal effort. I had a small team of about four people, and was responsible for generating all the nominal trajectories for those efforts. I did similar stuff for other lunar missions. The names of projects Surveyor and Prospector come to mind.

GE's N-body program was even more painful to use than NASA's. Built for interplanetary trajectory studies, it only accepted initial conditions in inertial, Sun-centered coordinates, with units of AU and AU/hr. Before we could even begin to use it for lunar work, I had to write a front end that accepted initial conditions relative to a rotating Earth. Later on, I added a computerized version of Bill Michael's celestial sphere, and even later, a wrapper that would iteratively refine the orbit.

There were a lot of groups using the nominal trajectories that we generated, including groups studying heat loads, lighting conditions, radiation dosage, etc.

A large part of my energies went into building quite a number of patched-conic approximations, programs to solve the complicated spherical trig, front-end and back-end processors, and "wrappers" for GE's N-body program.

My crowning achievement was a fully automated program that required only the barest minimum of inputs, such as the coordinates of the launch and landing sites, year of departure, lunar miss distance, and lighting conditions at both the Moon and Earth. The program would then seek out the optimal, minimum-energy trajectory that would meet the constraints.

There was a lot of inter-group rivalry at GE. One day a systems engineer casually asked me to study the problem of returning from the Moon to the Earth. As it turned out, the whole thing was a setup. This engineer wanted his own trajectory group, and hoped to get it by claiming that my group was not responsive. In addition to my group, he went to an outside vendor (STL, later TRW) and contracted with them to work the same problem. He was hoping we'd fail, so he could say, "Look at this: Crenshaw's group is so unresponsive I had to spend money with STL to get the results I needed."

Fortunately, thanks to my work at NASA, I knew exactly how to return from the Moon: Give the spacecraft a tangential velocity of about 600 f/s, relative to the Moon. From that desired end, it's fairly easy to figure out what sort of launch one needs at the Moon's surface. We generated quite a number of trajectories, both approximate and exact, that affected the Moon-to-Earth transfer. As far as I know, they were the first such studies ever done.

We completed the study and got results before STL did, and the sneaky systems engineer was soon sending out his résumé.

Later, I met a fellow from STL who had done similar work and I told him this story. He said, "Yep, I'm the one that got assigned to work the problem." He said he wasn't able to get the results on schedule, but that contract led to a nice career for him. That's great.

During this same period, I heard about a method due to Herrick for computing nearly parabolic orbits. The classical two-body theory upon which all approximate methods are based has three distinct solutions, depending on whether the orbit is elliptic, parabolic, or hyperbolic. When the orbit is not parabolic, but nearly so, the equations for the elliptic and hyperbolic cases all go singular. Unfortunately, that's exactly the kind of orbit involved in translunar missions. Herrick had developed what he called the "Unified Two-Body Theory." His equations worked just fine for the near-parabolic case. In fact, the closer to parabolic, the better they worked. Herrick's original formulation was not very useful because his functions involved power series depending on two arguments, both of which had units. But from Herrick's work it was a rather short step to new functions of only a single, non-dimensional argument. Between 1961 and 1963, I published quite a number of papers on these functions, most of them internal to GE and widely distributed within NASA. I converted every one of my approximate methods, N-body front ends, etc. to these functions. I wrote them up in an AIAA paper, "Unified Two-Body Functions" (*J. Spacecraft and Rockets* 3:6, June 1966, p. 955). Unfortunately for my ego, I never got much credit for the functions. For a time they were called the Lemon-Battin functions. Nowadays, the method is known as the method of Universal Variables.

In 1963 I worked at GE's Daytona Beach facility, studying the problem of how to abort from the lunar mission. At first glance, you might think nothing much can be done; once zooming towards the Moon, you're pretty much committed to continue. However, out near the Moon, the velocity relative to the Earth is rather low, and you have tons of available fuel for maneuvering, thanks to the need to enter and exit lunar orbit. Therefore, it's theoretically possible during some portions of the mission to simply turn around and head back for Earth. In other phases, you can lower the trip time quite a bit, by accelerating either towards the Moon (to hasten the swingby) or towards the Earth. In my studies, I discovered yet another class of returns that worked for aborts near the Earth. In these, you make no attempt to slow down, but simply deflect the velocity downwards enough to graze the Earth's atmosphere. Aerodynamic braking does the rest.

All these abort modes ended up designed into the missions and programmed into the Apollo flight computers. Two of them – accelerating towards the Moon on the way out, and towards the Earth on the way back – were used in Apollo 13 to reduce the life-critical flight time. If you watch the movie of the same name, you'll hear discussions of Fast Return and Free Return trajectories, both of which I had a hand in designing.

Some say that Apollo was primarily a political program. Of course it was also an extremely challenging engineering and scientific journey. What was the greatest challenge of Apollo?

Just getting there at all. When JFK made his famous speech urging us to go to the Moon and back, we all cheered and said, "Great, let's do it!" Then we looked at each other and said, "But how do we do it?" The technology simply wasn't there. You have to remember, in 1959 even the digital computer was a new thing. Fortran hadn't become widely used. Most of our early results were gotten using desk calculators and slide rules. Even the Kalman filter was new.

For knowledge of celestial mechanics, we relied heavily on what astronomers used to study comets and planets. They'd never even considered such problems as rendezvous and docking, of midcourse guidance, or of the kind of targeting that we did while designing the circumlunar trajectories. We were plowing new ground in every area, and we could only hope that our knowledge would increase fast enough to get the job done.

Just consider the problem of storing cryogenic propellants. While at NASA, I did a study of how to build insulated tanks for storing the cryogenics. My conclusion was, it couldn't be done. Even our best insulators weren't nearly good enough.

By the time Apollo actually flew, the cryo tanks in the lunar module (LM) were so good, they could sit at room temperature for weeks on end, without losing much of the lunar oxygen (LOX).

Or consider the problem of making spacesuits that could work in 200 degree heat or 40 Kelvin cold.

In so many areas like this, we were just inventing whole new technologies as we went along. It's a tribute to everyone's hard work that we pulled it off.

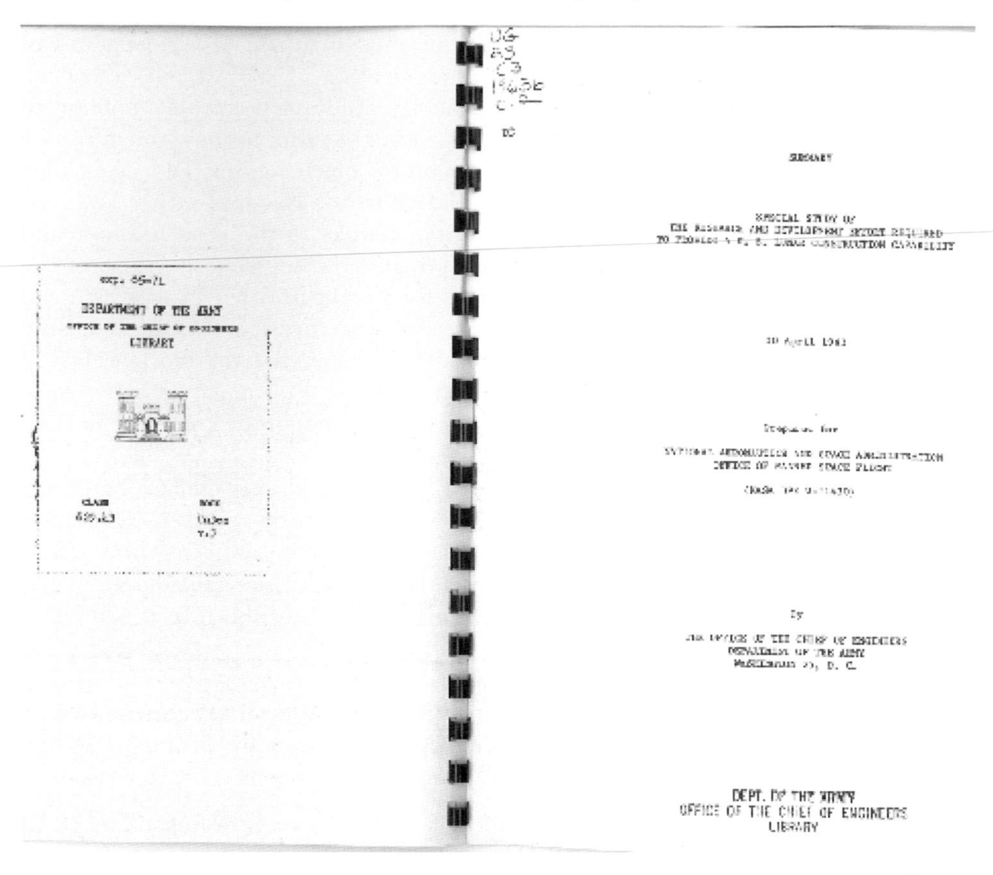

Fig. 2.17 Inside cover of Army Report, prepared for NASA, dated 30 April 1962. (Courtesy Al Smith and the U.S. Army Corps of Engineers)

In what ways were the people at NASA during Apollo special?

Stated simply, the engineers I worked with at NASA Langley were the best engineers I've ever worked with, bar none. Most of them had been there when the agency was the National Advisory Committee for Aeronautics (NACA), and many had been there during the war years (WWII, that is). Their work ethic was second to none. They didn't put on a lot of airs. They weren't pretentious or concerned with image. They didn't strut or pose, they didn't fight turf wars. They just rolled up their sleeves and got the job done. I sometimes wonder if NASA people can do that anymore.

It must have been tremendously exciting to be in the middle of the Apollo project. Was it very dynamic?

You bet. After being involved with Apollo, what do you do for an encore? I remember saying one day, "I can't believe they're paying me for this." I would have done it for nothing. If I'd had the money, I'd have paid them to let me do it.

Even if you set aside the thrill of helping us go into space, the work was still exciting. Everything we did was on the leading edge. Computers were brand new. So was Fortran. We were inventing the algorithms and developing the tools as we went along. Those are once-in-a-lifetime kinds of events.

In the 1960s where did you think we would be in 2010?

Certainly in permanent lunar colonies. We just assumed that was a given. We were all set to go to Mars too, and set up permanent bases there. The biggest disappointment of my career came when I realized that, having gone to the Moon – having realized the greatest achievement in the history of mankind – we had lost the drive to go back.

I do think that the hiatus was necessary, to some degree. We just weren't well enough equipped, either with the mathematics or the hardware, to sustain ourselves as a spacefaring nation. But we have all the tools we need now, but seem to have lost interest. Instead we chose to putter around in low Earth orbit, looking down instead of up, and watching spiders spin zero g webs. That's a pity.

Is settling the Moon important for civilization?

I certainly think so. I could offer the usual platitudes like "Because it's there," or "exploration is in our DNA." Both statements are true, but not likely to justify billions of dollars of funding. But two other issues trump every argument against it.

First, in the 1960s we thought the Earth was a pretty benign and stable place. I'm not sure why we thought that, in view of our experience with things like Krakatoa, Mt. Vesuvius, the San Francisco quake, etc. But we did.

Today we know that the universe can be a dangerous place. We know about black holes, about gamma-ray bursters, about the Chicxulub event that killed the dinosaurs, the Siberian traps event that caused an even more profound extinction, and the supervolcano under Yosemite. We saw what the Shoemaker-Levy fragments did to Jupiter, and we've learned more about the Earth-crossing asteroids that seem to turn up ever more frequently. Today we realize that the universe is a lot more violent place than we thought, and the stability the human race has seen during recorded history may be the exception, rather than the norm.

Today it would seem that having a foothold some place else in the Solar System might be important to the survival, not only of the human race, but all life.

Far more importantly, though, is what seems to be an innate human need for lofty goals. During the 20th century, we fought war after war. As horrible as those struggles were, they tended to bring out the best in us. It would be nice if the human race rose to the occasion when times were good, when peace reigned, and when we faced no crises. Ideally, we'd be able to spend our affluence and our leisure time to make discoveries and improve civilization.

Sadly, that doesn't seem to be the way it works, especially here in the USA. We don't seem to handle prosperity very well. Instead of using the leisure time to think deep thoughts and learn great new things, we seem to fall back into our baser natures, and fritter away the opportunities in self-indulgence and debauchery. The

Gay 90s, the Roaring 20s, and pre-Nazi Germany come to mind. On the other hand, when things are critical – during wars, the Great Depression, or natural disasters – our best traits seem to shine through. It's no accident that the veterans of WWII are called the Greatest Generation.

After the 9/11 attacks, we Americans banded together to help each other, and fight off a common and dangerous new enemy. For a time, we acted like the Americans of WWII.

That lasted about a week. Now we're back into watching "American Idol" and "Desperate Housewives," and getting high.

We seem to need lofty goals to stay vibrant and dynamic. Otherwise, we sink into what Teddy Roosevelt called "That gray twilight that knows neither victory nor defeat." For most of human history, the challenges have been about wars, plagues, and the like. But now, in this Nuclear Age, we need what William James called a "moral equivalent of war." Space can be such a goal. I've always been convinced that this is what JFK had in mind when he gave his challenge in the first place. With two nuclear powers glaring down our gunsights at each other, a great non-violent goal was a good thing.

How do we answer critics who say space is too expensive and that there are numerous problems on Earth to take care of first?

They said the same thing during the Apollo years. They said, "Why are we spending billions to bring back some stupid rocks from the Moon, when we could use them to end poverty and cure cancer."

Well, we did it their way. For all practical purposes, we ended the manned space program. We declared a War on Poverty and a War on Drugs. Have we ended poverty? Have we stopped using drugs? Have we cured cancer? Instead of achieving great things, we've used our billions to build bridges to nowhere, build ever-bigger SUVs, and pay CEOs' hundred-million-dollar bonuses. How is that better?

Fact is, if you wait around until you have no bills to pay or problems to solve, you will be frozen into immobility. Columbus discovered America during the height of the Spanish Inquisition. During the Apollo effort, this country was fighting a war in Vietnam and dealing with hippies, anti-war activists, the Symbionese Liberation Army, and the Weather Underground. If we had waited to solve such problems, we would still be waiting, and the problems would only have gotten worse.

What do you see as the major hurdles for our return to the Moon for permanent manned settlements? Is it technical, biological or political?

Technically, there are no major hurdles at all. We know how to do it now, and we have the hardware, the software, and the technologies to do it. Given the resolve, we could do it almost overnight. The problem lies in our seeming impotence in being able to muster up resolve for anything worthwhile.

How optimistic are you that President Bush's timeline for the return to the Moon will approximately be kept? The Obama Administration has just appointed an Administrator. Do you think they will support a manned return to the Moon?

Not very. When I first heard Bush declare his goals, I was naturally thrilled, but my excitement was tempered by concern that we'd be able to muster up, in these contentious times, the resolve to follow through. Now that Obama is in power, I worry not only about that resolve, but also about the danger that Democrats will scuttle the program simply because it was begun by a Republican.

I've been very encouraged by the rapid response by NASA and others, to help make the dream or reality. The outcome remains to be seen. Will we rise to the occasion again, or continue to argue over ridiculous political divisions.

When you envision the future, where do you see us in 50 years? In 100 years?

That's a tough question. It's easy enough to speculate about how things will be different, but I don't think we are capable of foreseeing the really big changes. 100 years ago was 1909. We had a very different culture, but not that different for the average citizen. We already had primitive versions of railroad trains, cars, airplanes, electricity and radio. Maxwell's equations were long since written. We even already had Jules Verne's concept of a lunar mission launched from Cape Canaveral. One could argue, however unconvincingly, that the technologies of today are mere detail improvements over those of 1909.

What we can't predict, and never could, are the astonishing breakthroughs that no one saw coming. Who, for example, could have predicted a CAT scan or an MRI?

Having said that, I'll offer my best guesses, knowing that they could be very wrong.

I think we can definitely assume permanent colonies on both the Moon and Mars. Perhaps even other places, like Titan or Europa. We'll have ion engines and maglev accelerators – maybe even Clarke's space elevator – making space travel less expensive. We'll have sophisticated robots with very high levels of artificial intelligence. Perhaps they'll be the explorers of Europa.

On the Moon, I think we'll find that sought-after water, which will make permanent colonies not only possible, but sustainable. Instead of rockets to land on the Moon and take off again, we'll have automated maglev accelerators. I think we'll not only be mining on the Moon, but we'll be digging mines to astonishing depths. We'll have rovers that can go for hundreds of miles, perhaps with robot drivers.

It remains to be seen if we ever find anything on the Moon valuable enough to ship it back to Earth. I've heard that Helium-3 may be the thing. Who knows?

Assuming that we don't still have Al Gore and Nancy Pelosi around to muck things up, I think it's close to certain that we'll have all-nuclear power plants (plus geothermal, etc.), and probably fusion reactors as well. The use of oil and gas for power will go the way of whale oil.

That changes everything, both here on Earth and in space. Fusion power gives you enough energy to level mountains and fill in valleys, should you choose to do so (here's hoping we don't). Given virtually limitless power, colonies on the Moon and Mars would have enough power to do pretty much anything they want, up to and including building the Tycho Hilton, New Las Vegas, and the Olympus Hyatt.

Even a Carnival Europa cruise ship, keeping warm while breeding more hydrogen from the unlimited sea.

On a more practical scale, plenty of power means large scale mining, smelting, chemical plants, factories, farms, everything you need to run a civilization.

But I also think that the Moon dwellers (lunians? lunatics?) won't care. As long as they can develop a maintainable presence, growing their own food (our plants love the lunar soil), making their own power, mining their own minerals, etc., they can thrive without us. And, of course, the Moon makes a not-bad waystation to the planets and beyond.

I expect that the habitats on the Moon, and perhaps Mars as well, will be underground, safe from cosmic rays and solar flares.

Permanent colonies on Mars are even more important than on the Moon, because you can't just stay for a week then come home again. The laws of physics require stay times measured in months, so the Martians will need to stay safe and comfortable there for long periods, perhaps forever.

I definitely think we're going to find water on Mars – heck, we've already found it. Probably primitive life, also. It's hard to imagine the impact that little discovery will have. Can you imagine? It's easy enough to speculate about life elsewhere in the universe, but actually finding it changes everything.

Will we have colonies on Titan? Europa? Could be. If not colonies, at least we will have explored them heavily, using sophisticated robotic probes.

What we can never predict are the big changes, the technological breakthroughs that no one saw coming. In 1909, who would have predicted the H-bomb? GPS and YouTube?

Scientifically, I think society is poised for breakthroughs, as long as we don't just slip back into that gray twilight. We seem to be on one of those cusps of discovery, much like the changes wrought by Einstein, Hubble, and the giants of quantum mechanics. Physicists nowadays are talking about string theory, extra dimensions, parallel universes, wormholes, and all manner of other bizarre concepts. If even one of them comes to fruition in practical usage, all bets are off.

The smart money says that things don't change much. As in 1909, people will still build houses, grow food, drive around, make love and war, and procreate. On the other hand, huge changes like antigravity, warp drive, and time travel aren't out of the question, either.

If we find out how to go back in time or zap over to Andromeda, I'm ready to go. How about you?

* * *

Such visionaries helped fuel the drive to a spacefaring society. Not only did they provide the goals, but they participated in a material way to make it happen. They were the foundation upon which society built its hopes and upon which the leaders of society drew their inspiration and courage to commit us to those goals.

Besides individuals there were groups that also created a framework from which we could map our way to the Moon and then Mars. Such groups offered anyone with an interest a way to become involved. Rocket science was not a prerequisite.

Fig. 2.18 Astronaut Eugene A. Cernan, commander, makes a short checkout of the Lunar Roving Vehicle (LRV) during the early part of the first Apollo 17 Extravehicular Activity (EVA-1) at the Taurus-Littrow landing site. This view of the "stripped down" LRV is prior to loading up. Equipment later loaded onto the LRV included the ground-controlled television assembly, the lunar communications relay unit, hi-gain antenna, low-gain antenna, aft tool pallet, lunar tools and scientific gear. This photograph was taken by scientist-astronaut Harrison H. Schmitt, lunar module pilot. The mountain in the right background is the east end of South Massif. While astronauts Cernan and Schmitt descended in the Lunar Module (LM) *Challenger* to explore the Moon, astronaut Ronald E. Evans, command module pilot, remained with the Command and Service Modules (CSM) *America* in lunar-orbit. (AS17-147-22526, 11 December 1972. Courtesy NASA)

The doors were open to all who wanted to learn and provide their proverbial two-cents. Meetings were held – if we had one launch of the Shuttle for every space meeting between 1990 and 2010 we would have had remarkable productivity in space. Sometimes the meetings were inspirational; often they were preaching to the converted. But they had their uses. They helped filter out ideas that would not work and thereby percolate useful concepts to the top. Such meetings were democratic – all could attend and participate, so the exchanges were useful, and sometimes colorful.

One organization that provided a venue for education and discourse was the National Space Society, which continues to this day to support a spacefaring society in many ways.

2.1 The National Space Society

The National Space Society (NSS) is an independent, educational, grassroots, non-profit organization dedicated to the creation of a spacefaring civilization. Founded as the National Space Institute (1974) and L5 Society (1975), which merged to form NSS in 1987, the NSS is widely acknowledged as the preeminent citizen's voice on space. Since the early days of space it has evolved into an organization that dedicates itself to furthering the goals of those who have inhabited the Moon, Mars, as well as the outposts on the asteroids and moons of the Solar System.

The National Space Society has a philosophy[11] that is reproduced next:

I. NSS Vision

The Vision of NSS is people living and working in thriving communities beyond the Earth, and the use of the vast resources of space for the dramatic betterment of humanity.

II. NSS Mission

The Mission of NSS is to promote social, economic, technological, and political change in order to expand civilization beyond Earth, to settle space and to use the resulting resources to build a hopeful and prosperous future for humanity.

Accordingly, we support steps toward this goal, including human spaceflight, commercial space development, space exploration, space applications, space resource utilization, robotic precursors, defense against asteroids, relevant science, and space settlement oriented education.

III. NSS Rationale

Survival — Of Human Species and Earth's Biosphere
It is the nature of every form of life, whether animal or plant, to strive to survive.

> Survival of the Human Species: The human species is encountering increased natural, man-made, and extraterrestrial threats, including disease, resource depletion, pollution, urban violence, terrorism, nuclear war, asteroids, and comets.
>
> Survival of Earth's Biosphere: Many forms of animal and plant life on Earth are suffering increased loss of population and quality habitat because of the growing presence of humans on planet Earth, via expansion, pollution, deforestation, fishing, farming, mining, and promotion of certain species of animals and plants.

[11] Reproduced with permission of the National Space Society.

Space technology provides both means to monitor threats to life on Earth and ways to help curtail them. Space industrialization and settlement provide safety valves to relieve the pressures that cause Earth-bound threats. They also provide escape routes in case of catastrophic man-made or extraterrestrial threats. Humanity has inherited the stewardship of the planet Earth. It will therefore need the vast resources of outer space to reverse the damage it has caused to the Earth's biosphere, and ultimately enhance all life on Earth.

Growth — Unlimited Room for Expansion
It is the nature of every form of life, whether animal or plant, to grow and multiply.

> New Habitats for Life: The human species, as well as all other animal and plant life on Earth, need room to grow and multiply. Earth has a finite supply of land, air, and water, for which humans, animals, and plants must compete. Of all Earth species, only humans have or can acquire and utilize the knowledge to create new habitats on other worlds or in space from the raw materials of moons and asteroids.

> New Frontiers for Humanity: To provide the human species with a new "frontier" for exploration and adventure, and to thought and expression, culture and art, and modes of government. The opening of "the New World" to western civilization brought about an unprecedented 500-year period of growth and experimentation in science, technology, literature, music, art, recreation, and government (including the development and gradual acceptance of democracy). The presence of a frontier led to the development of the "open society" founded on the principles of individual rights and freedoms. Many of these rights and freedoms are being placed under increasingly stringent limitations as human population grows and humanity moves towards a "closed society," where eventually everyone eats the same, speaks the same, and dresses the same. "Cultures that do not explore, die!"

Prosperity — Unlimited Resources
It is the nature of the human species to strive to improve the quality of its many lives and to provide a better future for its children.

> Improved Standards of Living: To provide humanity with the resources it needs to improve the quality of life for all humans on the planet Earth. The majority of humanity lives at an economic level that is far below that of the Western democracies. Outer space holds virtually limitless amounts of energy and raw materials, which can be harvested for use both on Earth and in space. Quality of life can be improved directly by utilization of these resources and also indirectly by moving hazardous and polluting industries and/or their waste products off planet Earth.

> Economic Opportunity: To provide every human individual with the opportunity to improve the well being of himself or herself, and his or her family. Vast new resources must be developed if all persons are to be given economic opportunities for themselves and their children even marginally equal to what many would consider a minimally tolerable standard of living.

Technological Development: To provide remote locations for the development, testing, and "perfection" of promising, but potentially hazardous technologies, such as biological experimentation; nuclear, fusion, chemical and antimatter power generation; and space propulsion. Such developmental facilities could be placed either in space or on other worlds far from both space settlements and unrelated facilities.

Curiosity — The Quest for Knowledge
It is the nature of the human species to learn more about its origins, its past, its fellow life forms, its environment, its limitations, and its possibilities for the future. Earth is but a tiny container of knowledge compared to the entire incredibly vast universe. "We are part of the universe, through our eyes, ears and minds, the universe may know itself."

IV. NSS Principles

These are the guiding principles of the NSS by which we will conduct our Mission in pursuit of our Vision.

Human Rights
NSS shall promote the principle of fundamental rights of every human being.

Ethics
NSS shall observe, practice, and promote ethical conduct.

Pragmatism
Within the bounds of these Principles, NSS shall promote and support any and all methods and practices that support achievement of our Vision.

V. NSS Beliefs

While we cannot say that the following are absolutely essential for space settlement we believe and support the following:

Individual Rights
NSS believes that space development and settlement will occur most efficiently, and humanity's prosperity will be best ensured, if every human being is given the freedom of thought and action.

Unrestricted Access to Space
NSS believes that space development and settlement will occur most efficiently, and humanity's survival and growth will be best ensured, if every human being is allowed the opportunity to travel, live, and/or work in outer space.

Personal Property Rights
NSS believes that space development and settlement will occur most efficiently, and humanity's survival and growth will be best ensured, if every human being is allowed the opportunity to own property in space and/or on other worlds.

Free Market Economics
NSS believes that space development and settlement will occur most efficiently, and humanity's prosperity will be best ensured, if the "free market" drivers of competition and profit are used.

Government Funding of High Risk R&D
NSS believes that space development and settlement will occur most efficiently, and humanity's prosperity will be best ensured, if national governments fund the research and development of space technologies deemed too "high risk" by their industries.

International Cooperation
NSS believes that space development and settlement will occur most efficiently, and humanity's survival and prosperity will be best ensured, if nations cooperate on space research and development, and leave competition to individual companies.

Democratic Values
NSS believes that humanity's growth and prosperity will be best ensured if the fundamentals of democracy are applied to and incorporated by space settlements.

Enhancement of Earth's Ecology
NSS believes that one of the goals and benefits of space development and settlement is to restore and enhance the biosphere of the planet Earth.

Protection of New Environments
NSS believes that space development and settlement should be pursued in a manner that safeguards alien life forms, natural wonders, and historical monuments.

* * *

Principles similar to those of the National Space Society were promoted by the American Institute of Aeronautics and Astronautics. These two leading space advocacy and professional societies, respectively, had a broad base of support to promote the efforts of a spacefaring nation.

2.2 The American Institute of Aeronautics and Astronautics

This professional organization for engineers, known as the AIAA, whose work is related to air and space, had technical committees in the early 21st century that focused on the issues related to the human return to the Moon. One of these committees, the AIAA Space Colonization Technical Committee (SCTC), created a document that outlined the AIAA view on the tasks that needed to be undertaken in order to implement the Bush vision.[12] It is presented below.

[12] "Robust Implementation of Lunar Settlements with Commercial and International Enterprise [Moon Base 2015]," An AIAA/SCTC Position Statement Prepared by the AIAA Space Colonization Technical Committee (SCTC) 9 January 2007, American Institute of Aeronautics and Astronautics. Courtesy AIAA.

Purpose

To implement the President's Vision for Space Exploration, the United States must commit to the early establishment of a Moon base by 2015. This Lunar base will be an ideal testbed for opening new frontiers to human exploration by maximally employing commercial and private products and services. The AIAA/SCTC recommends that specific research, development, test and evaluation (RDT&E) goals be implemented. This will be accomplished by establishment of the scientific and industrial capabilities of a permanent Lunar settlement and development of the commercial revenue sources on the Moon.

Introduction

On January 14, 2004, President Bush reaffirmed the United States' commitment to human space exploration by completing the International Space Station by 2010, returning to the Moon to stay, and then proceeding to more distant destinations. There exists an extensive database of research and technology development indicating that the logistical feasibility of space exploration will be greatly enhanced through space resource utilization (SRU). This AIAA/SCTC Position Statement emphasizes: human settlement of the Moon; development of Lunar observatories, energy and resources uses; and sustained, active encouragement of private and international enterprise.

Recommendations

The AIAA/SCTC recommends that the following actions be implemented by the Administration, U.S. Congress, and supporting Government agencies as appropriate:

The Early Period [The Present–2015]

- Implement a suite of orbital and lander precursor missions to the Moon to collect high- resolution data on: the Lunar environment; water, hydrogen and other resources to establish ground truth on resource distribution; and surface property characterization. The lander missions should demonstrate technologies and methods for establishing Lunar and space settlements.
- Establish and implement a strategic plan for use of space resources with substantial funding for SRU payloads, launch vehicles, robotic vehicles, landers for dedicated Lunar SRU missions, and Lunar surface testbeds.
- Develop an in-situ, self-sustaining infrastructure of solar energy production and storage derived mostly from Lunar materials, and wireless power distribution (power beaming) for both nuclear and solar energy transmission on the Moon.
- Deploy communications and navigation satellite system capability for cis- and trans-Lunar space to support Lunar development.
- Establish cost-effective crew and cargo space transportation systems with the capability to utilize Lunar-supplied propellants.

- Implement testbeds on the Earth, ISS and the Moon for human health issues related to long-term space flight, including tele-medicine, low-g environments, radiation and psychological issues.
- Implement testbeds for closed life support systems for sustaining a Lunar base.
- Implement testbeds for in-situ manufacturing systems for sustaining a Lunar base.
- Develop a human Lunar south pole base where valuable water and hydrogen, material, mineral resources can be explored/used and continuous sunlight can be utilized in a less extreme thermal environment.
- Develop technologies for Lunar surface and subsurface mining and excavation.
- Develop needed Lunar-specific resource production equipment to foster expansion of Lunar/terrestrial commerce.
- Initiate a Deuterium/Helium-3 fusion reactor development program to resolve the usefulness of Helium-3 from the Moon.
- Extend current advanced technology programs including electromagnetic, momentum transfer and other fuel-less launch technologies to establish capabilities that can be applied on the Moon.

Mid-Period [2015–2025]

- Develop solar, nuclear and other advanced energy systems to support Lunar base and orbital power needs.
- Deploy and operate a group (condominium) of observatory facilities on the Moon for observations of the Earth, the Sun, the Solar System and the Universe – providing a stable, nearly limitless aperture across the electromagnetic spectrum.
- Substantially expand the Lunar resource and space transportation infrastructures.
- Develop pressurized, crewed rovers and flight hopper technologies for Lunar operations.
- Establish large-scale manufacturing of Lunar base hardware elements from in-situ resources to enable among other things, construction capabilities for habitats, domes and mining machinery.
- Expand the production of Lunar water, hydrogen and oxygen to support transportation and life support.
- Construct planetary testbeds on the Moon, as needed, in preparation for future space exploration.
- Develop autonomous robotic mining and excavation technologies for the Moon.
- Develop Lunar construction technologies including landing-launch facilities, habitats, dome construction, building wall materials, roads, radiation shielding, free-form fabrication using regolith, Lunar concrete and inflatable structures.

Far-Period [2025–2050]

- Establish the first self-sustained, permanent Lunar settlement of approximately 1000 humans.

Further Recommendations

It is also recommended that the United States work with international partners to set precedent(s) through a constructive interpretation and evolution of applicable space law(s), including provisions for:

- Free-market rules and approaches to the exploration and development of space.
- Extension of international conventions on property and mineral rights to include assets in space based on U.S. and other historical precedents in the history of exploration.
- Extension of land management conventions and régimes to include provisions for homesteading.

Government agencies need to enable business development in the following areas to ensure sustainable viability of the exploration vision:

- Base and life support.
- Resource processing and manufacturing.
- Lunar communication systems.
- Lunar navigation systems.
- Lunar transportation systems.
- Space rescue capability similar to the Coast Guard.
- Methods for indemnifying business ventures from lawsuits based on fatalities or injuries.
- Government sponsored anchor-tenant production contracts.
- Government loan guarantees.

Conclusion

The AIAA/SCTC supports a strengthened space program through robust implementation of Lunar settlements with commercial and international enterprises as outlined in this position statement. The AIAA/SCTC encourages technology and business development to establish a Moon base by 2015 which will result in permanent Lunar settlement.

* * *

So we see that the seeds were planted over many decades in the 1900s and early 2000s in the hopes that nations would continue to marshal their resources and talents to enable humanity to become the spacefaring civilization it has become.

2.3 Quotes

- "Basic research is what I am doing when I don't know what I am doing." Wernher von Braun
- "To infinity and beyond." Buzz Lightyear

- "Engineering is not merely knowing and being knowledgeable, like a walking encyclopedia; engineering is not merely analysis; engineering is not merely the possession of the capacity to get elegant solutions to non-existent engineering problems; engineering is practicing the art of the organized forcing of technological change. ... Engineers operate at the interface between science and society." Gordon Brown
- "Circumstance put me in that particular role. That wasn't planned by anyone." Neil Armstrong
- "We can lick gravity, but sometimes the paperwork is overwhelming." Wernher von Braun
- "The greatest mistake you can make in life is to be continually fearing you will make one." Elbert Hubbard
- "I've been waiting all my life for this day!" (upon the launch of Sputnik) Sergei Korolev

3 Bootstrapping a lunar civilization

> *"It's like trying to describe what you feel when you're standing on the rim of the Grand Canyon or remembering your first love or the birth of your child. You have to be there to really know what it's like."*
>
> Harrison Schmitt

In this chapter we introduce and discuss the two financial aspects of the creation of a lunar settlement. The first is: *How do we fund the return to the Moon?* And the second aspect is: *How do we make it self-sustaining once we are there?*

One of the great challenges that faced our space pioneers was how to finance the settlement of the Moon. Initially, only rockets were available to carry mass into orbit from the Earth to the Moon. Near the turn of the 20th century, historical documents[1] tell us that Atlas and Delta rockets accounted for about 90% of U.S. commercial launches, each bringing about 8,000–10,000 lb into geosynchronous orbit. An Atlas launch cost about \$80 million and the Delta about \$60 million. The cost was therefore in the \$6,000–\$10,000/lb range. Even though costs dropped considerably in the subsequent decades, it was still very, very expensive to speed up the return to the Moon. It was generally believed that costs had to drop below \$1,000/lb for lunar development to become financially feasible.

The average cost to launch a Space Shuttle was about \$450 million per mission, although if we factor in development and related costs that number could be doubled. At liftoff, an orbiter and external tank carry over 800,000 gallons of the principal liquid propellants: hydrogen, oxygen, hydrazine, monomethylhydrazine, and nitrogen tetroxide. The total weight is about 1,600,000 lb. But with President Bush's speech of 2004, it was decided that return to the Moon we must, and if it took longer than anticipated due to cost and political roadblocks, then so be it.

The first decade of the 21st century saw continued growth in space-related commerce. Reports on global space commerce released in 2008 stated that space-derived revenue in 2007 exceeded \$250 billion, an 11% growth from the previous year. Of this total, commercial satellite products and services accounted for about 55%, an increase of 20% over the previous year. The U.S. government spent about 25% of the total global expenditures. Other spacefaring governments spent 6% of

---

[1] B. Iannotta, "Rockets Take Aim at Booming Market," *Aerospace America*, February 1998, 35–41.

that total. Commercial infrastructure expenditures were about 15% of the total. All these categories saw significant growth.

Commercial satellites were an outgrowth of initial government expenditures and R&D. Many proponents of a continued and growing effort in space pointed to the commercial satellite industry, and all of its derivative industries, as a model for success. Once the infrastructure and knowledge base was developed, private industry commercialized the technologies and created what was then the only profitable space business.

3.0.1 Why go to the Moon?

At the onset of humanity's return to the Moon, what were the views on how lunar settlements would evolve and what would be their economic lives? A 1992 study known as the Exploration Task Force Study[2] concluded the following about President Bush Sr.'s Space Exploration Initiative:

> "NASA's research in strong, light-weight metal alloys and plastics will support the efforts of such industries as car manufacturing, residential and commercial construction, and aircraft manufacturing. Miniaturization of electronic components will allow future designers to downsize many electronic systems. Sophisticated, portable, light-weight, health-monitoring and medical care equipment ... will become available in emergencies, at remote locations, and for local first aid squads. Long duration space missions will require partially closed life support systems in which air, water, and food must be conserved and recirculated, which could lead to advanced air purification devices for industry and medicine, water purification equipment for homes and industry, and vital new dietary information. Software ... can help manage terrestrial hazardous materials and waste. Automation and robotics ... can be applied to automobile assembly, undersea research, robotic manipulators, and vision systems. [The requirement for] high output, low weight, portable power supplies ... may yield a portable energy source for scientific, industrial, and military outposts at remote sites on Earth. Advances in the energy industry [may result], reducing our dependency on fossil fuels. In short, ... our lives [can be made] more comfortable and our environment more secure."

Others also agreed with the many benefits of space exploration. David Livingston, the host of a popular space show[3] in the early 21st century, expressed the following thoughts. "The money spent on space exploration is spent on Earth and in the most productive sectors of the economy. The money that is spent goes to manufacturing, research and development, salaries, benefits, insurance companies,

---

[2]R.McC. Adams, "Why Explore the Universe," 30th Aerospace Sciences Meeting & Exhibit, Reno, 6–9 January 1992.

[3]D. Livingston, "Is space exploration worth the cost?" *The Space Review*, 21 January 2008.

doctors, teachers, scientists, students, blue- and white-collar workers, and corporations and businesses both large and small. The money disperses throughout the economy in the same way as money spent on medical research, building houses, or any other activity we engage in with government or even private spending."

The debate at the time centered on potentially better uses for the "space money" that the government would spend. Of course, there are many worthy causes for government expenditures (although some say – with some justification – that the private sector is better suited to make such choices) that can be considered to be investments to benefit many people over a long period of time.

Our space show host also looks for such expenditures "to inspire others to do hard work, to go the next step, to push the envelope for the next level of advancement for all our benefit ... manned space is able to do it all." Thirty-four years after Apollo, 80% of people who were involved in science, engineering, and space-related fields and businesses said they were inspired and motivated because of our having gone to the Moon. "Thirty four years after all funding had stopped for the Apollo program, investment and wealth building ... was still going on as a result of our manned space exploration years earlier."

In the U.S., the return to the Moon was seen as a way to "replace the retiring generation of scientists and engineers inspired by Apollo."[4] Some say that the money spent on the space program is better spent directly on the education of scientists and engineers. But this is a misconception that was common in the United States at that time – more money spent on education led to a better education – even though the evidence proved otherwise. Also, students need something to inspire their efforts.

Safeguarding the species should Earth be destroyed – be it by a wayward asteroid or as a result of war – has always been a reason to place people extraterrestrially. Another related reason is to have a safe site for the scientific, technical and cultural information that has been created and upon which the survival of our civilization rests.[5] Such a secure facility on the Moon, tended by a permanent staff, would provide humans with the intellectual and other resources needed to rebuild a shattered people. "Unlike ancient manuscripts which have survived for centuries in unattended storage, the data collections of the future will require continual attention from trained staffs. Skilled individuals will be required not only to update the software and hardware, but to control the environment."

In a related effort to safeguard our Earth heritage, the *Svalbard Global Seed Vault* was created. It is to this day a secure seedbank located on the Norwegian island of Spitsbergen near the town of Longyearbyen in the remote Arctic Svalbard archipelago. The facility was established to preserve a wide variety of plant seeds from locations worldwide in an underground cavern. The Seed Vault holds duplicate samples of seeds held in genebanks worldwide. The Seed Vault provides insurance against the loss of seeds in gene banks, as well as a refuge for seeds in the case of large scale regional or global crises. The island of Spitsbergen is about 1,120 kilometers (700 miles) from the North Pole. The Seed Vault is managed under

[4]E. Sterner, "More than the Moon," washingtontimes.com, 11 April 2008.

[5]R. Shapiro, "A New Rationale for Returning to the Moon? Protecting Civilization With a Sanctuary," *Space Policy,* Vol. 25, 2009, pp. 1–5.

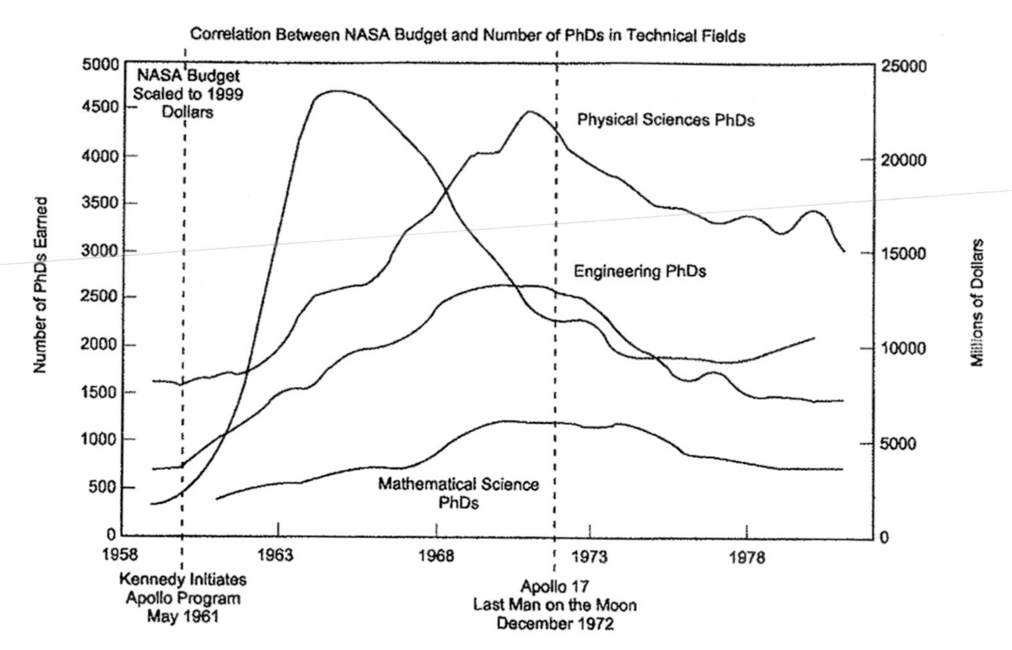

Fig. 3.1 Correlation between the NASA budget in 1999 dollars and the number of Ph.D.s that were granted in the physical sciences, engineering and mathematics. The curve labeled "NASA Budget Scaled to 1999 Dollars" peaks at about 1965, at the height of the Apollo funding. (Courtesy William Siegfried)

terms spelled out in a tripartite agreement between the Norwegian government, the Global Crop Diversity Trust (GCDT) and the Nordic Genetic Resource Center. It continues its work today.

Before we had settlements on the Moon and Mars, and before we created mining operations and outposts on major asteroids, all of the expenditures for space activities were spent on Earth. There was much debate regarding the multiplier[6] attached to NASA spending. An early study[7] utilized input-output analysis to investigate the impact of the U.S. space program on the economy. An extensive detailing of *all* of the industries that were involved in creating the hardware for NASA included *all* the interrelationships between *all* of the companies. Direct requirements generated by NASA expenditures resulted in indirect requirements in a cascading way as each contractor placed its orders with its vendors. "Specifically, in

[6] A multiplier is the net economic effect of a dollar's expenditure. As an example, suppose $100 was spent to procure an item. That $100 spent at one company results in the company also spending part of that $100 to buy the parts from a number of other companies that it needs to create the requisitioned item. Those companies, similarly, spend part of their share to purchase their raw materials. So that $100 can generate additional economic activity in a trickle-down fashion. The multiplier is the number that multiplies $100 to yield the total dollar amount of purchases.

[7] R.H. Bezdek and R.M. Wendling, "Sharing Out NASA's Spoils," *Nature,* Vol. 355, 9 January 1992, pp. 105–106.

1987, the NASA procurement budget generated $17.8 billion in total industry sales, had a 'multiplier effect' on the economy of 2.1, created 209,000 private-sector jobs, and $2.9 billion in business profits, and generated $5.6 billion in federal, state and local government tax revenues." These benefits cascaded throughout the United States even though the initial NASA expenditures were in a few states.

While the overall multiplier was 2.1, several states had multipliers greater than 10 – "for every dollar Indiana receives directly in space programme funds, it also receives $12 indirectly in business arising from the programme" – and certain industries have high multipliers, for example, electronic components had a multiplier of 5.9.

"Our findings are significant because we have for the first time estimated the benefits flowing from the second-, third-, and fourth rounds of industry purchases generated by NASA procurement expenditures. ... Many workers, industries and regions benefit substantially, and these benefits are much more widespread throughout the United States than has heretofore been realized. We believe our results imply that the economic benefits and costs of space exploration need to be reassessed." Indeed, that conclusion was always valid and needed repeating often as a reminder that the nation benefited in tremendous ways from the epic journey of humans into space.

Marc Cohen, a space architect, was a strong proponent of settling space for economic and social reasons.

3.0.2 An historical interview with Marc Cohen (June 2009)

Can you give us a one or two paragraph bio?

I was born between Hiroshima and Sputnik, which makes me a "baby-boomer" by default (I mean, nobody asked me). These two sets of events – World War II and the space race to the Moon – framed our generation's formative period. The Shoah and the Apollo 11 landing – the nadir and apex of human experience in the 20th century – demonstrate in the most primal way that people can use technology for evil or for good. We are the first generation to experience our existence so precisely bracketed and defined by such revolutionary technological events. Innovations in technology and widespread adoption of them always bring far-reaching societal impacts, whether it is the printing press, the steam engine, the telegraph, the automobile, the airplane or the television. However, our generation was the first for which the *ever increasing pace of change* became the norm, and with it all the stressors of 21st century life.

However, I must interject that being a baby boomer gave me the privilege of sitting with my father and brother in the right field upper grandstand at Yankee Stadium when Roger Maris hit his 61st home run, the highest-pressure accomplishment in the history of sport.

I grew up imagining that the NASA space program was the story of my life. Although I began with aircraft models, I built all the model kits I could find of spacecraft, and in this way came to know the Mercury, Gemini, Apollo, and Skylab in the intimate way only the biotactile digitizer (the hand) affords. From my years in elementary school, I wanted to build spacecraft. I also developed a strong interest in architecture, particularly vernacular architecture and the early Modern Movement. Later, my taste became more catholic in my fascination for architecture from many civilizations and historical periods. I came to recognize an evolving continuum from terrestrial architecture to orbital architecture that has informed much of my work in space human factors, habitability, and space architecture.

It looks as though the US will be without heavy launch capability beginning in 2010 when the Shuttle is retired. It seems almost unbelievable that we have gotten into this situation. What do you think will happen?

Au contraire, it is not "unbelievable" at all; I have no trouble believing it. From 1991 through 1994, I served in the Advanced Space Technology Office at NASA-Ames Research Center. Part of my job involved listening to the *weekly* heavy lift vehicle telecom and taking notes for all the people who had better things to do. This debate occurred about 40 to 50 times a year, pitting the shuttle-derived vehicle advocates (e.g., Shuttle-C) against the Evolved Expendable Launch Vehicle (EELV) advocates and all them against the resurrect the Saturn V advocates. These debates rehashed the same arguments that had much less to do with financial, programmatic, scientific, technical merit and much more to do with who was married to which launch vehicle, or which manufacturer a NASA manager or Center saw as "their" contractor or hoped would employ them after retiring from the civil service. The essentials of this debate remain unchanged and unresolved today as we see in the "shuttle derived" Ares I first stage versus EELVs. Unfortunately, for nearly two decades and perhaps longer, it has been far more important to the propulsion community, inside and outside NASA, who should provide and control the launch vehicle selection rather than meet the actual needs of the NASA space program. The military has largely solved this problem through the EELV program and some smaller launchers for military payloads, but within NASA, the arguing continues unabated.

Can you summarize for us what are the foundations of Space Architecture?

Yikes! How many pages are you willing to give me? I suggest that the readers look up the Millennium Charter of the AIAA Space Architecture Technical Committee and associated documents at www.spacearchitect.org. Click on Resources, and visit Publications for lots of free downloads.

OK, Ok, I will give you a quick rundown. Can I quote from the charter of the AIAA Space Architecture Technical Committee? (I say it more concisely there than I can extemporize).

"This charter defines Space Architecture broadly to encompass architectural design of living and working environments in space-related facilities, habitats, and vehicles. These environments include, but are not limited to space vehicles, stations, habitats and lunar and planetary bases; and earth-based control, experiment, launch, logistics, payload, simulation and test facilities. Earth analogs to space applications may include Antarctic, airborne, desert, high altitude, underground, undersea environments and closed ecological systems. Designing these forms of architecture presents a particular challenge: to ensure and support safety, habitability, human reliability, and crew productivity in the context of extreme and unforgiving environments."

Why is settling the Moon so important for civilization?

I hate to burst your inflatable, but actually, it is not important in my view. Aside from being the largest local piece of barren rock on which to land a spacecraft, it is virtually devoid of resources that we can exploit economically, operationally, technically, or most important: profitably. Yes, the regolith is about 50 percent oxides, and there is a benefit to the propellant mass biggest "gear ratio" to extract the oxidizer for ascent launch from the lunar surface and perhaps to supply a lunar orbit space station or perhaps one at a Lagrange point. However, the search for water – that is, hydrogen – remains elusive with the caveat that we just launched Lunar Reconnaissance Orbiter and Lunar Crater and Observing and Sensing Satellite (LCROSS), and I hope that it will prove me wrong.

What is important about the Moon – of paramount importance – is to use the Moon to our advantage as a practice, dress rehearsal, and testing site for design concepts, engineering solutions, manufacturing practices and techniques and operations that we will enable us to go to Mars and live to tell about it. I believe this approach is what Krafft Ehricke meant when – at the landmark 1984 conference at the National Academy of Science Lunar and Planetary Bases and Activities of the 21st Century – he argued, "If God wanted man to go to Mars, He would have given him a moon."[8]

What do you see as the major hurdles for our return to the Moon for permanent manned settlements? Are they technical, financial, physiological, psychological?

All of the above, but mostly the obstacles are political. As Tip O'Neill, Speaker of the House of Representatives said, *all politics are local.* Up to the present, the NASA space program is predominately a loose feudal confederation of local and regional enterprises tied to local NASA Centers and local contractors. Only when the space program becomes a *national* enterprise with the *national* interest standing above the local political "gimmees," "gotchas," and earmarks on the funding, the civil space program will always face these crippling impediments.

I notice that you are careful not to ask for any solutions to these problems. Perhaps, you can save that question for your next book.

How do we answer critics who say space is too expensive and that there are numerous problems on Earth to take care of first?

Are we still talking to those people? Seriously though, the answer is quite simple. The space program has generated vast economic and industrial growth and expansion for our country. These advances go far beyond Tang or Velcro. For example,

[8]The book of the conference proceedings gives the text as "If God wanted man to become a spacefaring species ...," but I was there and heard him say Mars.

to conduct the Apollo program successfully, NASA had to *invent* digital computers and electronics. Just think about the impact of digital devices throughout our country and the world and the immense information economy they enable.

Usually, the "Earth-firsters," the people who raise the "first fix the problems here on Earth" banner, are coming from the perspective of social and economic justice for up to now disenfranchised and disempowered communities. I understand that worldview deeply because, like my grandparents and parents, I have long been involved in progressive causes. My answer to them is that I believe there is less discrimination in employment in aerospace and all the closely allied high technology areas than in most other economic sectors. The reason is that, in this design and engineering work environment, more objective criteria exist to evaluate the quality and forward-leading character of an individual's work than in many other economic sectors subjective criteria dominate, such as superficial appearances, who is in the "old boys' club," and how well someone can exchange the correct social cues. Basically, a component, a subsystem, or a system either works according to the requirements or it does not, and it is eminently feasible to evaluate an employee on the relevant objective criteria.

We appear to have a number of very motivated competitors for the return to the Moon: China, Japan, Korea, Russia, and the European Space Agency, not to mention dozens of national space programs that are quite competent at placing objects in orbit. Do you think the U.S. and Americans in general take this seriously?

Does *anyone* take *anything* seriously *anymore* in the era of YouBook and TubeFace? How can a generation or nation of cyber-narcissists distinguish between the virtual worlds of World of Warcraft or Second Life and the challenges we need to overcome on the corporeal plane of existence?

I believe that the internationalist vision expressed in the first Star Trek series will prove prophetic. As humanity faces the vastness and mystery of the cosmos, it makes no sense whatsoever to take outworn and atavistic national rivalries with us on this ultimate journey. In becoming a spacefaring species, we must emphasize the things that unite us, not that divide us. The multinational crews on the International Space Station have begun to lead the way on this path to the stars.

The next step will be to make the next generation of lunar exploration even more truly *cosmo*politan by inviting all of the international space agencies to join in the lunar missions that the Altair lunar lander will enable. However, we should not be naïve that such international contributions will "reduce" costs for NASA if we incorporate them into Altair. On the contrary, the ISS experience shows that such hardware integration drives up costs. Instead, I expect that the international lunar collaboration will rely upon "clean interfaces" on the Apollo program model, so that each partner will deliver a complete unit of hardware for the surface systems, and not try to fit subsystems into modules built by other countries. There would then need to be a new Space Treaty stating that these modules do not remain "national" territory or property, but rather there is a collective ownership or system of shareholding in the whole lunar enterprise.

When you envision the future, where do you see us in 50 years? In 100 years? How do you see this evolving from the present?

As Yogi Berra said, "Prediction is difficult, especially about the future." However, I will go way out on a limb here and predict that the Chicago Cubs will win the World Series sometime in the next 100 years, but probably not until the World Baseball Classic eclipses both the Series and US Major League Baseball.

Oh, you were talking about space? I thought you meant the future of something important. Evolution is always painful. It always involves winners and losers, gloaters and whiners, the creation of new beginnings and the extinction of old favorites. We lose the National Space Transportation System (NSTS, the official name of the Space Shuttle Program) but we gain the Orion and its associated supporting elements.

Anyway, if you are looking for a banal assurance that we will have humans on Mars in 30 years or a city of 100,000 folks on the moon in 50 years, you are asking the wrong cynic. It will be *centuries* until space is no longer the most dangerous place we can go, until spaceflight is no longer the most expensive way to travel, and until anybody sees a return on investment based upon economic activities beyond low earth orbit.

In human spaceflight, I expect that we will see a blossoming of private launch and flight vehicles. Nobody should ask, "If these small companies can get a crewed capsule into orbit for such low cost, why is NASA having such a hard time?" In fact, NASA invested hundreds of billions of dollars in creating the hardware and software systems and technologies that the new crop of private adventurers can pluck almost from off the shelves of the space technology bazaar. However, it appears that nearly all of these startup space entrepreneurs can proffer their exceptionally low costs *only* by not investing in crew safety and mission assurance, reliability, and human system integration, which are all major cost drivers for NASA. It is therefore with deep regret that I submit that one or more of these small space startups will kill a crew. The consequences will be traumatic for the young companies, and it will drive up all of their insurance premiums to astronomical levels, severely impeding their ability to break the surly bonds of Earth.

We will see a crewed mission to an asteroid as a way of testing our flight hardware in deep space, outside the Earth–Moon system. This six- to ten-month mission could use the basic Orion and Altair hardware, perhaps with a downsized Altair descent module and a pressurized mission module to provide long duration habitation for the crew.

You have worked for NASA and for Northrop Grumman. Many say that private enterprise will be the motive force for space exploration and settlement, if NASA just gets out of the way. Given that you have been inside both, what is your view on the best way forward?

Here is the deal. The President and the Congress set the overall direction for the space program that NASA leads. NASA defines mission architectures, requirements, roadmaps, and all of those abstractions that are so useful for filling up PowerPoint slides. NASA really does set the direction and the requirements for new programs and procurements. Industry will mobilize to respond to any solicitation for which NASA has the funding and is willing to spend it.

Having said that, I can say there are some substantial differences in culture between NASA and industry, but also between companies, and between NASA and other national or state space agencies. What I could say about these differences in culture is largely speculative and subjective, so let us just say they exist and leave it at that.

Within NASA, it was my observation when I was in NASA that the agency is far too dependant upon thousands of in-house support contractors doing jobs that are inherently governmental and should be done by civil servants. The way you know that they are inherently governmental is that whenever NASA recompetes those contracts, the same people stay in the same support contract jobs, regardless of which offeror wins. The only thing that changes is that the low bidder eliminates a few of the highest paid positions, and cuts vacation and benefits for the employees they have "rehired." It should be obvious that if the same people continue doing the same work regardless of which corporate logo is on their paycheck, that they are doing inherently governmental work and that they should become NASA civil servants. That would greatly improve the cost-effectiveness, efficiency, and reliability of those support service functions because nobody would be draining a double overhead (government and corporate) plus profit from them. Also, NASA would receive better support because the workers would feel much more secure and can stand up for ethics in the work place and doing what is right rather than what is expedient to protect their job.

Given the recent financial meltdown worldwide, does the return to the Moon become more tenuous?

The Moon will still be there for a long time, so from Luna's perspective, it does not make sense to fuss over a few years or decades. However, for those of us who want to be involved in a new era of lunar exploration, it is a serious question – whether we will return astronauts to the Moon in our lifetime and go on to send them to Mars.

With the new President Obama in office for only six months, how do you see President Bush's vision evolving?

Again, there is that evolution word, so badly misused and so poorly understood. President Obama will have very little time for things to evolve of their own accord on almost any issue. It is much more likely that everything will be a crash course of drinking from the fire hose of new information. In fact, the new administration's priorities must be the wars in Iraq and Afghanistan, the potential conflicts with Iran and North Korea, health care reform, global warming legislation, energy policy, and didn't you mention the economy? Therefore, I imagine that the space program will start out low on the White House's list of priorities. I would not expect to see it get much attention for at least a year or more in terms of making real decisions about the Constellation Architecture and all the rest. That will give everyone in the space community high anxiety. President Obama seems to be sincere in his enthusiasm, but perhaps still a little naïve in terms of specific policies for space exploration.

If I can offer a design methodology comparison, the people in Washington will need to come to grips with the fact that President Bush's Vision for Space Exploration marks a more radical departure from the conventional approach than they realize. The following tables show this comparison. The point is that under the VSE, the Constellation Architecture and the vehicles act superior to the missions. Therefore, there are not actual requirements in the conventional sense – the missions get whatever capability the Constellation Architecture can give them. Whenever there is a change (usually a cut in capability) in the architecture or vehicle, it diminishes the missions we can accomplish, without any direct accountability to a mission manager or principal investigator who might hold them to their performance guidelines. It is possible that this approach can succeed, but all the delays in the Ares and Orion project reviews make us wonder.

Conventional hierarchy of a space program
1. Strategy: Scientific and Engineering Objectives
2. Mission
3. System Architecture
4. Systems
5. Technology Base – Development led and supported by NASA

Constellation program hierarchy
1. Strategy: Vision for Space Exploration
2. Constellation Architecture
3. Systems: Orion, Ares I, Ares V, Altair
4. Mission
5. Technology Base – None, all come hit-or-miss from industry and academia

* * *

3.1 Three-sector model for lunar economy

Commercial activities must always be put into a larger social context. Economic issues can almost never be considered in isolation. They must be studied as one, albeit major, dimension of the human experience. This has always been true on Earth, and without question has been part of the social order on the Moon.

In the early 21st century we viewed the lunar economy at any stage of its existence, whether pioneering or well-established, as three strongly coupled and overlapping sectors: (i) self-sustaining operations; (ii) production for export to Earth and Earth orbit;[9] and (iii) production for export to Mars and environs. In the

[9]The region between the Earth and the orbit of the Moon is called *cislunar* space.

Fig. 3.2 As commerce develops on the Moon, tracts of the lunar surface will be dedicated to various industries such as lunar oxygen production, communications and Helium-3 production. (Artwork done for NASA by Pat Rawlings, of SAIC. S95-01562, February 1995. Courtesy NASA)

present – in 2169 – we see that there is also a fourth sector, the outer Solar System.

Dual-use technologies[10] had a significant impact on the lunar economy because sectors (ii) and (iii) were by far the largest fraction of the gross lunar product (GLP). An historical document outlining the early ideas on the leveraging of dual-use technologies as a financial model for lunar development is provided in Section 3.2 and in a set of tables in Section 3.9.

Many lunar exports are placed on long-haul transports to the budding settlements and outposts on Saturn's and Jupiter's moons, as well as facilities that orbit Neptune and Uranus. These are today's frontiers. They would not be possible had we not advanced the technology of robotic *in situ* resource utilization. We are now able to send robotic construction crews to a moon or asteroid and begin to prepare a significant local infrastructure. Whatever cannot be made from local materials awaits the initial astronaut missions for infrastructure completion. If we had to rely on astronauts to build our infrastructures we would be hardly off the Moon today.

Examples of early commercial activities on the Moon included, in particular, *in situ* resource utilization as well as processes that gain benefit from low gravity and,

---

[10] H. Benaroya, "Economically Viable Lunar Development and Settlement," *Journal of the British Interplanetary Society*, Vol. 50, pp. 323–324, 1997.

H. Benaroya, "Economic and Technical Issues for Lunar Development," *Journal of Aerospace Engineering*, Vol. 11, No. 4, October 1998, pp. 111–118.

therefore, a smaller gravity-well than Earth's, hard vacuum, temperature extremes, and isolation from Earth "noise." In addition, it was recognized that dual-use technologies would be a powerful vehicle for partially financing and supporting the development of lunar bases.

By dual-use we mean technologies that have applications in at least two different economic sectors, such as materials technology that has a military and a civilian purpose. Promising dual-use technologies for the second-generation lunar base were self-repairing systems, low-gravity and microgravity technology, micro- and nano-devices, robotic manipulators, and instrumentation.

While many of the issues that needed study and resolution for the permanent return to the Moon were technical, many others were not. These included the related issues of the environment and territoriality, the questions of ownership, and the eventual and unavoidable independence of colonies are but a few that we return to in later chapters. These issues, along with physiological and psychological concerns and uncertainties, in the end governed the pace and style of lunar and Martian economic and social development.

For the purpose of understanding the financial aspects of lunar settlement, we place ourselves in the scenario that existed during the transition from the first lunar base to the second generation lunar base. The first lunar base was comprised of several space station type modules on foundations that were linked and shielded. Some sketches of these are provided in Chapter 5.

The first base was really a pioneering footing, establishing a baseline for survival, including resource recovery. Once established, science and commerce evolved rapidly. Commercial activities were primitive and comprised of small-scale feasibility and demonstration projects. Population was in the range of 5 to 20 people with a broad spectrum of skills and with initial six month tours of duty that evolved to one or two years.

This is what led to the second generation lunar base that was comprised of more advanced structures of a variety of classes, such as hybrid stiffened-inflatable concepts. The ability to construct such structures of varying sizes led to structures that were differentiated according to function – for housing, agriculture, manufacturing, science, storage and maintenance. Larger power generation facilities were required, as were broader transportation capabilities. This second generation lunar base expanded commercial activities beyond those of the first lunar base. At that point, once survival seemed reasonably assured, the focus went to creating self-sufficiency and then profitability.

Essential to profitability was the ability to export materials and products to Earth, and to use the Moon as a testbed and staging area for the colonization of Mars. Our lunar population grew quickly to the range of 100 to 300 people with more specialized skills, many of whom became permanent residents. Astronauts landed on Mars in 2034 and by 2041 we had a permanent settlement. The site on Mars was similar to the first lunar base and was on the same scale.

When the lunar settlement evolved beyond sustenance, items mined from the lunar surface, as well as those manufactured on the Moon, were exported to Earth and low-Earth orbit (LEO). Due to the low lunar gravity, it now costs less to place satellites in orbit in LEO from the Moon than from the Earth since the infrastructure is already in existence on the Moon – thus the early drive to manufacture such components on the lunar surface.

Once a manufacturing and mining capability existed on the Moon, it was possible to support the settlement of Mars. In addition, many future Martians were trained on the Moon and lived and trained there for years before moving on to Mars.

Fig. 3.3 This painting was used as a visual at an April 1988 Houston-hosted conference titled "Lunar Bases and Space Strategies of the 21st Century." A deep drill team shown in the lower central portion of the image obtains cores for petrological studies of the floor units of the young, 30-kilometers, 4200-meter crater, Aristarchus. The pea of Aristarchus is a few kilometers to the south of the drill rig. This perspective from the crater floor shows the prominent slump terraces of the crater walls and the solidified impact melt rivulets which flowed down the steep inner wall immediately after the crater was formed. Because the crater is very "young" the rivulets and the volcanic-like features and cooling cracks of the impact melt floor unit are only slightly muted by meteorite erosion and ejecta blanketing. The drilling activities are taking place at 23.7 north and 47.5 degrees west. (The painting was accomplished as a joint effort by Pat Rawlings and Doug McLeod of Eagle Engineering. It is one of a series of paintings done on subcontract to, and under the technical and scientific direction of, Lockheed Engineering and Management Services Company. The work was sponsored by the NASA Johnson Space Center. S88-33127, 7 April 1988. Courtesy NASA)

In Section 3.2, some activities that were envisioned as growth sectors for the lunar economy are listed and discussed. A few of these appeared far-fetched in the early 21st century when GPS, cell phones, hand-held computers, massive data transfer rates were relatively new technologies.

3.1.1 Relation to Mars development: Moon-first strategy

For the decade prior to the Bush speech of 2004, the intellectual battle within the space community was whether the first goal of settlement should be the Moon or Mars. The Mars-First contingent deemed the Moon to be, at best, a diversion from the real goal of colonizing Mars. The Moon-First-on-the-way-to-Mars group, while also supporting the eventuality, desirability, and perhaps the dominance of a Martian civilization, believed that the rational choice for a first settlement was the Moon.

They saw the clear benefits of initially settling a planetary body three days from Earth versus one that is about a year away. From any perspective except public relations, the clear and rational way for Man into space was via the Moon, as we know today. At that time, bypassing the Moon ignored the critical technical and physiological issues that were then as still unresolved. Pretending that the existing technology need not be tested extensively before being sent on a yearlong mission to Mars with humans bordered on wilful ignorance.

As we reasoned, and as the Bush speech directed, the way to send humans to Mars was to first create a significant presence on the Moon. Such settlements had multiple benefits. We learned about the risk and reliability of engineered and human systems in the space and low gravity environment. All of these studies required data with a significant time history and had to be performed on another planetary body with less-than-Earth gravity, the Moon. How fortunate we were and are to have a moon in such close proximity to Earth. It is worth repeating Krafft Ehricke's statement: "If God had wanted man to explore space he would have given us a moon."

One of the very challenging aspects of becoming a spacefaring civilization was to learn how to manage super-projects, that is, very large projects spanning decades and vast distances. We learned how to do that on a smaller scale during Apollo, and did it well. But the return to the Moon was ten, maybe a hundred, times more difficult. Thousands of elements of the super-project had to be coordinated and ready at specific instants of time. Imagine thousands of strands of rope that must be spun at the right rates so that they intersect at specific locations and times to create a weave of incredible complexity, with almost no room for error. That is the weave that began to be spun after Bush's speech of 2004.

Having colonies on the Moon permitted us to develop manufacturing and construction capabilities. The inhabitants of the Moon created resources required for manned trips to Mars. In-situ resources were utilized in support of Mars expeditions, and the technologies developed on the Moon were sent to Mars in advance of manned flights to begin to synthesize oxygen, hydrogen, and to build structural shells that were completed by the manned expeditions. The low lunar gravitational field made the Moon an ideal source, not only to fly manufactured goods into low Earth orbit, but also to fly products to Mars and eventually much of the Solar System.

From the financial perspective, there were significant economic benefits to operating on the Moon first and using some of these profits to go on to Mars. The following section provides a discussion of economically viable activities on the Moon and in lunar orbit that only hints at the economic power of the lunar civilization. Part of the gross lunar product (GLP) was related to Mars development because Mars became a major market for local industries.

The Moon flourished, continuing to attract permanent settlers and investors within a century. It became a viable and a desirable second home for humanity. These pioneers became the fount for the human settlement of the Solar System.

Sections 3.2.1 and 3.2.2 are taken verbatim from studies in the late 1990s and early 2000s that discussed the ideas used to create a financial basis for privately funding part of the costs for the manned return to the Moon. They have been left in their original context.

3.2 Dual-use technologies as a financial model for lunar development

The defining question in the late 20th century and early 21st century was: *How do we structure projects and technological developments that are so expensive, and generally require such long time scales, so that investors will finance them?* This question applies to many large-scale expensive endeavors, including lunar development. However, other than space, such projects were widely viewed to be in the public interest, and therefore justifiably funded by government. Examples included airports, highways, environmental cleanup facilities, the military, and the space exploration of the 1960s.

Compared to these closed-ended projects with well-understood economic and social benefits, lunar development was viewed open-ended. The vastness of the enterprise made it difficult to detail its benefits to the public. (It was challenging to justify the multi-generation benefits of colonizing a planet over a period of one to two hundred years.) The settlement of the Moon was also very expensive when viewed as one large project. Attempts by proponents at justification by using cost comparisons to other aggregate national expenditures, such as the amount of money spent by Americans to see movies,[11] did not enhance public desire for funding lunar development. Therefore, it became necessary to identify segments of the larger endeavor that could be pointed to as being beneficial activities – independently of the fact that they also supported the space program. Worthy is, of course, in the eye of the beholder, so that a variety of activities had to be identified.

This effort to capitalize on the intermediate products and output of a long-term effort to settle the Moon took hold in the early 21st century. There were always "spin-offs" from the space program. These were generally not known to the public except – incorrectly – in the cases of items such as Tang[12] and Velcro.[13] But as we discuss in this book, many technologies originally developed for the space program have found application in the Earth-based economy.

---

[11] According to *Entertainment Weekly*, in 2004 Americans spent $24.1 billion on home video rentals and $9.2 billion at theaters. The NASA budget for that year was $15.2 billion.

[12] "Tang is a sweet and tangy, orange-flavored, non-carbonated soft drink from the United States. Named after the tangerine, General Foods Corporation marketed it in powdered form in 1959. It was initially intended as a breakfast drink, but sales were poor until NASA began using it on Gemini flights in 1965, which was heavily advertised. Since that time, it has been associated with the U.S. manned spaceflight program, so much so that an urban legend emerged that Tang was invented for the space program." [http://en.wikipedia.org/wiki/Tang_(drink)]

[13] The Velcro® brand hook and loop fastener was invented in 1941 and named for the French words "velour" and "crochet." These fasteners are ubiquitous – replacing shoe laces, to anchoring equipment on NASA's Space Shuttle.

On the Moon today, there are whole companies that license and market goods and technologies – that we create here for local use and consumption – to Earth, and less so to Mars. Technology after all is the solution to a need. Human needs are in many ways similar on Earth, the Moon and Mars.

The two sections that follow are a late 20th century view[14] on the use of dual-use technologies to assist in lunar settlement and development.

3.2.1 The new paradigm (circa 2000)

The essential framework proposed here for supporting and managing long-term and expensive projects is to substructure the projects into smaller independent and profitable units. This is true regardless of the particular goal of the larger project. In the same way, the path to the Moon will be supported by scores of existing and newly created independent businesses. These businesses will be such that the whole is larger than the sum of its parts. The whole will get us to the Moon, and investors can support any or all of the parts that get us there.

While this may seem to be an obvious solution to the problem of funding the return to the Moon, there are many hidden difficulties in such an approach. For example, one needs to be certain that the whole project is not held together by a weak link, or a series of weak links. For robustness and reliability, a parallelism of technological capabilities is necessary. In addition, an independent entity must be created that can pull together all the pieces that we will need in order to land on and develop the Moon. At this point, it is reasonable to stipulate that R&D efforts in propulsion and rocketry will require major government subsidies, although entrepreneurial rocket companies are becoming successful. Most other efforts can be justified to private investors.

One must address the question of how to coordinate such a disparate group of business enterprises and, at the appropriate time, actually embark to the Moon with its first settlers. Ideally, a leadership group must be in place to properly coordinate the design and manufacturing efforts of the various organizations that are supporting lunar development. While it is not necessary to own or acquire these businesses for this paradigm to be effective, it is important that a central organization have the large picture in mind and have the resources, intellectual and otherwise, to ensure the development of the necessary technologies to support lunar development. These resources can only be developed in a financially viable sense – meaning that we need investor interest rather than government taxation. Of course, government acts on behalf of the population and, therefore, will become a customer of any space enterprise.

A *Lunar Development Corporation* (LDC) can be created as an independent company that will work towards the above goals. It will include essential professional talent, such as management, science and engineering, financial, and legal teams. These teams will act as venture capitalists, coordinating activities around

[14]H. Benaroya, "Economically Viable Lunar Development and Settlement," *Journal of the British Interplanetary Society*, Vol. 50, pp. 323–324, 1997.

H. Benaroya, "Economic and Technical Issues for Lunar Development," *Journal of Aerospace Engineering*, Vol. 11, No. 4, October 1998, pp. 111–118.

each venture, attracting capital for start-up endeavors that cannot be accomplished by existing industries, and attracting investors for existing companies that have a role in lunar development. All of these activities are viewed as for-profit. Part of the profits will be used to repay investors, and the remaining funds used to create the financial strength needed to initiate and support a return to the Moon. As the LDC grows, it may be appropriate to include debt financing (bank) in certain instances rather than relying solely on equity financing (venture capital).

It is preferable that the LDC be a for-profit corporation. An issue yet to be addressed is whether the LDC is to be privately held, and therefore with no fiduciary responsibility except to its owners, or publicly held with all of the requisite fiduciary duties. It may be that the LDC starts out privately held and evolves into a public corporation.

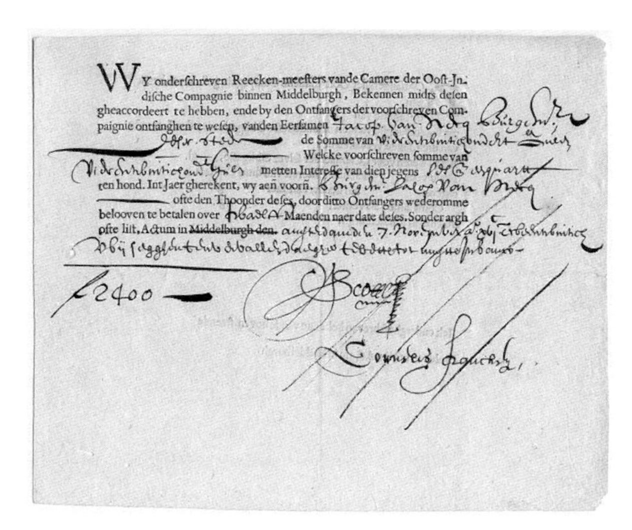
WY onderschreven Reecken-meesters vande Camere der Oost-Indische Compagnie binnen Middelburgh, Bekennen midts desen gheaccordeert te hebben, ende by den Ontfangers der voorschreven Compaignie ontfanghen te wesen, vanden Eersamen [illegible] de Somme van [illegible] Welcke voorschreven somme van [illegible] metten Interesse van dien jegens [illegible] ten hond. Int Jaer gherekent, wy aen voorn. [illegible] ofte den Thoonder deses, door ditto Ontfangers wederomme belooven te betalen over [illegible] Maenden naer date deses. Sonder argh ofte list. Actum in ~~Middelburgh den.~~ [illegible] 7. Nov[illegible] [illegible]

f 2400

Fig. 3.4 A bond issued by the Dutch East India Company, dating from 7 November 1623, for the amount of 2,400 florins. While the LDC is not envisioned as monopolistic – as were the East India Companies of Holland and England – it would operate in a similar way.

It is clear from our general description that the LDC must be created with start-up funds to hire the teams needed to begin planning. As part of the start-up, a business plan must identify preliminary activities that will be used to attract investors. Investors providing such start-up funds are expected to receive reasonable rates of return. Given the magnitude of the proposed venture of lunar development, a 20-year time period is expected between the creation of the LDC and the first year of lunar colonization. It may be technically feasible to go back to the Moon in ten years if the taxpayers provide full funding. This is unlikely to happen. Therefore, the proposed approach, which is a bootstrapping method, is needed. Thus, the two-decade time frame.

The primary goal of the LDC is to ensure that the right technologies are available when needed during the twenty-year path to lunar development. In addition, the LDC will attempt to do this with dual-use technologies that are profitable for

Earth applications as well as lunar needs. In this way, financing and profits can be reasonably expected. However, although our ultimate goal – the initiation of lunar development – is twenty years away, our individual dual-use ventures will provide reasonable rates of return. Technologies selected for development must be prioritized according to a critical path method as well as an economic benefits model for the technology.

Once we are assured of the necessary technologies, we need to create the path to lunar development.

3.2.2 Paths to lunar development (circa 2000 – continued)

Existing industries that are not traditionally considered aerospace have a significant role in lunar development. Certainly, these industries will be the backbone of early development. However, in planning a project that will not complete its first phase of operation before about two decades, new industries are anticipated for which early stages of technical development are necessary. The technological issues for lunar development are relatively well understood. Of course, there will be debates on technical options. For example, should the prototypical lunar structure be inflatable, a truss, or within a lava tube? Nevertheless, once a choice is made, the technical issues can be addressed, even if this means that new technologies need to be developed.

The long-term biomedical issues are less well-understood and require continued investigation. In order for the proposed paradigm to be successful, a significant percentage (dollar wise) of the technologies must be dual-use, meaning that they not only have a role in lunar development but also have a more immediate civilian or other application.

As examples, the following are promising dual-use technological arenas for investment:

- *Self-repairing systems* can be used to help safeguard systems that are hit by micrometeorites in space and on the Moon. Such systems can also be utilized in monitoring and repairing micro-cracks in aircraft fuselages and other mechanical components.
- *Materials development and processing* are among the most economically valuable scientific and engineering activities because of the importance of new materials in our world. The same is true for materials developed for space applications.
- *Low gravity and microgravity technology* will be developed as a result of our experience with the space station. Such capabilities, whether for fluid mechanical applications or materials processing and handling in such conditions, will become extremely important for the practicalities of spacefaring. The mechanical equipment developed for the extreme space environment will be useful in extreme environments on Earth such as the deep-sea and the polar ice caps.
- *Robotic manipulators* have very broad applications. In space and on the Moon they could be extremely useful in minimizing the workload of the astronaut construction corps. They also can be used in delicate manipulations such as medical procedures and hazardous material handling on Earth and in space.

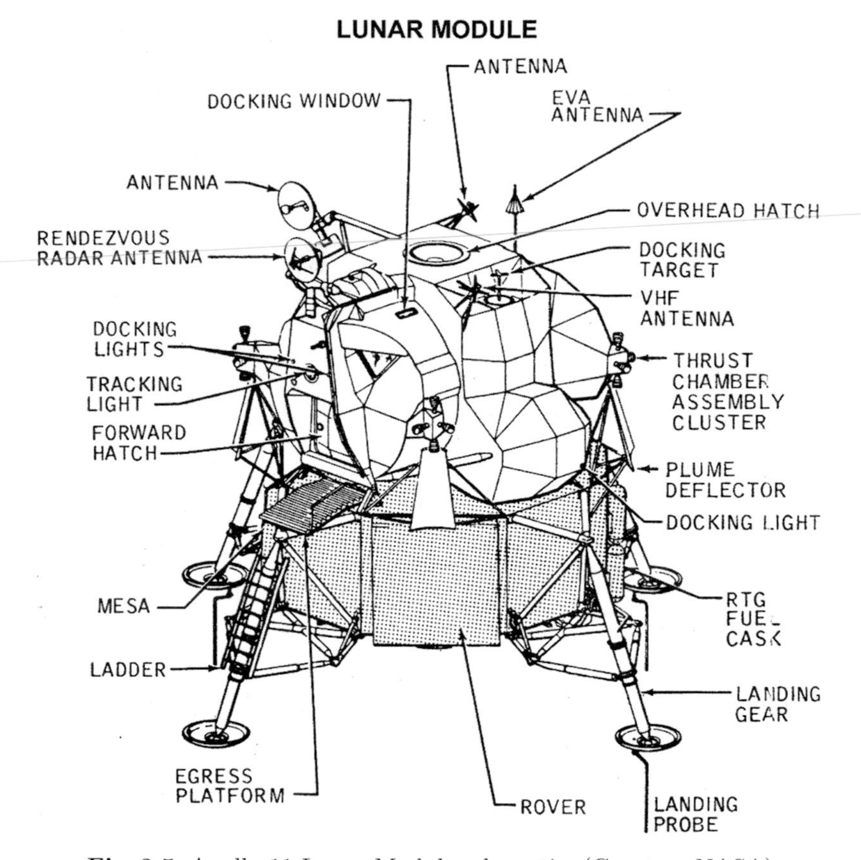

Fig. 3.5 Apollo 11 Lunar Module schematic. (Courtesy NASA)

- *Instrumentation* is an industry with broad applications. Two possible areas for investment are the monitoring of material integrity and fluid flow control. Both have significant dual-use possibilities.
- *Micro- and nano-devices* are those of a size that approach the smallest of scales – the molecular and atomic scales. Potential applications include the tiniest computers, and the strongest and lightest materials. Such devices would revolutionize manufacturing, electronics, materials, and medical procedures. The applications to the space and terrestrial economies would have immense impact.

The above list is only a sample of possibilities.[15] Once very focused and specific studies are initiated, this list would grow by multiples of ten or a hundred and it will become clear which technologies have the highest potential for dual-use and, therefore, rate of return. These technologies will form the backbone for technology development. Some of these profits will help support the less profitable, but necessary, technologies that must also be developed.

[15] A more complete list is provided at the end of this chapter.

In this way, the necessary resources can be accumulated for early lunar development. The LDC will evolve as an organization that resembles a venture capital holding company, with the overarching ultimate goal of lunar development. Intermediate goals include appropriate technology development and the earning of profits for investors and for the LDC. The long-term goal is an expedition to the Moon, to stay. Because of that long-term goal, some technology development undertaken may not lead to profits but, rather, losses.

In creating a path to lunar development, various options will exist, each leading to different sequences of technology development. The path to the Moon is not unique. There may be hundreds of alternate paths that lead to the same end. However, cost and time will differ. Also, there is the important issue of whether a particular component should be developed in year 1 or in a more advanced form in year 15. Decisions need to be made on the acceptability of certain older technologies in lunar development. What we see are hundreds of paths to the same end, but with significantly different technology and financing stops along the way.

Nevertheless, at certain points in time, a commitment has to be made to use acceptable technologies so that plans can be finalized and manufacturing set into motion for a return to the Moon.

During the last phase of this twenty-year plan, the LDC will begin to contract certain needs for the return to the Moon. Among a host of needs, the LDC will contract for launch dates, bays on the space station for systems integration, equipment, as well as the initiation of training for its astronaut/settlers who have been recruited from the best of those eager to become the pioneers of the 21st century.

In addition, at this stage, the LDC will begin to negotiate contracts for access by companies and governments to facilities on the fledgling lunar base that is to be constructed by the LDC. Interested parties will include government, industry and research organizations. Issues of reliability and insurance will become part of the negotiations. Numerous details need to be considered and engineered, from crew size and mix to the lunar base functions. Planning-beyond-twenty-years must be initiated so that the long-term vision ensures the economic viability of this lunar colony. Possible early lunar-based enterprises include tourism and sports, energy production, pharmaceuticals development, medical trauma centers, materials extraction, defense, sites for R&D facilities, as well as a spaceport between Earth and the rest of the Solar System. Science facilities will be well-represented on the Moon, especially astronomical observatories.

Not all of the issues raised and discussed can be resolved *a priori*; rather, it will become clear which pose major difficulties and which will be easily handled by the evolving technologies of the next twenty years. The importance of getting started cannot be overstated!

* * *

3.2.3 Afterthought

As we discussed in the last chapter, many dual-use companies were created in the early-to-mid 21st century. Some flourished into the 22nd century. All were part of the foundation for the permanent human settlement of the Moon and Mars. NASA played a critical role – the creation of the transportation infrastructure, with major assistance from the Japanese space agency JAXA, the European space agency ESA and other major space agencies. Of course, the early ideas on how an LDC would work had to evolve and adjust to changing perspectives, unanticipated technologies, the shock of catastrophic failures, political crises, and economic fluctuations. But in the end the essential idea worked as hoped – private enterprise was energetic and took advantage of opportunities for the development of space technologies, and governments provided the infrastructural underpinnings.

3.3 On the chances and limits of lunar development (2008)

This section is a summary of a study and model development by H.-H. Koelle during the mid-2008s.[16] The text is an edited version that summarizes his thoughts and work on lunar development. Here the different stages of the evolution of a lunar human presence are described. The three stages are grouped by population level, and estimates are given of the *in situ* resources mined and processed and the costs associated with different sectors of the lunar economy. In Section 3.5, additional studies by H.-H. Koelle on the lunar economy are summarized.

The models used mathematical representations of economic, logistic, transportation and other activities that can be represented by an equation. Any mathematical model is based on an understanding of the underlying behavior. Mathematics is an efficient way to represent that understanding, and can be only as good as the underlying understanding. There are two fundamental modeling uncertainties, one is the form of the equation used and the other the parameter values used in the equation.

There are many different classes of equations used for mathematical models in physics, engineering, econometrics, biology, and any body of knowledge that can be quantified. Because of the efficiency of mathematics and due to the predictive powers of the equations, most disciplines attempt to use mathematical models if they are so amenable. The social sciences, including psychology and sociology, have gone beyond utilizing statistics to organize data and now use mathematical models to represent the behavior of groups and the dynamics of the behavior of individuals.

How accurately a mathematical model can predict an outcome depends on the accuracy of the underlying assumptions. Those assumptions are critical.

3.3.1 Lunar development model

Lunar development is an open-ended and worldwide activity. The underlying principle is that we "learn by doing." From the very beginning the future lunar population, whether it be small or large, must organize its efforts with a priority on survival. This includes the production of oxygen gas for the replenishment of losses of air and water experienced during lunar settlement operations. In addition, the lunar population must produce services and products that can be marketed in the

[16]Courtesy H.-H. Koelle.

Fig. 3.6 Moon landing map. Legend: Light downward-pointing triangles: Surveyor. Dark upward-pointing triangle: Luna. Light upward-pointing triangle: Apollo. (Courtesy NASA)

long-term. Examples include the supply of energy to Earth, the mining of rare minerals for export, propellants, research facilities, waste deposits and lunar tourism.

It is very difficult at this time to make predictions of what can be done and to what level of production, but it is clear that there will be surprises and products that no one has thought of as yet. Most people on the Moon will be employed in providing housing, life support, and local transportation services, but there will also be explorers and customers. What we can do at this time is to compile the available information and insights to create mathematical models of what can be expected.

Evolutionary development scenarios can be broken down into the following steps:

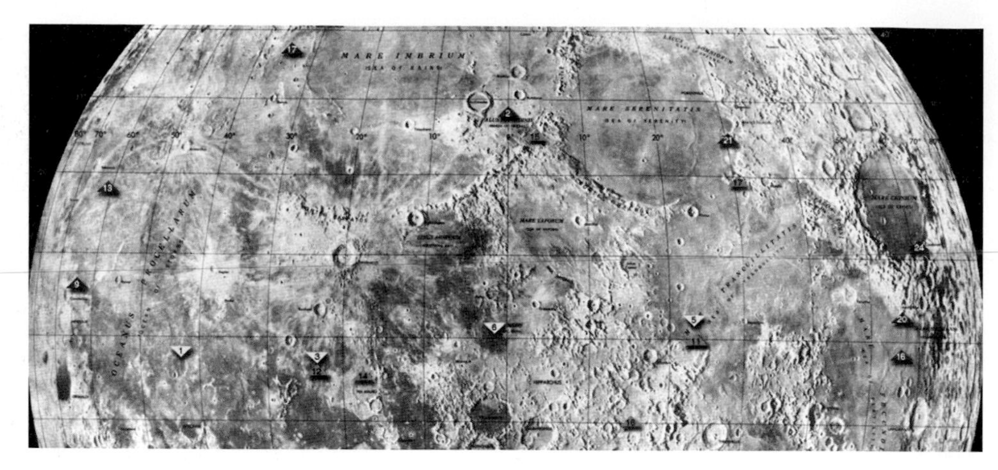

Fig. 3.7 Moon landing map – cropped version for closeup. Light downward-pointing triangles: Surveyor. Dark upward-pointing triangle: Luna. Light upward-pointing triangle: Apollo. (Courtesy NASA)

1. Lunar Outpost (up to 10 people)
2. Lunar Laboratory (up to 100 people)
3. Lunar Settlement (up to 1,000 people).

The next several sections discuss these steps in detail.

3.3.2 Lunar outpost

Currently, efforts at NASA are concentrated on developing capabilities for returning to the Moon by the year 2020, as directed by the Bush Administration, with a budget authority of Congress in 2005 promising resources through Fiscal Year 2018. Total expenditures over the thirteen years, as expected in 2005, were on the order of 108 billion dollars.

We expect that this phase of the program – replicating the Apollo successes of the last century – will only be the beginning. Present plans envision sending two to three crewed missions annually to the Moon for several years and the establishment of a small outpost that can temporarily house four astronauts for periods of up to several months. Including cargo missions providing equipment and supplies, this phase would require three to four launches annually from the Earth at a cost of approximately 4 billion dollars per year.

These tentative plans have not been firmed up as yet, and the new Obama administration must find public support. In this context, it must be mentioned that other spacefaring nations have also expressed interest in human exploration of the Moon, either as partners or within their national programs. The target date for the establishment of a lunar outpost that has been mentioned by those planning this phase is the year 2025.

As things stand today, this limited human presence of four astronauts on the lunar surface, even if performed annually several times lasting several years, will not suffice to explore the Moon, and even less so to achieve the overall objectives of the Bush U.S. space policy.

3.3.3 Lunar laboratory

The announced objective of the 2004 U.S. space policy is to establish a permanent human presence in space beyond low Earth orbit, on the "Moon, Mars and beyond." Thus, by definition, a small lunar outpost, temporarily occupied by four people, can be only an interim step of a few years.

An expansion of the outpost into a modest research and development laboratory would be the next logical step, which could take one or more decades, depending on long-term global economical and political developments. The usefulness and priority of such a lunar installation must be assessed periodically. Past studies indicate that this phase of lunar development requires a new logistics approach to make it attractive and affordable. This critical point will be discussed first.

Past Space Transportation Systems (STS) employing propulsion systems using chemical propellants have been utilized in the Apollo program landing humans on the Moon beginning in 1969 through 1972. Currently, expendable launch vehicles have been selected by NASA and are under development to re-establish this capability. However, their cost would be high in an extended program because they are not reusable. Thus they would not permit the establishment and maintenance of an adequate lunar installation that is able to demonstrate sustained operations of humans on the Moon in preparation for interplanetary expeditions, such as landing a crew on Mars.

Consequently, this question must be asked: Could improved space transportation systems do better than the current selected concept? Advanced propulsion systems are not likely in the next few decades. Higher *specific impulses*[17] than the 470 seconds available with the LOX/LH2 (liquid oxygen/liquid hydrogen) combination are possible with chemical propulsion systems but are not considered a realistic alternative. The only realistic option is to change from expendable vehicles to reusable vehicles, as demonstrated by the air transportation industry over the last century.

Experience with reusable rocket vehicles has been accumulated since the late 1960s. It began with the rocket airplane X-15, and since 1982 with the payload stage of the Space Shuttle. Rocket casings of Shuttle boosters have also been recovered and refurbished for reuse. Thus, multiple flights of launch vehicles or subsystems have been demonstrated and can be applied systematically to the larger ballistic launch vehicles of the next generation. Winged launch vehicles are also possible, but these would be limited in size to about 1,000 metric tons launch mass, and relatively small compact payloads. Lunar transportation systems require that large crews and voluminous cargo must be transported.

A reusable three-stage ballistic launch vehicle for the leg "Earth-lunar orbit-Earth," in combination with a reusable single stage lunar shuttle for the leg "lunar

[17]The specific impulse is a measure, usually in units of seconds, of the efficiency with which a rocket engine utilizes its propellants, and is equal to the number of pounds of thrust produced per pound of propellant burned per second.

Fig. 3.8 The Space Shuttle was the world's first reusable spacecraft, and the first spacecraft in history that could carry large satellites both to and from orbit. The Shuttle launched like a rocket, maneuvered in Earth orbit like a spacecraft and landed like an airplane. Each of the three Space Shuttle orbiters – Discovery, Atlantis and Endeavour – was designed to fly at least 100 missions. Altogether they have flown a combined total of one-fourth of that. The Shuttle had a length of 184 ft, the orbiter: 122 ft; height of orbiter on runway: 57 ft; wingspan of 78 ft, a liftoff weight of 4.5 million pounds; an orbit of 115 to 400 statute miles; and a velocity of 17,321 mph. The image shows the components of the Space Shuttle system: orbiter, external tank, and solid rocket boosters. (Courtesy NASA)

orbit-lunar base-lunar orbit," and supported by a lunar orbit space operations center (traffic node), appears to be the best solution at this time for providing the logistic support of the lunar installations currently envisioned and in the planning stage.

It is generally agreed that it will take up to ten years to develop and test a new space transportation system. It will also take probably ten years to establish an initial lunar base that could grow into a lunar settlement. Exploring the growth potential of lunar installations and lunar space transportation systems takes a planning process that is complicated by several interdependencies. Program optimization is difficult and time consuming. It will take many trade studies[18] before a realistic balance for a specific scenario can be achieved. This task requires suitable planning tools if the purpose of the analysis is to explore the upper limits of lunar development with respect to performance and financing. A first cut at this problem can be accomplished by employing simulation models[19] that are compatible with respect to input and output.[20] Inputs and outputs for the simulation are discussed below.

3.3.4 Lunar settlement

A lunar settlement established during this century (21st) is a potential goal in line with the new space policy of the United States of America. It may be a dream, but can be modeled and analyzed using our engineering methodologies.[21] In such an endeavor the available and foreseeable technologies have practical but yet unknown limits that could be either technical and/or financial. In general, however, past experience indicates that if a specific technology is known well enough, then its practical limits can be determined for a new application. Thus it appears opportune and interesting enough to explore the limits of lunar development in this century.

There are two formal simulation models available as practical tools for this analysis: LUBSIM (LUnar Base SIMulation) and TRASIM (TRAnsportation SIMulation). These computer codes have been developed and extensively employed during the last two decades at the Institute of Aeronautics and Astronautics of the Berlin Institute of Technology. Nearly one thousand parameters and more than one hundred nonlinear equations are involved in these models, allowing us to estimate the performance and costs associated with such an undertaking on the Moon. Since mass is the key parameter that establishes the cost of taking an object from Earth to the Moon, simulation programs that are used to mathematically model lunar

[18] A trade study or trade-off study is the activity of a multidisciplinary team to identify the most balanced technical solutions among a set of proposed viable solutions. These viable solutions are judged by their satisfaction of a series of measures or cost functions. These measures describe the desirable characteristics of a solution. They may be conflicting or even mutually exclusive. Trade studies are commonly used in the design of aerospace and automotive vehicles and the software selection process to find the configuration that best meets conflicting performance requirements.

[19] A simulation is a computer-based study of how a very complex system behaves. For example, there are simulations of the economy, of a manufacturing process, of a transportation system, and of the settlement of the Moon. These computer programs are used to solve thousands of mathematical equations for the parameters that are used to define the system.

[20] An input is the value of a variable that is known, for example a thrust or a dimension, and an output is something that is determined by the model, such as a velocity or a deflection. The details of such an analysis are available in a separate note entitled "A Perspective of the Space Exploration Vision: Lunar Development," pages 7–11, Info 11/2008, *Lunar Base Quarterly*, Vol. 16, No. 2, April 2008.

[21] Engineering methodologies essentially encompass the way engineers define and solve problems. These are rational methods by which tasks are organized for analysis, design and construction by engineers in dozens of engineering disciplines.

settlement and space transportation convert all activities into an equivalent mass for computational purposes.

A good starting point would be to assume that there will eventually be a justification for 1,000 lunar astronauts and extensive facilities to support them. More than half a century has passed since the first human space launches started with Gagarin, and our goal here is to estimate what it takes to achieve the Bush Vision. Furthermore, it is assumed that an international consortium of government agencies and commercial entities would be in charge and provide the necessary annual resources.

The analysis begins with separate models of the Lunar Settlement and the supporting Space Transportation System. Cargo and personnel required for lunar base operations must be provided annually by the space transportation system selected.

Both systems – the Settlement and the Transportation – represent fairly ambitious projects and establish an initial frame of reference for a development period of ten years for a settlement that will operate for fifty years. They are complemented in such a way that lunar propellants produced would suffice to refuel a lunar shuttle for the ascent to lunar orbit within a period of ten operational years.

Preparatory steps for simulating lunar development, with this hypothetical goal of a permanent lunar installation in mind, results in the following initial estimates of logistics requirements, in units of millions of kilograms per year, on average:

Lunar regolith processed	30
Beneficiated minerals mined	10
Share of gaseous products	0.6

Beneficiation of minerals is the process of treating the ore to make it more suitable for smelting. The development costs of the initial lunar facility for the first year is estimated to be $12 billion.

After many simulations of such a lunar settlement life cycle the following tentative characteristics of this scenario were obtained. This information will be used to conceive an adequate space transportation system.

A lunar base has many individual facilities depending on the concept and purpose of the lunar base. These facilities undergo interrelated changes during their life cycle. They have to be designed in such a way that the system effectiveness will improve with time. The model presented in this document allows for the construction and staffing of twenty individual facilities: strip mining, beneficiation, chemical processing, mechanical processing, fabrication shop, assembly, laboratories and scientific equipment, construction equipment, gas processing and liquefaction, storage for rocket propellants, power plant system on lunar surface, lunar dump, lunar spaceport and equipment, central storage (other than for rocket propellants), central workshop for maintenance, repair and facility extensions, central carpool and surface transportation facilities, control center for all lunar facilities and activities, housing and offices – including health and recreation facilities, life support and recycling, and a lunar solar power satellite in space serving the lunar facilities.

Cost categories of recurrent expenditures (as percent of total) can be grouped as follows:

Equipment delivered	34.5%
Consumables delivered	10.5%
Training lunar personnel	6.3%
Salaries lunar personnel	18.2%
Earth support personnel	30.5%
Total	100%

Another way of grouping how costs are accrued is through the following cost centers,[22] including development share:[23]

Export products	9.6%
Laboratory services	42.4%
Propellants and port services	32.9%
Internal consumption	15.1%
Total	100%

3.3.5 Model of a lunar settlement in summary

The following summarizes the key assumptions and results of the simulation:

1. A 10-year development followed by a 50-year operational lunar program requires average annual deliveries of cargo (facilities, equipment, supplies) of about 1,080 metric tons and during 50 years or a total of 54,000 metric tons, beginning with a few hundred metric tons during the first years, of up to nearly 2,000 metric tons per year towards the end of the 50-year life cycle.
2. The average lunar population that can live and work in the lunar settlement will start out with about 40 astronauts. The entire lunar population will approach 1,000 in operational year fifty, with an average annual level of 510 people during the assumed fifty-year life cycle. At the end of the fifty-year cycle, there would be room for 250 people on the Moon as paying customers engaged in research and development or tourism.
3. The production capacity of the lunar facilities is laid out to gradually produce up to 1,300 metric tons of oxygen propellant per year, or an annual average of about 570 metric tons. These will be used for the ascent leg of the reusable lunar shuttle to lunar orbit.
4. The production system scenario assumes a potential of 60 metric tons per year of export of rare materials or valuable research products.
5. With the assumptions made, annual costs per lunar inhabitant have been estimated to be about $5,600 million/person-year – total lunar settlement life cycle expenditures of $143 billion (2005) or $2.38 billion annually during a 10 plus 50 year life cycle. This total cost includes the $47 billion cost of the lunar propellants required by the shuttle. [The cost of producing lunar propellants

[22] A cost center is any unit of activity, group of employees or machines, or line of products, for example, isolated or arranged in order to allocate and assign costs more easily.

[23] The development share is the percent of total cost for a group of items.

could also be transferred to the cost of the logistics system if total transportation costs are to be compared and thereby reduce direct settlement cost to $96 billion.]

With a total of 27,000 people living and working on the Moon for about a year, the specific[24] cost of operating a lunar settlement in a 50 year scenario would amount to $3.555 million/person-year, or $68,376/week per person.

3.3.6 The logistics system serving a lunar settlement

The following parameters govern the transportation costs of a major space program, and their quantitative values need to be estimated and then used in the simulation model equations. They are:

Annual delivery of cargo mass and space crews
Launch vehicle payload capability per mission
Duration of operational life-cycle
Annual launch rate
Mission reliability
Vehicle losses
Turnaround time of a specific launch vehicle between two launches
Turnaround time of launch positions between two launches
Number of reuses of launch vehicle subsystems
Unscheduled replacement of components due to random failures
Maintenance and repair work hours required per mission
Learning curves for production
Learning curves operation
Level of effort of sustained engineering and product improvement
Cost of direct human labor in reference year.

In analyzing the logistics support system of the conceived lunar settlement, the following assumptions have been made:

1. Maximum operational life cycle of the space systems analyzed will be 50 years after the 10 years of development. If this phase of lunar development is initiated in the year 2020, the life-cycle time frame will be 2020 to 2080. If Mars expeditions have a higher priority it could well be 20 years later (2040 to 2100).
2. Positive effects of other possible space programs such as Interplanetary Expeditions or Space Solar Power Systems have not been taken into consideration.

[24] The term "specific" in commerce denotes a fixed cost per unit, for example number, weight, or volume. In the physical sciences, it generally denotes a physical constant or physical property in the form of a ratio to a standard unit, for example, as the ratio of the quantity in the substance to the quantity in an equal volume of a standard substance, as water or air, or as a quantity per unit length, area, volume, or mass.

3. Vehicle size and capacity clearly affect the maximum size piece that can be delivered to the Moon – the larger the pieces that can be delivered, the smaller will be the human workload of assembling and maintaining lunar facilities. It must also be remembered that the cost of human labor per hour in space is one to two orders of magnitude higher than on Earth! Two orders of magnitude is a factor of 100. This calls for a sizable launch vehicle with a large payload mass and payload volume. However, there will also be limits in the size of what is delivered to the Moon due to handling and surface transportation difficulties.
4. The turnaround time of the same launch vehicle between two launches will strongly influence how many vehicles must be available at the launch site. It is assumed that 2 months would be achievable in the beginning and that technical improvements and operational experience will lead eventually to leveling off at about 30 days.
5. Another important variable will be the number of reuses of a single launch vehicle. This is still debated and considered by many as a potential bottleneck. The vehicle concept chosen is based on ballistic reentry and dry and soft landing (using rocket power) on the ground. Either there is a suitable island downrange or an artificial island (comprised of retired tankers or aircraft carriers) for the first stage. The number of reuses per vehicle will be different for the subsystems comprising the launch vehicle.
6. The lunar orbit space operations center (LLO-SOC) would be placed at its destination during the first year of operation. It serves as a transportation node for the stationing of lunar shuttles, as a propellant depot and for the rotation of personnel and cargo.
7. The lunar shuttle is normally attached to the LLO-SOC or stored at the lunar spaceport. Its design life is assumed to be limited to 20 years due to the harsh space environment. Vehicles are retired on the Moon after 20 years of operation and used for spare parts or other purposes.

Performance and costs of the logistics system

The following table summarizes the key logistics categories, their simulation-based estimated costs (in millions of 2005 dollars) and their percent of total cost.

Launch vehicle & lunar shuttle development	23,309 (12%)
Launch vehicle & lunar shuttle production	54,666 (28%)
Crew capsule and cabin & SOC development	9,875 (5%)
Crew capsule and cabin & SOC production	18,075 (9%)
Ground support operations – personnel	53,760 (28%)
Ground support operations – spares	35,280 (18%)
Sum	$M 194,965 (100%)
Lunar propellants and port services costs	$M 47,050
Total logistics costs	$M 242,015 (124%)

During a 60-year program life cycle, total expenditures associated with the creation of a lunar facility – as outlined earlier – are estimated to be approximately

$143 billion – and it needs to be supported by a logistics system that is estimated to require expenditures of approximately $195 billion. This adds up to $338 billion, or an average of $5.6 billion per year. These are the "should" costs of a competently managed program that would not be constantly changed by the customer![25] Thus, these cost estimates assume no major mistakes and no changes in plans or policies. Strategic reserves, government oversight and contractor fees are not included.

In this program scenario a total of up to 27,100 people would spend a year on the Moon at a cost of about $338 billion or $12.5 million per lunar man-year. 54.1 million kilograms of imports would be delivered from the Earth at an average specific cost of $2,060 per kg. Round-trip costs are estimated to be about $3,530 million per person, excluding lunar propellants.

If the $47 billion cost of lunar propellant production is transferred from the books of the lunar settlement to that of the logistics system, these funds would increase logistics costs by 24% from $195 billion to $242 billion, leading to corresponding increases in specific transportation costs of $2,556 per kg and $4,380 million per person, respectively.

3.3.7 Financial prospects of lunar tourism

In the scenario outlined above there would be room for allowing tourists to visit the Moon for a week or more. It is of interest to determine the expected costs in a specific scenario compatible with the lunar settlement analyzed. Obviously safety of the transportation system is the most important criterion to assess the prospects of lunar tourism. This must first be demonstrated for many years before allowing large-scale tourism. Thus, the first tourist can board the spaceship perhaps during the second decade of operation of the lunar space transportation system described above. In this scenario it will be assumed that the second decade will see the first tourists, with their number increasing with time and costs reduced simultaneously.

It is assumed that there would be an average of 30 tourists on one flight to the lunar spaceport and back. They would be prepared on the ground for one week, have a travel time of three days to the Moon, stay there seven days, and require four days for return. They will remain at the landing site for debriefing[26] for another week. A visit to the Moon would thus require a four-week vacation. This model can be analyzed on an annual basis or for selected time periods. The table below indicates the trend for the decades considered, with data at operational years 15, 25, 35 and 45 considered to be typical.

[25]This was a problem with the Space Shuttles and the International Space Station. Political conflicts led to numerous redesigns of these spacecraft, leading to sub-optimal systems at larger costs.

[26]Debriefing could include medical tests as well as the gathering of feedback on the experience as part of an effort for improvement.

Operational year	Flight frequency	Tourist flights/ decade	Tourist trips/ decade	Total sales $B
10 – 20	0.25/mo	40	1,200	8
21 – 30	0.5/mo	60	1,800	10
31 – 40	1/mo	120	3,600	17
41 – 50	2/mo	240	7,200	30
Sum		460	13,800	65

This number of tourists – 13,800 – would require nearly half of the available transportation capacity, leaving the other half for the regular crew. That would mean a doubling of the average stay time for these astronauts, which might be acceptable. Program costs to the operator would be reduced by about $65 B (or 27%) as a result of the income received from tourists. The alternative would be to charter extra flights for transporting the tourists, resulting in higher program costs.

In summary, a tourist trip to the Moon costing $5 M (2005 value) appears attractive when compared to the current week-long trip to the International Space Station in low Earth orbit that has been sold for $20 M.

* * *

The view of the analysis above from the vantage point of 2169 is that it was prescient and accurate. While many variables in the mathematical model had to be estimated based on very limited experience at the time, the expertise of the analyst led to quite accurate predictions of how lunar settlement and tourism eventually evolved.

In the next section we summarize some of the industries that have become profitable as a result of almost a century and a half of lunar settlement and economic development. This is followed by Section 3.5 that summarizes the view from 2008.

3.4 Possible economically viable activities on the Moon and in lunar orbit

Some of the commercially-viable activities on the lunar surface (one-sixth g) and elsewhere in the Solar System are listed below. Given that the lunar infrastructure is well developed, it is less expensive to place objects into low Earth orbit from the Moon than it is from the Earth's surface. Therefore, materials that may be needed in LEO, or for other space exploration activities – such as asteroid mining or Mars development – are supplied from the lunar surface, if possible. This fact alone justifies early proponents of settling the Moon first, before Mars. The use of the Moon and its orbital space has become the base for our expansion to Mars and the rest of the Solar System. Our space stations and space elevators comprise an infrastructure that can move massive amounts of material as well as hundreds of people a day to and from lunar orbit.

The list that follows is meant to provide the reader with only a glimpse of the kinds of commercial activities we find on the Moon today, two hundred years after the first man landed on the Moon.

Remote sensing of Earth and the asteroids for minerals and precious materials, of Earth's environment to monitor weather, for pollution studies and land use, and by news organizations, is a valuable service. Satellites are manufactured on the Moon and placed into LEO. The cost savings here has driven much of the satellite industry to the Moon. The Moon, with its advanced telescopes, now monitors all bodies in the Solar System for potential sites for settlement and robotic exploration.

Medical research requiring hard vacuum and/or low gravity and/or isolation takes advantage of the lunar environment. Medical treatments of massive injuries are now routinely done on the Moon. Low gravity and microgravity are beneficial for treating burn and trauma patients, those with massively degenerative diseases and is very helpful to those with serious injuries that require extensive rehabilitation. We have facilities for the growth of human organs and cartilage. It was discovered that the low gravity environment permitted greater control of the processes involved. There are laboratories for immune systems research. The pharmaceutical industry has set up major research and production facilities under the lunar surface as soon as it was feasible. Not only was there a major need to understand and control how medicines work in low gravity, the low gravity and hard vacuum environment led to the development of new methods of pharmaceutical delivery systems, including nanotechnology-based systems.

Materials research on the Moon has led to a renaissance in advanced materials development. Examples include single crystal components such as fan blades and engine cores. Improved sedimentation and cooling processes have led to materials with reduced mass and increased strength. Low gravity casting processes have led to glasses and ceramics that are widely used in lunar construction. In particular, perfect optical systems have been manufactured and used in scientific research equipment as well as telescopes of fantastic resolving power – one thousand times that of the Hubble. Many of the processes developed for materials production take advantage of the extreme lunar temperatures.

***In situ* and meteor/asteroid materials utilization** are the largest sectors of the lunar economy. Large quantities of Helium-3 are mined and processed into fuel for nuclear fusion energy systems. Such systems are now ubiquitous on Earth, the Moon and Mars, and all of our spacecraft and our robotic systems. Some of these power systems are as small as a shoebox. Due to the large quantities of Helium-3 on the Moon, energy costs have been cut by two orders of magnitude – one hundred times less expensive than energy costs at the turn of the 22nd century. Petroleum products are no longer used for energy, only for certain types of product manufacturing. Elements such as oxygen and hydrogen are utilized to synthesize liquid rocket propellants. And the regolith, when we include the resources we are able to recover from the asteroids, has most of the elements we need to survive and prosper on the Moon.

Fig. 3.9 A lunar mining operation. (Drawing by Pat Rawlings for NASA. Courtesy NASA) See Plate 5 in color section.

Solar power generation continues to be a major sector of the lunar economy. It preceded fusion power, and during the late 21st century solar power was beamed to Earth, generating large revenues for the Moon. Such revenues were valuable since we were still importing many items from Earth – our *in situ* resource utilization processes had not yet met their present high returns. The lunar surface receives as much solar power as does the Earth, without any atmospheric obstructions. A power grid was placed strategically on the lunar surface within a decade of the first permanent settlement so that 50–60% of the grid is always in the sun. Energy generated this way exceeded the needs of the early base and the surface grid was linked to orbital transmitters for beaming to Earth. Today solar power is still used in some of the more remote regions of the Moon where there are no fusion reactors. Similarly, solar power is beamed back to isolated regions on Earth.

Board and chip manufacturing was an early business activity on the Moon. As soon as a basic infrastructure was erected, such manufacturing began, taking advantage of the hard vacuum and lower gravity and resulting in advanced computing devices. This, in conjunction with the manufacturing of electronics for LEO satellites, is a significant sector of the lunar economy.

Hard vacuum processes have long been used for the manufacture of "perfect" materials. Low-pressure materials-synthesis processes are common in applications where film uniformity over large areas is desired. For example, Chemical Vapor Deposition (CVD) usually involves placing the object to be coated in a large vacuum chamber at low pressure, heating the surface, then introducing process gases that decompose on the surface, creating the film. Many of the costs are due to the vacuum chambers and pumping equipment. Since a hard vacuum is readily available on the Moon, CVD is cheaper and easier. The hard vacuum on the Moon results in fewer impurities in the films, and this helps in some electronic materials applications where dopant levels need to be highly controlled. There are a whole host of advanced technologies for fabrication of microstructures or thin films by molecular beam epitaxy or using electron beams. All of these require high vacuum. Another area is non-agglomerated nanopowder synthesis, where we make large quantities of powder with particle sizes in the tens of nanometers, where the particles are not strongly bonded together (that is, agglomerated). There are many new applications for these nanomaterials. These powders are easy to manufacture on the Moon where vacuum exists.

Astronomical research has blossomed on the far side of the Moon where telescopes and equipment are effectively shielded by the Moon from Earth noise. Naval Research Laboratory astronomers in 2008 created the concept of the Dark Ages Lunar Interferometer (DALI), a telescope that is based on the Moon that studies the young Universe during the first 100 million years of its existence. "Dark Ages" refers to that early time in which the Universe was unlit by any star. With no atmosphere, and shielding from the Earth, the far side of the Moon is a nearly ideal environment for a sensitive Dark Ages telescope. Scientists and engineers investigated novel antenna construction, methods to deploy the antennas, and electronics that can survive in the harsh lunar environment. Research and development for the lunar telescope lasted over a decade before it was deployed. Much was learned about the very young universe. Phased array optical telescopes have been on the far side of the Moon for over one hundred years, yielding fantastic images as well as a wealth of data.

Near-Earth asteroid tracking has been performed in conjunction with astronomical research using unmanned and remotely controlled telescopes. It was recognized in the early 21st century that asteroids were not only the stuff of science fiction. The region in our Solar System between Mars and Jupiter, called the Asteroid Belt or Main Belt, contains millions of asteroids ranging widely in size from Ceres, which at 940 km in diameter is about one-quarter the diameter of our Moon, to bodies that are less than 1 km across. By 2008 there were more than 90,000 numbered asteroids. Today we have them all numbered and tracked. Of primary concern are the asteroids that pass through the Solar System. They are difficult to detect. Three had to be deflected since the tracking telescopes were placed on the Moon in the late 21st century.

Human physiology and psychology in the space environment have always been important research areas. No single factor in manned space exploration is more important than how humans will survive, physically and mentally, in this severe environment. The Moon has provided us with a testbed for better understanding and for the development of coping tools. The first teams of people on the Moon struggled with the new environment – psychological factors weighed heavily. When couples started to settle in large numbers, many of the difficulties faded away, although some new ones appeared.

Plant biology and growth are critical aspects of food production and human survival in a permanent lunar facility. Living in space, the Moon and Mars is possible only if humans are self-sufficient. Plants not only provide us with sustenance, but are also our air filtration backbone.

Entertainment and tourism resorts and amusement parks along with hotels and restaurants can attract huge amounts of capital. Hilton Hotels has shown this with its surface lunar hotel, as has Marriott Hotels. The key has been to demonstrate that such facilities are reliable systems with very low risk. As of 2169 there are four large hotels on – and under – the lunar surface. Each is unique, providing lodging for a different kind of tourist. One is more geared to shoppers, providing unique items that have been created *in situ*. Another is for the active crowd, highlighting low gravity sports. The third organizes treks to the historic Apollo sites. And the fourth is favored by honeyMooners and includes "special accommodations."

Facilities for artists are always sought after. Artists have always been a part of the exploration of new lands. This was true when the "New World" on Earth was discovered and colonized. It was also true when the "New Worlds" in the Solar System were visited and settled. Today we have quite a few artists on the Moon and on Mars – some visit for months at a time and some have settled. Special lofts are available for those who experiment with new vistas and environments. Three-dimensional techniques have led to paintings that could not have been created under Earth's gravity.

Competitive sports have been redefined for the low and micro-gravity environment. Completely new sports have been created – and known ones reinvented – for the Moon and later Mars. Not only do tourists visit to engage in low gravity sports, but professional activity is increasing. We now have two decently-sized stadiums on the Moon. It is clear that low g changes the dynamic functioning of our bodies and their sports abilities in profound ways.

Educational and research facilities for researchers and students take up almost two-thirds of the habitable facilities on the Moon. Once the early settlements of the mid-to-late 21st century stabilized and could be self-sustaining, planetary scientists and medical doctors, as well as a variety of researchers, set up their labs on the Moon to broaden their research activities. Numerous industries set up shop here to explore commercial opportunities given the unique lunar environment. Of course, many of these ventures succeeded and became profitable and grew. From the beginning, these scientists required an infrastructure to support them and their work. When the first families moved to the Moon in 2084, with the first births beginning in 2099, the need for educational institutions and facilities arose. As more children are born, schools are created at all educational levels. However, much of the early education is home-based.

Operations management has always been a part of the lunar enterprise. But as the number of residents grew, the infrastructure became more complicated and the need for large-scale management and maintenance grew. This need created new business opportunities. Private specialists have a large role in the management of the lunar base.

Support services are also privately run. Maintenance, food production, waste disposal, medical services, recycling and salvaging, and some aspects of security are primarily commercial enterprises. One of the difficulties that arose as the lunar settlement grew from one to many and became larger with time was how to set up the legal infrastructure for the commercial operations. Governance required much attention. And even so, significant problems arose.

* * *

There are many commercial activities on the Moon that are quite profitable and keep our economy growing. As an aside, I want to mention something about education and sports.

I have two children – both born on the Moon, a boy in 2142 and a girl in 2143. Both are very healthy. Like me, neither has visited Earth although both are training to be able to do so in the next few years. My husband and I home-schooled both kids until they entered high school even though the lower grades are available publicly. We felt that we could offer more to their education in those early years than could the school.

High school is another story. Due to the high percentage of people with advanced education, and since education is highly regarded, teaching is one of the most prestigious professions on the Moon and the teachers are exceptionally talented. For my Earth-based readers, you may know that college on Earth is very different from lunar college, which is always a part-time activity since everyone has to become economically productive after the age of sixteen. Regardless of your interests you need to contribute to the lunar economy. In addition, you will take a variety of courses in specific and general subjects. Everyone must study emergency procedures and first aid – we all have survival training. It takes the average person about ten years to complete professional or vocational training.

My son, who is 27 years old – we measure our years according to Earth time – finished engineering training last year. He also studied human physiology – he may be interested in medicine at some point. My daughter, who is 26 years old, is finishing her training as a master ISRU mechanic. She was always taking things apart when she was very young and then trying to put them back together again. During her college years she worked at the central manufacturing and maintenance facility in our city.

Both kids are very involved in sports. My son became proficient in three-dimensional football. My daughter is an avid swimmer. As you know, our swimming pools are rotating toroidal cylinders – 50 feet in diameter – where the water – of depth 10 feet – is pushed against the inside walls by the centrifugal forces. The walls are made of a special glass so that we can look into the pool. Sports and physical activity are very important to all the people of the Moon.

My daughter is involved with the use of lunar resources. The ISRU in her title refers to the part of her job that utilizes regolith and converts it into the elements needed to manufacture machine parts. She works with the chemical and mechanical processes that are used to take raw regolith and separate out elements such as magnesium. Some of our structures and machines are designed with ISRU-based self-repair capabilities. This is especially important for our buried structures where access is very limited.

Lunar resources are our lifeblood. We survive on the Moon because we are able to fully utilize the elements found in the regolith and the rocks.

3.4.1 Lunar resources as a driver to lunar development

Proponents of the manned return to the Moon in the early 21st century suggested that the bountiful natural resources on the Moon could economically justify that return. There is abundant oxygen, about 42% of the weight of lunar rocks and soils is chemically bound oxygen. These materials also contain considerable silicon, iron, calcium, aluminum, magnesium, and titanium, which can be extracted as a by-product of oxygen extraction. In addition, helium, hydrogen, nitrogen, and carbon can be found in the lunar regolith. The chart in Figure 3.10 shows how the most abundant elements are distributed on the lunar surface.

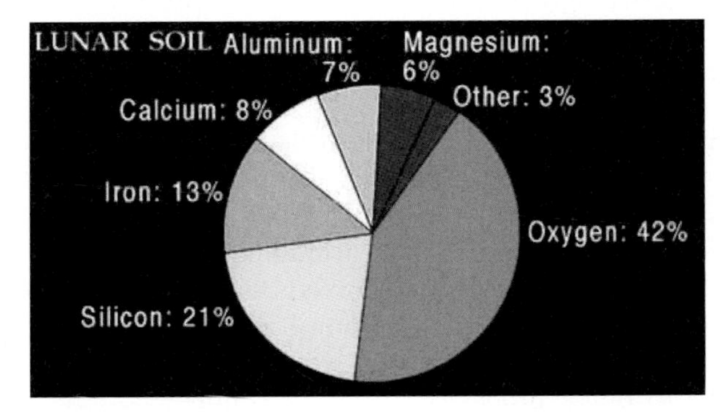

Fig. 3.10 The top six constituents of the regolith.

All of this suggested to the proponents of the return to the Moon in the early 21st century that many important components could be extracted, resulting in oxygen- and hydrogen-based rocket fuels, both for Earth–Moon operations as well as for ships going to Mars. The existence of various metallic ores also suggested a multitude of other uses.

The lunar resource of highest potential was Helium-3, a light isotope of helium and potentially a fuel for nuclear fusion reactors. Unfortunately, these reactors had not been engineered at that time and, had Congress not cut off funds for the Princeton Tokamak fusion research facility, they might have become technically viable by 2040, rather than the actual date of their first use in 2070.

But as discussed elsewhere, *in situ* resource utilization was really the backbone of the lunar economy. There was no other way to go if we were to become self-sufficient in the Solar System. Mars did not offer the same economically viable rationales for would-be colonists, at least in the early stages of settlement. It was too far from Earth and the added time and cost of transporting any Martian products to Earth did not make sense in the early stages of spacefaring.

Section 3.5 presents an economic analysis of lunar products and services written in the early 21st century.

3.5 Prices and sales potential of lunar products and services (2008)

This section is completely based on the work of H.-H. Koelle (August 2008).[27]

3.5.1 Introduction

In the initial phase of establishing a lunar settlement, costs and prices of lunar products and services will not play a major role. Functionality and crew safety will be more important. Self-sufficiency is an important objective, but will probably not be achieved for a long time, if at all. In the long term, however, as commercial activities develop, costs and prices will become important.

The development trend of prices, after a permanent presence has been achieved, would be influenced strongly on the price policies adopted. The effects of different policies will be demonstrated in this analysis using a few selected examples. The following policy options are conceivable:

- **First option**: Lunar propellants are furnished by an owner consortium without charge.
- **Second option**: Lunar propellants are furnished at the production cost for local uses other than propulsion (FOB Moon[28]).
- **Third option**: Lunar propellants are priced including their Earth–Moon transportation costs, but without the burden of the cost of propellants needed to transport.
- **Fourth option**: Lunar propellants are sold commercially on the basis of full transportation costs including lunar propellant costs.

The simplest model of the computer code LUBSIM assumes that no space transportation costs are charged to the Lunar Settlement operator. They are considered to be an investment in the development of space travel by the owner consortium. Resulting specific costs of lunar products and services, such as the cost of lunar-produced oxygen propellants, would cover only cost items resulting from local lunar operations. This case is of interest only as a point of departure providing a lower limit. It is a simplified development model, attempting to keep the cost of lunar products at the lowest possible level in order to stimulate the development of a market.

The next step of estimating the real specific costs of crew and cargo missions requires the input of their specific transportation costs. The computer model TRASIM must be employed. It provides all of the required transportation data and is compatible with the LUBSIM model. However, the cost of propellants produced on the Moon remains a cost center of the lunar organization and is not taken into

[27] Courtesy H.-H. Koelle.

[28] Indicating "FOB Moon" means that the seller pays for transportation of the goods to the port of shipment, plus loading costs. The buyer pays cost of freight transport, insurance, unloading, and transportation from the arrival port to the final destination. The passing of risks occurs when the goods pass the ship's hatch at the port of shipment.

account as a transportation cost. In a full cost model, the cost of lunar propellants is taken into account when calculating market prices in a commercial scenario.

Costs and prices are not the same. Costs are expenditures required to offer a service or product. In contrast, prices are dictated by the market in a free market economy. In the space program, prices have been established for launch services and commercial satellites. There is currently no active lunar program and thus there are no lunar services or products available. That will change after the return of astronauts to the Moon, scheduled for 2020. A market will develop slowly. Laboratory services, rare minerals, lunar art, and lunar tourism can be envisioned in the long term. Their costs will strongly depend on the size of the lunar installation, the mix of products offered and the annual sales volume. An attempt will be made to estimate the market potential of a few near-term items.

3.5.2 Relative importance of lunar products and services

It is obvious that the costs of lunar services and products will depend on a large number of variables and will be critical for their development and usage on the Moon. These potentials can be estimated only for a specific development scenario and a specified lunar base concept. Initial surveys have identified typical products and services a lunar installation might have to offer.

Lists presented below have been developed by participants of the Lunar Development Forum.[29] An attempt was made by this group to determine the relative importance of these services and products for an initial lunar facility currently planned to be established after 2020.[30] They are listed here in order of relevance as decided upon by experts.

Near-term products of a lunar installation are: oxygen for life support and liquid oxygen for propulsion, thermal and electrical power, construction materials, raw materials, hydrogen, metallic products, feedstock (beneficiated minerals), food, electric materials, ceramic products, technical gases other than oxygen and hydrogen, pharmaceuticals, and nuclear fuels (Helium-3).

Near-term services provided by a lunar installation are: launch services for space transportation systems, process engineering development services, equipment engineering, maintenance and repair of space transportation systems, knowledge derived from science on the Moon, knowledge derived from the science of the Moon, training services for other space projects, tele-education, space observation and the protection of Earth, tele-entertainment, materials engineering development, tourism, health care of special ailments, waste storage services, and administrative services.

---

[29]The Lunar Development Forum is an informal group of people from all over the Earth who are interested in the development of space travel. They observe and participate in public discussions of current and future activities to return to the Moon and beyond. The Forum was created by H.-H. Koelle in 1992 under the auspices of the International Academy of Astronautics. A *Lunar Base Quarterly* (LBQ) was published by the group for many years.

[30]From Work Package 7, Lunar Base Quarterly 2005.

3.5.3 Estimating prices of lunar products and services

The most important manufactured product of a lunar facility would probably be lunar oxygen because it is essential for the survival of the crew. The cost of producing lunar oxygen for life support and propulsion purposes depends primarily on the pricing strategy, and on the number and volume of other lunar products and services. All of them together have to support the expenses of the enterprise. Thus, specific product costs will be different in each scenario and each year of the life cycle.

First option: Only those costs originating from establishing and operating the lunar facilities enter the cost calculations of products and services. In this case, all logistics costs are not taken into account. More precisely, these are considered to constitute the globally supported investments for space travel in the interest of securing the future of mankind. In this model, the two program partners – the *Lunar Settlement Corporation* and the *United Space Lines* – are jointly financed by an international consortium, and they agree to support each other without an exchange of funds. The budgets and accounting systems of these organizations are kept separate in the interest of full transparency.

Second option: In addition to the direct settlement costs, as defined in the first option, transportation costs are now taken into account, with the exception of the cost of lunar propellant used by the lunar shuttle.

Third option: In addition to those costs defined in the second option, the costs of lunar propellant used by the lunar shuttle are included. They are transferred from the Lunar Settlement accounts to the account of the logistic enterprise accounts of United Space.

Fourth option: Both organizations (and their possible daughters) are separate enterprises of different owners and will exchange funds and pay for services and products they deliver to the other partner. This is the last step in the evolution of the full scenario, which is several decades away and impossible to specify in detail now.

The LUBSIM model provides very detailed cost output information for many conceivable products and services. The following itemized costs are these outputs, available on an annual basis per unit mass (kg) or per man- or person-year (MY), respectively:

Raw material ($/kg)
Feedstock ($/kg)
Fabricated products ($/kg)
Assemblies ($/kg)
Oxygen propellants ($/kg)
Construction materials ($/kg)
Laboratory services (M$/MY)

Workshop services (M$/MY)
Port services ($/kg)
Road construction services ($/kg)
Surface transportation services ($/kg/km)
Habitat services (M$/MY)
Recycling services ($/kg)
Control system services (M$/MY)
Power plant output ($/Mwh)
Lunar produced consumables ($/kg)
Lunar produced spares ($/kg)
Lunar facility extensions ($/kg)
Lunar produced supplies ($/kg)
Average lunar products for self-supply ($/kg)
Average lunar exports ($/kg)
Total lunar products ($/kg).

For commercial missions, the users would have to be charged lunar propellant cost. Each passenger mission requires 20 Mg of lunar propellants at a specific cost of 3.3 M$/Mg.[31] In this scenario, with 34 people on board in a single mission, the surcharge would be 20 Mg $\times$ 3.3 $M/Mg/34 = 1.94 $M/passenger. Figures 3.11 and 3.12 depict cost trends for transportation and oxygen.

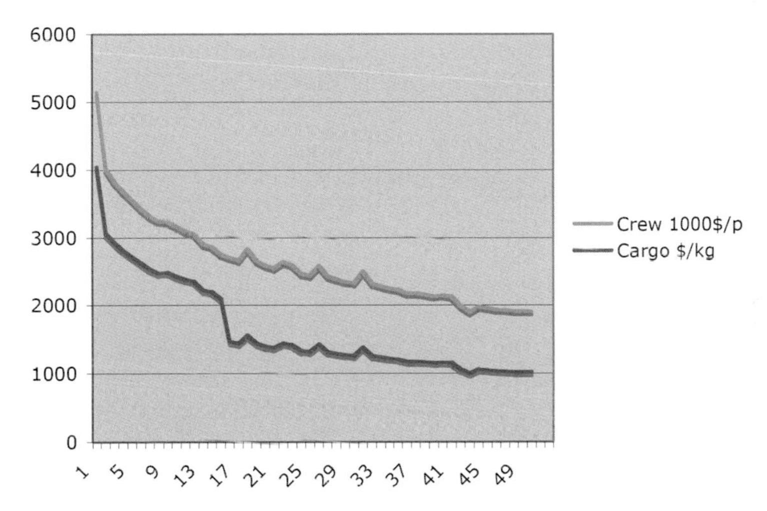

Fig. 3.11 Specific transportation costs for the standard case excluding cost of lunar propellants. The two curves depict crew transportation costs in units of thousand dollars per person, and cargo transportation costs in dollars per kg. The horizontal scale equals the number of years since initiation of the transportation system.

[31]In metric units Mg stands for mega-gram, or 1 metric ton, which equals 1,000 kg.

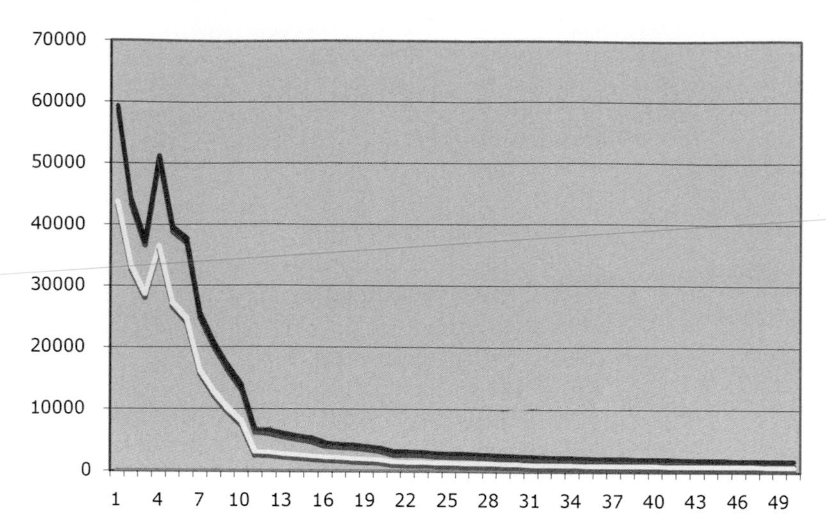

Fig. 3.12 Specific cost of lunar oxygen ($/kg) including and excluding transportation cost. The horizontal scale equals the number of years since the initiation of the transportation system.

3.5.4 Prospects of selling lunar products and services

As the lunar settlement grows and becomes more cost-effective, opportunities will arise to sell services and products. Due to this expectation it is extremely desirable to identify those services and products that may achieve commercial significance.

Laboratory spaces for lease

The current program model uses the number of lab spaces available for direct mission activities as a major driver of growth. In the first decade there is little working space for scientists and developers. The few laboratory spaces available will be assigned to members of public research organizations to accomplish high priority tasks such as the search for valuable minerals and experiments. Beginning about year 15, more space becomes available in the laboratories and commercially-oriented organizations will have a chance to lease space for company employees who need to utilize the lunar environment. It would certainly also be helpful and desirable for company image to have a staff member on the Moon exploring new opportunities. A cost of $15 million to $30 million per lunar person-year would be affordable and found desirable by quite a number of commercial entities.

Mission spaces in lunar laboratories can be used by public research institutions or commercial entities. Public users would reimburse costs through their national R&D budgets; commercial users would have to pay rent at the annual rate determined.

Lunar exports

The LUBSIM model provides detailed output of cost information of many conceivable products and services. In the category of lunar base exports, numerical values for the following items are obtained from the simulation:

Raw material ($/kg)
Feedstock ($/kg)
Fabricated products ($/kg)
Assemblies ($/kg)
Oxygen propellants ($/kg)
Construction materials ($/kg)
Average exports ($/kg).

Raw materials could be rare minerals, for example. One could also think of souvenirs or pieces of art that are created on the Moon by the lunar population. However, exports will be limited because of the relatively high transportation costs and the lack of transportation capacity.

The return cargo of ferries arriving with supplies and equipment can be anywhere between 0 Mg and up to 10 Mg. In the model analyzed here, it is assumed that the average return cargo would be 5 Mg. One can derive a model that describes a plausible scenario.

Due to the relatively high production and transportation costs, sales of lunar export products in the Earth market will be modest and hard to project. However, it is expected that quite a large number of people on Earth would be interested to pay several thousand dollars for a piece of the Moon, even if it is just 100 grams. Thus, these potential sales could be significant and cannot be disregarded.

* * *

Many of the above projections of potential lunar commercial activities came true. As the manufacturing and transportation infrastructure grew, costs declined significantly, as predicted in the charts above, and exports became cheaper to transport and therefore their quantities skyrocketed.

Commercial activities can only flourish within a legal framework that is very specific about the rules of ownership and relationships between entities. Space law had – and continues to have – a difficult evolution. Because of the diversity of people on the Moon from numerous countries with different legal and political histories, and with a multitude of customs, the creation of a unified legal structure is only crawling into being. It has been difficult, but the current system is satisfactory – for now.

3.6 Space law

Laws are necessary in republican[32] democracies. These laws are a framework of what is "allowed" and what is not. At the turn of the 21st century and almost till

[32] A republican form of government is a state in which the supreme power rests in the body of citizens entitled to vote for – and exercised by – representatives chosen directly or indirectly by them.

the 22nd century, there was no body of laws that defined even the most basic of activities that were permitted in space or on the Moon or Mars. The issue of ownership was especially contentious. Thoughtful legal minds came down on opposite sides of whether individuals and corporations – non-governmental entities – were allowed to stake a claim to land on the Moon, Mars or on asteroids. Given that neither government nor individuals can support the settlement of space without attendant rights, there need to be incentives that balance the immense costs and time needed for this effort to succeed.

Various models of ownership have been proposed.[33] Some groups proposed property rights as a powerful force for exploration and the efficient development of the resources of space. Property rights were viewed as the only economic incentive to the space entrepreneur. Then there was the question of how extensive those rights should be. Would they be lease rights or easements? Other models based ownership as contingent on a sustained physical presence by a group of people. This limited claims on more property than could have been developed. A popular option was to set aside some of the extraterrestrial land for a time to also allow less developed countries a chance to have their citizens place claims.

A model known as the "land claim recognition" plan,[34] was based on laws where governments would recognize private extraterrestrial property claims on the condition that the claimant had established a settlement there. This plan limited holders to property areas of the size of the State of Alaska. This size was viewed as sufficient to establish an economically feasible enterprise, and if sold, the property would generate a profit even after the massive startup costs.

The debate on property rights were vague because the Outer Space Treaty of 1967 was vague. It stood as the prevailing basis for those rights. The treaty forbade claims of national sovereignty in space, but it did not mention private citizens. So some believed that the Treaty did not preclude private ownership. However, the Treaty did hold nations responsible for the actions of their citizens in space. Some legal thought viewed this as denying the possibility of private ownership. It came down to legal interpretation.

As long as space settlement and development were viewed as decades away, there was little impetus for setting up the legal framework. But this lack of legal framework hindered development and became part of the projected cost of doing business in space. It was a Catch-22.

Therefore, the legal framework inched forward as needs arose. Technological capability leaped forward permitting commercial entities to create a presence on the Moon and forcing the legal framework to catch up. Progress was not smooth, rather, it was jumpy with starts and stops. There are numerous legal grey zones to this day.

---

[33] This summary is culled from an on-line article in *The Boston Globe*, "My Space – If we want to explore space, maybe we should sell it off to the highest bidders," D. Bennet, 18 May 2008.

[34] A. Wasser, D. Jobes, "Property Rights, and International Law: Could a Lunar Settlement Claim the Lunar Real Estate it Needs to Survive?," *Journal of Air Law and Commerce*, Winter, 2008.

3.7 And water

On 22 October 2008, India's first mission to the Moon, Chandrayaan-1,[35] was launched to gather information about the Moon's origin and the development and the evolution of terrestrial planets in the early Solar System. The Moon Mineralogy Mapper (M^3) was one of two instruments that NASA contributed. M^3 was a state-of-the-art imaging spectrometer that provided the first map of the entire lunar surface at high spatial and spectral resolution, revealing the minerals of which it is made.

On 24 September 2009, NASA scientists announced that the M^3 reported observations of water molecules in the regolith, as well as Hydroxyl, a molecule consisting of one oxygen atom and one hydrogen atom. Data from the Visual and Infrared Mapping Spectrometer, or VIMS, on NASA's Cassini spacecraft, and the High-Resolution Infrared Imaging Spectrometer on NASA's Epoxi spacecraft contributed to confirmation of the finding. "Water ice on the Moon has been something of a holy grail for lunar scientists for a very long time," said Jim Green, director of the Planetary Science Division at NASA Headquarters in Washington. "This surprising finding has come about through the ingenuity, perseverance and international cooperation between NASA and the India Space Research Organization."

"For silicate bodies, such features are typically attributed to water and hydroxyl-bearing materials," said Carle Pieters, M^3's principal investigator. "When we say 'water on the Moon,' we are not talking about lakes, oceans or even puddles. Water on the Moon means molecules of water and hydroxyl that interact with molecules of rock and dust specifically in the top millimeters of the Moon's surface."

The M^3 team found water molecules and hydroxyl at diverse areas of the sunlit region of the Moon's surface, but the water signature appeared stronger at the Moon's higher latitudes. Water molecules and hydroxyl previously were suspected in data from a Cassini flyby of the Moon in 1999, but the findings were not published until [the report of this discovery].[36]

There was great excitement in India about the discovery. " 'It's very satisfying,' said Dr. Mylswamy Annadurai, the project director of Chandrayaan-1 at the Indian Space Research Organization in Bangalore. [The discovery] will also provide a significant boost for India as it tries to catch up with China in what many see as a 21st century space race."[37]

As firm evidence of water on the Moon, NASA released a report[38] of the analysis of the LCROSS[39] impact data. A Centaur upper stage rocket impacted the crater Cabeus near the Moon's south pole while being viewed by the LCROSS spectrometers – only water matched the four minutes of spectral data! Other measures were taken to be certain that the data was not contaminated. LCROSS was launched on 18 June 2009 as a companion mission to the Lunar Reconnaissance Orbiter. After separating from the orbiter, the LCROSS spacecraft held on to the spent Centaur

[35]Meaning "Lunar Craft" in ancient Sanskrit.

[36]http://www.nasa.gov/topics/moonmars/features/moon20090924.html

[37]Timesonline.co.uk, 24 September 2009.

[38]On 13 November 2009 on the NASA website.

[39]Lunar CRater Observation and Sensing Satellite.

upper stage rocket of the launch vehicle, and traveled through a complex set of orbital maneuvers for about 113 days, to bring it to the impact trajectory.

Even looking back at this discovery from the year 2169 is very exciting. At the time, there were doubts that water molecules would be found on the Moon. Once discovered, a major hurdle for lunar development was removed. This was clearly a turning point as important as man's first steps on the Moon.

3.8 Summary

With the launch of the first artificial satellite in 1957, humanity started to explore space and its resources. The first manned excursion to another celestial body was successfully completed in July 1969 with the crew of Apollo 11. Lunar exploration was de-emphasized after 1972. With the new space policy announced in 2004, the United States of America set out again to extend human presence beyond low Earth orbit to nearest-neighbor Moon, eventually to Mars and then beyond. Today, in 2169, the door is wide open for the utilization of the resources of our Solar System with the means available to spacefaring nations and private groups.

A person from the early 21st century would be astounded at our rapid progress. Much of what was speculated then has been achieved. Beyond the technologies, there are societies on the Moon and Mars and outposts all over the Solar System. Human civilization has evolved into a spacefaring species, with expanded customs and mores befitting its larger presence and its more sophisticated view of the Universe and its place within.

3.9 Tables of Dual-Use Technologies

The following pages present dozens of technologies[40] that support the Return to the Moon, while having significant terrestrial applications. Each such technology has been supported by investors individually, who then did not have to wait for the Moon to be settled to receive their return on investment.

[40]The tables shown here are an evolution of similar tables created by S.J. Hoffman, D.L. Kaplan, Eds, *Human Exploration of Mars: The Reference Mission of the NASA Mars Exploration Study Team*, July 1997.

Dual-Use Technologies: Communications/Information Systems

Terrestrial Application	Technology	Space/Lunar Application
• Communications • High-Definition TV Broadcast • Business Video Conferencing	• Ka Band or Higher	• Telepresence: Vision and Video Data • Interferometers: Raw Data Transmission
• Entertainment Industry • Commercial Aviation • Powerplant Operations • Manufacturing Operations	• Machine-Human Interface	• Control Stations • System Management
• Communications • Archiving • Computer Operating Systems	• Data-Compression Information Processing • Large-Scale Data Management Systems	• Interferometers: Raw Data Transmission Information Processing • System Management, Expert Data • Archiving/Neural Nets

Fig. 3.13 Dual-use technologies for communications and information systems.

Dual-Use Technologies: In-Situ Resource Utilization

Terrestrial Application	**Technology**	**Space/Lunar Application**
• Mineral Analysis, Yield Estimation - Deep Mine Vein Location and Tracking • Wall and Cell Integrity	• Advanced Sensors	• Mineral Analysis, Yield Estimation of Surface Mineral Analysis, and Resource Location
• Deep Mine Robotic Operations for - Mining, Beneficiating, - Removal	• Advanced Robotic Mining	• Surface Mine Robotic Operations for - Mining, - Beneficiating, - Removal
• Improved Automated Processing: Increased Efficiency	• Automated Processing Technology	• Remote, Low-Maintenance Processing
• Reliable, Low-Pollution Personal Transmission • Regenerable Energy Economies • Small, Decentralized Power Systems for Remote or Third World Applications	• Alternative, Regenerable Energy Economies - Methane/O2, H2/O2	• ISRU-Based Engines • Regenerable Energies • High-Density Energy Storage
• Environmentally Safe Energy Production	• Space-Based Energy Generation and Transmission	• Surface Power Generation and Beaming

Fig. 3.14 Dual-use technologies for *in situ* resource utilization.

Dual-Use Technologies: Surface Mobility - Vehicles

Terrestrial Application	**Technology**	**Space/Lunar Application**
• All-Terrain Vehicles for - Research (Volcanoes), - Oil Exploration • Automobiles	• Mobility	• Surface Transportation for - Humans, - Science Equipment, - Maintenance and Inspection
• Reactor Servicing / Hazardous Applications • Military	• Robotics and Vision Systems	• Teleoperated Robotic Systems
• Earth Observation, Weather, Research	• Super-Pressure Balloons (110,000 ft - Earth Equivalent)	• Mars Global Explorations
• Efficient, Long-Term Operations with Low-Maintenance • Machines in Arctic/Antarctic Environments	• Tribology	• Surface Vehicles: Drive Mechanisms, Robotic Arms, Mechanisms
• Helicopters, Autos	• Variable Speed Transmissions	• Surface Vehicles
• Automated, Efficient Construction Equipment • Military	• Multipurpose Construction Vehicle Systems and Mechanisms	• Robotic Construction and Set-up Equipment

Fig. 3.15 Dual-use technologies for surface mobility – vehicles.

Dual-Use Technologies: Surface Mobility - Suits

Terrestrial Application	**Technology**	**Space/Lunar Application**
• Hazardous Materials Cleanup • Fire Fighting Protection • Underwater Equipment • Homes • Aircraft	• Lightweight, Superinsulated Materials	• Surface Suits: Thermal Protection • Surface Facilities: Thermal Protection
• Robotic Assisted Systems • Orthopedic Devices for Mobility of Impaired Persons • Human Power Enhancement	• Robotics • Mobility Enhancement Devices and Manipulators	• Robotic Assisted Suit Systems • Human Power Enhancement
• Hazardous Materials Cleanup • Fire Fighting Protection • Underwater Equipment	• Dust Protection, Seals, Abrasive Resistant Materials	• Surface Suits: Outer Garment
• Hazardous Materials Cleanup • Underwater Breathing Gear	• Lightweight Hi-Reliability Life Support	• Portable Life Support for Surface Suits • Backup Life Support Systems
• Remote Health Monitoring	• Portable Biomedical Sensors and Health Evaluation Systems	• Surface EVA Crew Member Health Monitoring
• Hypo-Hyper Thermal Treatments • Fire Fighting Protection and Underwater Equipment • Arctic/Antarctic Undergarments	• Small, Efficient, Portable, Cooling and Heating Systems	• Surface Suits: Thermal Control Systems • Rovers: Thermal Control Systems

Fig. 3.16 Dual-use technologies for surface mobility – suits.

Dual-Use Technologies: Human Support

Terrestrial Application	Technology	Space/Lunar Application
• Stored Food: - US Army, NSF Polar Programs, Isolated Construction Sites	• Long-Life Food Systems with High Nutrition and Efficient Packaging	• Efficient Logistics: for - Planetary Bases, - Long Spaceflights, - Space Stations
• Improved Health Care • Sports Medicine - Cardiovascular Safety • Osteoporosis - Immune Systems • Isolated Confined Environments / Polar Operations • Noninvasive Health Assessments • Military	• Physiological Understanding of the Human Chronobiology • Understanding of Psychosocial Issues • Instrumentation Miniaturization	• Countermeasures for Long-Duration and / or Micro-g Space Missions • Health Management and Care • Systems / Structural Monitroing and Self-Repair
• Health Care • Disaster Response • Military	• Long-Term Blood Storage	• Health Care for Long-Duration Space Missions
• Office Buildings ("Sick Building" Syndrome) • Manufacturing Plants • Homes	• Environmental Monitoring and Management	• Environmental Control for: - Spacecraft Cabins, - Planetary Habitats, - Pressurized Rovers
• Contamination Cleanup • Waste Processing • Homes	• Waste Processing • Water Purification	• Closed Water Cycles for: - Spacecraft Cabins, - Planetary Habitats, - Pressurized Rovers
• Long-Life Clothes • Work Clothes in Hazardous Environments • Military	• Advanced Materials / Fabrics	• Reduced Logistics Through Long-Life, Easy-Care Clothes, etc. • Fire-Proof / Low-Outgassing Clothes • Building Material for Inflatable Structures
• Efficient Food Production	• Advanced Understanding of Food Production / Hydroponics	• Reduced Logistics Through Local Food Production for: - Spacecraft Cabins, Planetary Habitats

Fig. 3.17 Dual-use technologies for human support.

Dual-Use Technologies: Power

Terrestrial Application	Technology	Space/Lunar Application
• Batteries for: - Autos, - Remote Operations for DOD, NSF Polar Programs	• High-Density Energy Storage • Alternate Energy Storage (Flywheels)	• Reduced Logistics for Planetary Bases • High Reliability, Low-Maintenance Power Systems • Spaceship Power Storage
• Clean Energy From Space	• Beamed Power Transmission	• Orbital Power to Surface Base • Surface Power Transmission to Remote Assets
• Remote Operations for: - DOD, NSF Polar Programs	• Small Nuclear Power Systems	• Surface Base Power • Pressurized Surface Rover • Interplanetary Transfer Vehicle
• Remote Operations for: - DOD, NSF Polar Programs • High Efficiency Auto Engines	• High Efficiency, High Reliability, Low-Maintenance Heat-to-Electric Conversion Engines	• Energy Conversion for Planetary Bases: - Low Servicing Hours, - Little or No Logistics

Fig. 3.18 Dual-use technologies for power.

Dual-Use Technologies: Science and Science Equipment

Terrestrial Application	Technology	Space/Lunar Application
• Energy Resource Exploration • Environmental Monitoring, Policing	• Spectroscopy: - Gamma Ray, - Laser, - Other	• Geo-Chem Mapping • Resource Yield Estimating • Planetary Mining Operation Planning
• Undersea Exploration • Hazardous Environment Assessments and Remediation	• Telescience	• Remote Planetary Exploration
• Environmental Monitoring • Medicine	• Image Processing: - Compression Technique, - Storage, - Transmission, - Image Enhancements	• Communication of Science Data • Correlation of Interferometer Data
• Improved Health Care • Sports Medicine - Cardiovascular • Osteoporosis - Immune Systems • Isolated Confined Environments / Polar Operations • Noninvasive Health Assessments	• Physiological Understanding of Humans • Instrumentation Miniaturization	• Countermeasures for Long-Duration and / or Micro-g Space Missions • Health Management and Care

Fig. 3.19 Dual-use technologies for science and science equipment.

Dual-Use Technologies: Structures and Materials

Terrestrial Application	Technology	Space/Lunar Application
• Vehicles • Fuel-Efficient Aircraft • Modular Construction, Homes	• Composite Materials: - Hard, - Soft • Advanced Alloys, High-Temperature	• Cryogenic Tanks • Habitat Enclosures • Pressurized Rover Enclosures • Space Transit Vehicle Structures
• Aircraft Fuel Tanks • Home Insulation	• Superinsulation • Coatings	• Cryogenic Tanks • Habitable Volumes
• Large Structures, High-Rise Buildings, Bridges • Commercial Aircraft: - Improved Safety, - Lower Maintenance	• Smart Structures • Imbedded Sensors / Actuators	• Space Transit Vehicle Structures • Planetary Habitat Enclosures • Surface Power Systems • Rover Suspensions

Fig. 3.20 Dual-use technologies for structures and materials.

Dual-Use Technologies: Operations and Maintenance

Terrestrial Application	Technology	Space/Lunar Application
• Military • Systems and Structures Health Monitoring • Inventory Management	• Task Partitioning • Reliability & Quality Assurance in Long-Term, Hazardous Environments • System Health Management and Failure Prevention Through AI and Expert Systems, Neural Nets	• Systems and Structures Health Monitoring • Inventory Management • Self-Repairing Technologies • Logistics Improvement

Fig. 3.21 Dual-use technologies for operations and maintenance.

4 Extraterrestrial tourism

"I fully expect that NASA will send me back to the Moon ... and if they don't ... why, then I'll have to do it myself."

Pete Conrad

The first tourists on the Moon were the astronauts. Everyone was a tourist. It took a lot of willpower to focus on one's work rather than look around in awe at the extraordinary sights. Of course, eventually, the view became more routine and, regardless, the astronauts were very disciplined and work came first.

Several generations of my family visited the Moon before my parents permanently moved to the Moon in 2115. My great grandparents visited for the first time in 2059. They were involved with the construction of the Earth and later lunar space elevators. The initial strands of the Earth elevator were placed in orbit in 2046. It was brought up to full capacity within four years. The initial strands of the lunar elevator were placed in orbit in 2049.

The first time my great grandparents visited the Moon, the low gravity was not a problem for them since they were already used to the microgravity of their orbital habitats. But my parents who came straight to the Moon from Earth had a lot of adjusting to do. They often described to me how difficult it was for them to walk – it was more like hopping and the gait was very different than a walk on Earth. Eventually their bodies incorporated the new skill into their subconsciouses and they "hopped" all over the place. I was born on the Moon, so it all came naturally to me.

I always had a laugh – not out loud of course – when tourists visited the Moon. Many tourists were eager to see the wondrous lunar sights and forwent their orientation period. These visitors could be seen bouncing off walls and ceilings. It was painful to watch.

The other important component of the three-day orientation period was to learn how to perform bodily functions and how to shower. Low gravity changes the fluid mechanics within the human body – the blood flow is very different – but the mechanics of fluids outside the human body is also different. Visitors were more patient with that particular orientation.

Today, tourism generates one-third of the lunar domestic product. There are hotels, support services, and souvenirs. Some of the tourists become enamored with the low gravity effects on their body – they can maneuver as they did when they were teenagers – and look into retiring on the Moon.

4.1 Earth orbital tourism

By the end of 2008, six space tourists flew to and from the *International Space Station* on *Soyuz* spacecraft arranging their trips through the space tourism company *Space Adventures*. They were,

1. Dennis Tito (American): 28 April – 6 May 2001
2. Mark Shuttleworth (South African/British): 25 April – 5 May 2002
3. Gregory Olsen (American): 1 October – 11 October 2005
4. Anousheh Ansari (Iranian/American): 18 September – 29 September 2006
5. Charles Simonyi (Hungarian/American): 7 April – 21 April 2007
6. Richard Garriott (American): 12 October – 23 October 2008.

Others subsequently went to the ISS as well. When *Virgin Galactic* began commercial operations from the *Mojave Spaceport* in California and then from *Spaceport America* in New Mexico, the number of space tourists jumped. Eventually prices came down, more spacecraft were built and there was a considerable wait to get into space. Supply and demand was beginning to take hold and space tourism, while not commonplace, was safe enough to recede into the background noise of the up-and-coming spacefaring population.

Interorbital Systems made public in 2008 the design of its manned orbital launch vehicle, *Neptune*, shown in Figure 4.1. It passed a major milestone by completing the propellant tank construction of its *Sea Star* microsatellite launch vehicle. Sea

Fig. 4.1 Neptune was the first of a new generation of low-cost and highly reliable manned orbital launch vehicles that came into operation in 2009. It was designed for minimum cost and maximum reliability. Unnecessarily expensive, complex, failure-prone, and sometimes performance-limiting systems such as wings, ignition systems, and turbopumps had been eliminated from the design. (Courtesy Interorbital Systems)

Star was a subscale version of – and testbed for – the Neptune six-passenger orbital tourism ship. Both vehicles employed a novel modular, pressure-fed, two-stage-and-a-half-to-orbit configuration.

These are just examples of the large commercial investments in space tourist spacecraft, orbital hotels, lunar hotels, and – for the pioneering tourist – Martian tourist outposts. Eventually, people came to the Moon for extended stays. Some of those with scientific or engineering skills searched for work in the research labs. College students from Earth took academic years abroad – on the Moon. Those early days were very exciting!

4.2 Lunar tourism

The first real tourists – those who traveled here to vacation – arrived about five years after the first outpost. It was very difficult. There was not much of anything – tight quarters, packaged foods – not really much to do except walk on the surface and look around, and not get in the way of the engineers building the base and its facilities and the scientists setting up experiments and laboratories.

In a paper delivered at the 1967 American Astronautical Society meeting, Barron Hilton delivered the first vision by a major hotelier of a lunar hotel.[1] Hilton was serious about the prospect of such a hotel, and described how it might look. "[The] entrance to the Lunar Hilton will be on the surface of the Moon, but most of the Hilton will be situated beneath the surface – say 20 to 30 feet – to establish constant temperature controls and a more workable hotel area. The experiments of Surveyor 3 seem to indicate that excavations on the Moon are possible and that the Moon soil might be used for construction.

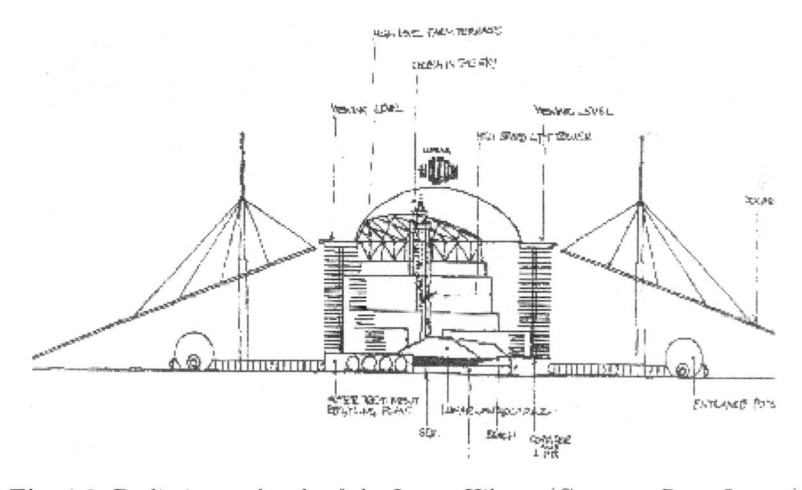

Fig. 4.2 Preliminary sketch of the Lunar Hilton. (Courtesy Peter Inston)

[1] B. Hilton, "Hotels in Space," AAS Conference Proceedings, Preprint AAS, 1967, pp. 67–126.

The Hilton will have three levels. At the bottom mechanical equipment will be housed. The center level will consist of two 400 ft guest corridors crossing in the middle core. These corridors will contain 100 guest rooms. The top level will be used for public space. ... The rooms will be large, with carpets and drapes and plants. ... There will be wall-to-wall television for programs from Earth and for views of outer space. ... Water, oxygen, weightlessness – such problems are being studied. If they cannot be solved, we cannot have a space Hilton."

In late 1999, a report came out[2] that the British-based Hilton chain had "hired London architect Peter Inston to design a $3 billion glass-domed hotel with 5,000 pressurized guest rooms, private baths, galactic views, and a lunar beach and artificial sea." In an update, "Inston, who is taking a year off to focus on the project, [predicts that] 'Lots of people would love to go on holiday to the Moon, especially if they could stay in luxury surroundings.' Insisting 'this is a serious exercise,' Inston acknowledges, 'we may be a little early, but it will happen.' "[3]

Fig. 4.3 Exterior view of the Lunar Hilton. Included in this drawing are the Earth in the background, a rocket taking off for parts unknown, tourist- or scientist- astronauts walking on the surface, as well as a meteoroid. (Courtesy Peter Inston)

From a biological point of view, having relatively untrained people here, civilians so to speak, was valuable. Physiology was and is the Achilles heel of a permanent human presence on the Moon. At that time, little was known about how humans would survive on the Moon. There is the low gravity, the constant bombardment by various types of radiation. Civilians spent about a year to get into shape and to learn what they would need to survive here for a week. And it was very expensive, about five times the average annual salary in the United States.

Today we have several large hotels, largely buried under the surface. One of the early concerns was how to deal with claustrophobia. Given the spartan facilities,

[2] J. Dash, "Out-of-this-world hotels: Architects develop cosmic designs," *The Des Moines Register*, 5 September 1999.

[3] S. Kaltenheuser, "Moonstruck: The Lunar Hilton Update," lodgingmagazine.com/9807/reflect.htm.

Fig. 4.4 Interior atrium of the Lunar Hilton. There is an undeveloped area where the cratered surface is visible. The Earth can be viewed through the clear cover at the top. (Courtesy Peter Inston)

we always lost a few who did not stay for the whole week and had to get back to the open spaces of Earth. It takes getting used to, except for the *lunites*, who have the advantage of being born here and are used to the reality of lunar living. However, lunites have a real problem if they go to Earth. Especially difficult is the six times larger gravity. But on a psychological level, having all that space can be disconcerting and frightening if you are used to close quarters. We lunites frequently need to find close quarters on Earth to help us cope. The problem is a variant of agoraphobia.

Hilton had the first hotel on the Moon. It was a surface structure, with little foundation. When it was built the infrastructure was minimal, and digging was very challenging. In fact, it is still very challenging. We'll discuss those challenges in another chapter. The first Hilton on the Moon was based on the 20th-century design of Peter Inston. It is a dramatic structure, looking much like a pyramid. Inside is an atrium with shops and cafes, some of the lunar surface is exposed, and there are glass expanses allowing us to look out at the black sky, and sometimes to Earth. When it was initially built, the glassy material used was not up to the shielding job it was given. After several piercings by micrometeorites it would shatter and fail. Fortunately, there were only minor injuries. Our fail-safe anti-decompression pumps always gave people a few minutes to escape into the pressurized portion of the structure. We still have problems with such expanses.

Some suggested that we replace the clear views offered in some of the older lunar structures by flat panel televisions. Eventually that is what happened. Today's technology is quite remarkable by comparison to the flat panel technology of the 21st century. The panels are super-thin now and can be just rolled up and carried away. A panel of area 900 square feet (100 square meters) can be rolled into a tube of one inch diameter and then folded with no damage. The resolution is quite re-

Fig. 4.5 At the Lunar Marriott, people are flying at the left of the image and there is surface activity at the lower left. (Image created by Paul DiMare for Popular Mechanics – popularmechanics.com. Courtesy Popular Mechanics) See Plate 6 in color section.

markable with the nano-optical elements in use today. Seeing such a panel unfurled and placed on a wall is wonderful, especially once powered up and providing a view of, well, anything. Windows are hardly used today. There are those who are not happy unless they are really looking at the sight itself, but they have to be satisfied with relatively small windows.

We now move to an interview carried out with an early proponent of space tourism and who was interested in the financing of space settlements based on space tourism.

4.2.1 An historical interview with Patrick Collins (May 2008)

Can you give us a one or two paragraph bio?

Born in England in 1952. BA in Natural Sciences & Economics from Cambridge University 1976. Ph.D. from Imperial College, London, on the Economics of Satellite Solar Power Stations (SPS) 1985. Worked as Consultant to ESA on SPS 1979-83. Lecturer and then Senior Lecturer in Economics at Imperial College. Guest Researcher at ISAS, NAL, RCAST, NASDA in Japan.[4] Visiting Professor

[4]In October 2003, the Japan Aerospace Exploration Agency (JAXA) was established as an independent administrative institution, integrating the Institute of Space and Astro-

at Hosei University, now Professor of Economics at Azabu University. Co-founded www.spacefuture.com in 1997.

You were the first person to do a complete feasibility study of space tourism. What drew you to this line of study?

At university during the 1970s, the first "energy crisis" was a big subject, and I got interested in Peter Glaser's proposal to deliver continuous solar power from space to Earth. My main conclusion was that launch cost to low Earth orbit is the key issue.

If it's low enough, anything is possible in space. If it isn't, nothing is. In order to reduce this, reusable launch vehicles are needed – and also a large new launch market. Demand for satellites is just far too small. The demand for tourism seemed to be the ideal market. Professor Makoto Nagatomo in Japan had reached the same conclusion, and we became friends and colleagues in researching both.

Did you have to contend with skepticism?

Yes, all the time! There is almost no research funding for passenger travel – although every government says that their space policy is to encourage commercial activities in space. A dreadful failure of policy.

How important do you think the Virgin Galactic enterprise is to space tourism?

Virgin Galactic has been very valuable – not least because of the company's high public profile. I have my fingers crossed they will be successful.

When you envision the future, where do you see us in 50 years? In 100 years?

If serious investment in orbital passenger vehicles starts soon, within 50 years there could be cities on the Moon, and within 100 years domed cities. The Earth's environment could be restored to become a "biotope." The seas could be pristine again, with much of the energy and minerals people use coming from outside the Earth.

Why is settling the Moon so important for civilization?

The Moon could become a very popular tourism destination. two-week trips would be very convenient. Developing this industry will raise humans' consciousness about our place in space.

How do we answer critics who say space is too expensive and that there are numerous problems on Earth to take care of first?

Controlled space flight was achieved 66 years ago. If rocket engineers had been allowed to continue this work for nonmilitary purposes, suborbital passenger flight

---

nautical Science (ISAS), the National Space Development Agency of Japan (NASDA) and the National Aerospace Laboratory of Japan (NAL). The Research Center for Advanced Science and Technology (RCAST) is part of the University of Tokyo, Japan.

services could have started in 1950 – so we've wasted 60 years so far! No joke. But this also shows how easy it is: just a few percent of today's space budgets would enable suborbital flights to grow into a popular low-cost leisure industry worldwide, and would lead on to the development of orbital vehicles. This could reduce launch costs to 1% of Expendable Launch Vehicles (ELV) launch costs – more than repaying any public subsidy. Thus it would be more expensive to carry on using ELVs.

From a different angle, Marco and Cristina Bernasconi published a paper in 1997 explaining that if we do not start to use space resources soon, life on Earth will get less and less civilized.[5] Events since then have shown them to be tragically prescient. We cannot afford not to use space resources.

What do you see as the major hurdles for our return to the Moon for permanent manned settlements?

We need to develop tourism in low Earth orbit (LEO) as soon as possible. In energy terms LEO is "Half-Way to Anywhere," so lunar travel and tourism will then become easy. In addition, demand for various materials from orbiting hotels could provide a first market for large-scale exports from the lunar surface.[6]

How optimistic are you that President Bush's timeline for the return to the Moon will approximately be kept?

It seems unlikely to me. Developing reusable passenger space vehicles is a much better route, whether privately or publicly funded.

* * *

4.2.2 Lunar sports

We have all experienced how physical activity in one-sixth g differs from counterparts in one g. We can justifiably argue that the first lunar sport was walking! What a challenge that is for Earthlings visiting the Moon for the first time. The low gravity throws off all of the internal and automatic mechanisms of the body and mind for balance and motion that hundreds of thousands of years of evolution helped develop. The act of walking, something that becomes natural after a few years on the planet Earth, requires active thought and re-learning on the Moon. Not only do the biophysical processes of the body need to adapt to the low gravity, but the mechanics of motion and balance need to be brought up to speed.

The low gravity environment on the Moon gave rise to numerous adaptations of Earth sports as well as new lunar sports that could not work on Earth because of the six-fold gravity. The most interesting of the lunar sports were accidentally discovered. For example, *human twirl*, the purpose of which is to rotate above the

[5] www.spacefuture.org/archive/why_implementing_the_space_option_is_necessary_for_society.shtml

[6] www.spacefuture.org/archive/tourism_in_low_earth_orbit_the_trigger_for_commercial_development.shtml

ground as many times as possible before landing on one's feet. The initiation of the twirl occurs when a person takes three rapid steps and "trips" on a fixed block on the ground. I think you can guess how this sport was born.

Lunar sports have a long and discontinuous history that began during Apollo. Alan Shepard actually hit two golf balls on the Apollo 14 mission of 1971, and they are still on the Moon, as he said in a 1991 interview on the Academy of Achievement website for students. He said he was looking for a way to demonstrate what the Moon's lack of atmosphere and much smaller gravitational force would mean for a familiar Earthbound activity. "Being a golfer," he said, "I thought if I could just get a club up there, and get it going through the ball at the same speed, that it would go six times as far as it would have gone here on Earth."

Fig. 4.6 View shows the javelin and golf ball used by Astronaut Alan B. Shepard Jr., Apollo 14 commander, during the mission's second extravehicular activity on 6 February 1971. Just to the left of center lies the javelin, with the golf ball just below it, almost perpendicular to it. Dark colored trails are the results of tracks made by the lunar overshoes of the astronauts and the wheels of the Modularized Equipment Transporter. (This photograph was taken through the right window of the Lunar Module, looking northwest. NASA Photo ID: AS14-66-9337. Film Type: 70mm. Date Taken: 6 February 1971. Courtesy NASA)

So with NASA's permission, Shepard designed a club head to fit on the handle of the device the astronauts used to scoop up dust samples. (The collapsible club was brought back to Earth and became the property of the United States Golf Association.) Before the flight, he practiced using it in a space suit and made a deal that if the mission went well, "then the last thing I was going to do, before climbing up the ladder to come home, was to whack these two golf balls."

"It was a one-handed 6 iron because it was very clumsy with our suits," he said in an interview in 1994. "The first one I shanked. The ball came off the handle and it rolled into a crater 40 yards away. The next one I hit pretty flush. Here it would have gone 30 yards, but because there's no atmosphere there, it went about 200 yards." [And also because of the low gravity.]

In 2008, Ken Harvey, No. 57 of the Washington Redskins, had proposed[7] a game called *Float Ball.* It would combine elements of basketball and football into a sort of free-for-all. While the proposal suggested playing in a zero-gravity plane, the Moon was also suggested. On the Moon, players would operate under one-sixth g allowing them to jump six times as high as on Earth. Under such conditions players would have to be able to navigate the "playing field" so as not to hit the walls, obstacles and one another while herding weightless balls of various colors to either end.

"There's a bonus," said Harvey to an attentive audience of National Aeronautics and Space Administration engineers, technicians and scientists at the Goddard Space Flight Center, "where you have to pick up a person holding a certain ball and throw them through a hoop as a sort of extra point."

Another popular sport that had been slated for the Moon and Mars was skateboarding. The skate park would have fully 3-D surfaces on which to skate, including the ceiling. "With new carbon nanotube construction the ceiling of the skate park could be see-through or made opaque by simply a change of the electric current."[8]

One elaborate study suggested a variety of zero-gravity (really microgravity) sports centers.[9] These would be attached to orbital space hotels. Some were eventually adapted to low gravity locations such as the Moon and Mars. The proposals included a gymnasium for individual and group use, a swimming pool for individual use as well as for other full-scale team sports such as water polo, and a large stadium also suitable for full-scale team sports. One hope of the proposers was to use the facilities for a space-based Olympics. The first lunar Winter Olympics occurred in 2094, one hundred years after this study was published in 1994.

Water sports in very low gravity poses a challenge to the designer of a swimming pool as well as the swimmer. In low gravity, water will float in spherical blobs. And without gravity, there will be no buoyancy for the swimmer to rely on to be pulled to the surface of the water. One could just sit there in the middle of a large sphere of water relying on one's swimming skills for motion.

For the swimming experience that we have on Earth, there exists a rotating swimming pool. The pool structure is a pressurized circular cylinder made of a clear material and rotating to create artificial gravity. One can "dive up" from one side of the pool to the other side.

The erection of zero-gravity stadiums allowed team sports to become fully three-dimensional. Players have the ability to "fly" between different parts of the stadium volume. Games such as three-dimension soccer and American football are very

[7] "For Ex-NFL Star, a Dream of Sports in Space," M. Brick, *The New York Times*, 30 October 2008.

[8] L. Winslow, "Skateboarding in the Lunar Colony," EzineArticles.com. 13 April 2007.

[9] P. Collins, T. Fukuoka, and T. Nishimura, "Zero-gravity Sports Centers," *Proceedings of Space 94*, ASCE, 1994.

Fig. 4.7 Pole-vaulting records of more than 30 m at an International Lunar Games event might be established in the next century. The Moon's one-sixth gravity would be an excellent environment for athletic competitions that are hampered by Earth's stifling gravity. This visionary concept of NASA's Advanced Concepts Office looks far into the future at the potential for lunar development and tourism. *(Leap of Faith* by Pat Rawlings for NASA 1 July 1995. Courtesy NASA)

popular. One player is allowed to throw a team-mate from one location to another in order to gain advantage. These football players do not look much like those on Earth. The skills and body type needed on the Moon are very different than those required in Earth's gravity.

4.2.3 Also sex?

Alex Comfort's original *Joy of Sex*, published in 1979, became the lovemaking guide to intimate discovery and experimentation for a generation of adults. It is still popular today, but on the Moon the mechanics of lovemaking is, let us say, very different. In space and on the Moon, we can report on *The Joy of Low g Sex.* As in sports, gravity both helps and makes more difficult the physical act of love. On the Moon in low gravity, dynamic motion can result in unexpected trajectories and injuries.

There was no first-hand evidence of sex in space during the early days of space travel, although there were numerous "urban legends." However, clearly the idea of human intimacy in zero g in space or low g on the Moon and Mars was deliberated in great detail. "The bottom line is that sex in space will probably take some practice and hard work at first. Since people are very creative, I have no doubt that it will make for a wonderful otherworldly experience."[10] But a cautionary note is made about conceiving in space. Conception "may be dangerous to the mother and baby. Based on animal experiments, we know that fetal development is affected in space. Bones, muscles (including the heart), and neurology, will simply not develop properly without Earth gravity. We also know that human hormones and even sperm motility are affected by the lack of gravity. Radiation is a serious problem too, even in Earth orbit where [Earth's] magnetic field protects us somewhat."[11]

But the reader will be pleased to learn that "sex in space may be the ideal exercise to prevent muscle atrophy. During sex, the heart beats faster, muscles in the buttocks, abdomen, chest, and even the face, hands, and feet, contract and spasm. The act of holding on to each other in zero gravity would take a lot of energy too. The resulting vigorous flexing and muscle spasms make space sex a better exercise than stationary bike riding or any of the other exercises that astronauts and cosmonauts [have historically done] to keep in shape. But you would have to be diligent and have sex every day, maybe even twice a day to gain the benefits.

"Studies have found that having sex regularly can keep a person healthy because the activity somehow helps the immune system stay strong. This will be especially important in space where zero g suppresses immune function. Sex may truly be the way to counteract many health problems. This is why it is so important that the space agencies study sex-related issues."

Of course, today – with births occurring on the Moon – we can report that there has been sex in space, and on the Moon, and on Mars (we guess)! And pretty much wherever people live and visit.

Maybe we can trace the first sanctioned sexual activity in space to the Japanese firm First Advantage and the U.S. spaceflight company Rocketplane Global who announced in 2008 that they would host weddings in space for about $2.3 million. A NASA spokesman at that time stated that "we don't study sexuality in space, and we don't have any studies ongoing with that."[12]

We are not sure that we believe them.

4.3 Quotes

- "Space flights are merely an escape, a fleeing away from oneself, because it is easier to go to Mars or to the Moon than it is to penetrate one's own being." Carl Gustav Jung
- "Space is the breath of art." Frank Lloyd Wright
- "There are so many benefits to be derived from space exploration and exploitation; why not take what seems to me the only chance of escaping what is otherwise the sure destruction of all that humanity has struggled to achieve for 50,000 years?" Isaac Asimov, speech at Rutgers University

[10] L.S. Woodmansee, *Sex in Space*, CG Publishing, 2006.

[11] L.S. Woodmansee, "Sex in Space: Imagine the Possibilities," *Ad Astra* online, 4 August 2006.

[12] J. Bryner, "For Better or Worse, Sex in Space is Inevitable," Space.com, 7 July 2008.

- "For me the single overarching goal of human space flight is the human settlement of the Solar System ... no greater purpose is possible." Michael Griffin
- "The most important thing we can do is inspire young minds and to advance the kind of science, math and technology education that will help youngsters take us to the next phase of space travel." John Glenn
- "I'm proud to be an American, I'll tell you. What a program and what a place and what an experience." Charlie Duke, Apollo 16 Lunar Module Pilot, saluting the U.S. flag on the surface of the Moon, 21 April 1972

5 Lunar bases

"It's a brilliant surface in that sunlight. The horizon seems quite close to you because the curvature is so much more pronounced than here on Earth. It's an interesting place to be. I recommend it."

Neil Armstrong

When we think of cities on the Moon and Mars, outposts on asteroids and the moons of the planets, what we see in our mind's eye are structures. These structures are unique to their environments, just as are structures on Earth. And when people started to imagine the possibility of settling these final frontiers what they envisioned were structures and cities.

Just as cities on Earth are centers of commerce and art, and structures house all of our creative endeavors, cities and structures are places where we live and work but also are creative ends of their own merit. Structures are a reflection of our civilization. They are engineered as a culmination of science and aesthetics, serving many functions for people. On Earth they house us and are a place for our belongings. On the Moon, they also keep us alive.

Some of the early documents we have recovered and restored have given us a view of how structural concepts evolved from the functional to the beautiful. Structures are designed at a minimum to satisfactorily function in the environment where they will be built.

There are environmental forces to contend with. On Earth, there are wind and rain, sometimes earthquakes and other times ocean waves. These are the environmental forces. In addition there are the "loads" inside, those due to people and equipment that are sometimes stationary and other times in motion.

On the Moon there are environmental characteristics that on the whole are very different than those found on Earth. There is no wind since there is no atmosphere. The lack of atmosphere leads to a number of serious human survival issues that the structure needs to counter by way of its design. The first is the need for an atmosphere within the structure, pressurized to an acceptable level for humans. While the atmosphere on Earth protects inhabitants from galactic and solar radiation as well as meteoroids, the lack of one around the Moon means that lunar structures – at least those on the surface – will need to be designed with shielding against these dangers.

These issues and others are discussed in this chapter in more detail. Some historical concepts for lunar structures are depicted in the images. Other more recent concepts were created with computer drawing tools. All concepts evoke the essence of the human spirit of exploration, the thirst for new adventures and to know more about our universe and thus ourselves.

5.1 Lunar architecture

As humanity prepared for the settlement of the Moon, individuals recognized that everything that civilization had learned over thousands of years would have to be brought to bear on this greatest of challenges. It would not be enough for scientists and engineers to build rockets for travel and structures for habitation. We were planning for a manned and permanent presence on the Moon, and then Mars. It was an urban environment that we were looking to create.

Humans are complex. Given life support, we can survive almost anything during a round trip to the Moon, as we did with Apollo. Close quarters and no recreation or distraction were not a problem. The color of the interior of the space capsule was not a concern. But then that was a trip of eight to twelve days. The following table shows different time durations for three of the Apollo trips to the Moon.

(h-min-s)	Trip duration	Time on Moon	EVA[1]duration
Apollo 11	195-18-35	21-38-21	2-31-0
Apollo 12	244-36-24	31-31-0	7-45-0
Apollo 17	301-51-59	74-59-40	22-4-0

As groups on Earth moved forward with plans to settle the Moon, considerations of the "other issues" that needed attention grew in importance. Humans in close quarters will tend to be stressed and will require some way to relax. The interior environment needs to be a psychologically positive experience.

A study[2] began to consider how some of these habitation issues could be addressed. "Space Architecture is the theory and practice of designing and building the human environment in outer space." Clearly an all-encompassing definition.

How does space architecture – a discipline born in the space age – look at the human environment? "Many considerations familiar to terrestrial architects – productivity, privacy, assembly, aesthetics, place identity, sensations, view, mood, safety, utilities, and adaptive use, to name just a few – are increasingly relevant to the design of habitable environments for outer space. In addition, living in space brings into sharp focus some new considerations increasingly important to architecture on Earth; sustainability, material recycling, regenerable life support."

This study also explored the roles that artists could play as members of human space project teams. The artist's aesthetics were a valuable resource in the design of facilities for long-duration habitation. It is the rare person that can envision what

[1]Extravehicular activity.

[2]Report of the IAA Commission VI Study Group 6.9 – *Space Architecture: Tools for the 21st Century*, December 2008.

Fig. 5.1 The deployment of the United States flag on the surface of the Moon is captured on film during the first Apollo 11 lunar landing mission. Here, astronaut Neil A. Armstrong, commander, stands on the left at the flag's staff. Astronaut Edwin E. Aldrin, Jr., lunar module pilot, is also pictured. The picture was taken from film exposed by the 16mm Data Acquisition Camera which was mounted in the Lunar Module. While astronauts Armstrong and Aldrin descended in the Lunar Module "Eagle" to explore the Sea of Tranquility region of the Moon, astronaut Michael Collins, command module pilot, remained with the Command and Service Modules "Columbia" in lunar orbit. (S69-40308, 20 July 1969. Courtesy NASA)

it means to live in very close quarters for long periods of time. A prisoner lives that life – even though the prisoner was recruited to that life involuntarily. For the lunar facility, colors can greatly enhance the psychological state of occupants. The windowless lunar structure initially had "virtual" windows composed of flat-screens, giving the impression that the outside was not so far away. Artists created powerful sceneries – not necessarily lunar scapes – that provided a supportive ambiance to the inhabitants.

Space architecture is not only for architects. Engineers also practice space architecture when they integrate human factors into structural designs. The needs of the human occupant of the structural system is incorporated into the technical analysis. While there are many goals for a human lunar settlement, a primary goal must be the wellbeing of the people. All else flows from a satisfied population.

Brent Sherwood, an originator of space architecture, put lunar architecture into the larger framework of architecture as "that group of disciplines that gives vision to the human environment."[3] Referring to earlier remote and hostile environments such as the submarine and Antarctic, human factors became critical to successful lunar habitation design. Such a facility had to "encourage and support a conflict-

[3]B. Sherwood, "Lunar Architecture and Urbanism, 2nd Edition," 05ICES-79. The earlier paper by this author was of the same title and presented at Lunar Bases and Space Activities of the 21st Century, Houston, 5–7 April 1988.

free and pleasant environment" for those within. "Human living is an exceedingly complex activity that requires much more than passably engineered accommodation because it includes all we do: working, resting, playing and growing. ... Architecture has as its purpose the creation of facilities that foster human living."

5.1.1 An historical interview with Brent Sherwood (March 2009)

Can you give us a one or two paragraph bio?

I am a space architect, interested in space construction techniques, design principles, and eventual urbanism, and in technology and science in general. My advanced degrees are in architecture from Yale, and aerospace engineering from the University of Maryland.

I lead Strategic Planning and Project Formulation for the seven businesses of NASA's Jet Propulsion Laboratory in Pasadena, California. Before that I was at the Boeing Company for 17 years, working on human and robotic planetary mission concepts, International Space Station manufacturing engineering, the international Sea Launch joint venture, and commercial space business development.

I have provided leadership in the international community of space architects, through the American Institute of Aeronautics and Astronautics, the International Academy of Astronautics, and independent publications. I have produced over 40 papers and book chapters on the exploration, development, and settlement of space, and coedited the book *Out of This World: The New Field of Space Architecture.*

It looks as though the U.S. will be without heavy launch capability beginning in 2010 when the Shuttle is retired. It seems almost unbelievable that we have gotten into this situation. What do you think will happen?

I expect we will "weather" a gap for a few years in U.S. human orbital access, relying on our Russian friends for rides to and from the International Space Station.

Can you summarize for us what are the foundations of Space Architecture?

If I understand the question, the field of space architecture *per se* originated officially when NASA hired a small number of architects and industrial designers in the early 1970s to participate in the design of Skylab, recognizing it would be supporting crews for long-duration missions. Individual space architects today got their start in a variety of ways, but share a common attraction both to designing the human environment and doing so for space. More broadly, space architecture is "merely" an extension of the design of human environment into the new domain of orbital and planet-surface space flight. In this it shares characteristics with "naval architecture."

Why is settling the Moon so important for civilization?

It is not, necessarily. In the 1960s, space was the "final" frontier. Today, molecular biology and molecular engineering hold that honor, and are more immediately relevant, promising, and dangerous for civilization than space flight of any kind, let alone lunar settlement. The reason planetary settlement is important is that it is inherently interesting and challenging, and is a likely element of learning to use offworld material and energy resources, which may become important for humanity and Earth life (assuming we even survive the next century or two on Earth) and would be essential if humanity is ever to become an interplanetary or interstellar species.

What do you see as the major hurdles for our return to the Moon for permanent manned settlements? Are they technical, financial, physiological, psychological?

The major hurdles are not psychological or technical as far as *in situ* activities go. They are principally psychological in terms of national will to make this a priority, which shows no signs of occurring. Deriving from that lack is a crippling financial hurdle. No one knows whether physiological limiters will come into play (reduced-gravity deconditioning, or long-term damage from natural radiation).

How do we answer critics who say space is too expensive and that there are numerous problems on Earth to take care of first?

Space is too expensive. Part of this is inherent; but part is an artifact of how the military-industrial complex has evolved its contracting environment. The argument that Earth's numerous problems need to be fixed first is specious, because it presumes that great challenges must be approached serially, and that Earth's numerous problems are in fact solvable in any final way. As an argument for inaction, it makes no sense.

We appear to have a number of very motivated competitors for the return to the Moon: China, Japan, Korea, Russia, and the European Space Agency, not to mention dozens of national space programs that are quite competent at placing objects in orbit. Do you think the U.S. and Americans in general take this seriously?

No, nor should they. First, the barriers to entry in space are tremendous: the current crop of suborbital entrepreneurs are far away from achieving orbital flight; the new orbital player China is far away from routine, complex in-orbit operations; the attainment (or re-attainment) of lunar orbital and surface operations is beyond even the U.S. and Russia presently. It is one thing to send a robotic probe to lunar orbit, and quite another to "land a man on the Moon and return him safely to the Earth." Second, we should encourage all the "competitors" we can, for in the end these will all become collaborators. Indeed, the most efficient global progress would be made if the "competing" national programs were to swallow their pride, embrace interdependence, and divvy up the many elements of a complete settlement architecture among them, so that the architecture could become collectively affordable. Else it is not.

When you envision the future, where do you see us in 50 years? In 100 years? How do you see this evolving from the present?

I believe human social and political behavior is essentially constant (i.e., it does not evolve appreciably). That means actual progress will continue to be less than linearly-increasing compared to our desires. However, actual progress will instead, simultaneously, be exponential in dimensions we are not paying attention to, predicting, or interested in for the subject at hand. I believe the only hope for significant space settlement is for progress toward the genetic-nanotechnology-robotic singularity to occur. If it does, truly vital problems like overpopulation, access to potable water and sufficient nutrition, and environmental devastation can be mitigated. Then, humanity as a species could regain the semblance of discretionary investment and leisure that would permit space settlement to re-emerge as one direction for human achievement.

You have worked for NASA and for Boeing. Many say that private enterprise will be the motive force for space exploration and settlement, if NASA just gets out of the way. Given that you have been inside both, what is your view on the best way forward?

It makes no sense to pretend that government investment of any kind is "the problem," even if the principal investor is a federal agency driven by old glories and bureaucracy. The resources required to enable lunar settlement are vast, far exceeding the capacity or likely interest of private capitalists: you would need Elon Musk plus Paul Allen, raised to a significant exponent. I have yet to see a prospective business case that convinces me there is an inherent source of economic value on the Moon to justify enormous investment on Earth to get it or grow it. I believe in the long run – the very long run – extraterrestrial sources of solar energy and rare elements may prove valuable, but this is a completely untested hypothesis.

Given the recent financial meltdown worldwide, does the return to the Moon become more tenuous?

Yes. I think the most likely use of any increased NASA funding in the short term is Earth science.

With the new President Obama having taken office, how do you see President Bush's vision evolving?

Human exploration should fully utilize the huge investment in the ISS as a laboratory to learn high-leverage things like how VASMIR[4] performs, how variable gravity levels affect physiology, etc., before we invest in a lunar lander. And when we do invest in a lunar lander, it should be reusable, designed to take advantage of propellants manufactured *in situ*.

Addendum:

My responses will leave you with the impression that I am "down" on lunar settlement and human spaceflight. I am not. I personally remain highly motivated and inspired by it. But I believe it is farther away today than it was in 1992 in the

[4]Variable Specific Impulse Magnetoplasma Rocket.

denouement of the first Space Exploration Initiative. It is simply not essential to any of the pressing problems we face as one among many nations, on a poisoned, warming Earth; wracked by violence, intolerance, and adherence to myths; and pressing as we are on other, higher-leverage technological frontiers.

The philosophy I've believed for many years (story below) is that the most realistic way to chart a course into the future is to look at what will be natural extensions for a civilization. Viewed that way, the development of Los Angeles around the automobile was a "natural act" in the mid-20th century because American society was primed to combine its romance of the wide-open spaces of the west, its (young) tradition of individual freedoms, and its evolution of horse-powered transportation into internal-combustion-engine-powered automobiles. So although the network of freeways and horizon-to-horizon grid might at first look shocking to a stagecoach driver from a century earlier, it was in fact a natural outgrowth of the civilization of the time. And now if we were to ask, "What is the replacement cost of Los Angeles?" the answer is both moot and (paradoxically) astronomically unaffordable. The lesson is that gradualism and incremental achievement can accomplish astounding feats. But it requires patience.

Applied to the analogy of space settlement, we have to ask, "For what kind of civilization will multi-year expeditions to Mars be a 'natural act'?" I believe it is not us, and this is the fatal flaw of the urging of a Zubrin.[5] I believe the answer, instead, is "a civilization that is already routinely operating away from Earth, in an alien and lethal planetary environment, with all the logistics, technological, financial, industrial, and attitudinal resources that requires." As you know Ehricke argued, "If God had intended man to explore space, He would have given him a Moon." For such a civilization, exploratory expeditions to Mars would be highly risky, a real stretch, but reasonable. For one thing, the societal resources to enable such expeditions would have to be able to be contained to something like 0.5% of the society's economic productivity.

Now take one more step back to ask, "For what kind of civilization would routine operations away from Earth, in an alien and lethal planetary environment (e.g., the Moon), be a 'natural act'?" I believe it is still not us, and this is the fatal flaw of efforts like the Space Exploration Initiative (SEI) in 1989–92 and the Vision for Space Exploration (VSE) in 2004-x. I believe the answer, instead, is "a civilization that is already routinely operating in Earth orbit, exchanging passengers, crew, and cargo with no particular challenge." You can see this is almost us. Russia is closer by this standard than is the U.S., and nobody else comes close. The Shuttle was

[5]Robert Zubrin (born 19 April 1952), an American aerospace engineer and author, best known for his advocacy of manned Mars exploration. He was the driving force behind Mars Direct – a proposal intended to produce significant reductions in the cost and complexity of such a mission. The key idea was to use the Martian atmosphere to produce oxygen, water, and rocket propellant for the surface stay and return journey. A modified version of the plan was subsequently adopted by NASA as their "design reference mission."

Disappointed with the lack of interest from government in Mars exploration, and after the success of his book **The Case for Mars** as well as leadership experience at the National Space Society, Zubrin formed the Mars Society in 1998, an international organization advocating a manned Mars mission as a goal, by private funding if possible.

intended to embody it when coupled with the ISS, but as we all know, the Shuttle is still technologically developmental and economically disappointing. Flying the Shuttle is not without challenge even today – too much pucker factor each time.

One might argue that since we are at the cusp of this criterion, however, we are indeed "ready." I counter-argue that to "push" to the next level when not fully comfortable at the precursor level takes enormous, deep, and sustained national or international will. Any martial artist or musician will concur that skipping a level is sometimes possible, but always dicey. That is why Project Apollo was an anomaly, and why it could not be sustained upon reaching its primary goal. It was "out of step" with the natural order of things – an unnatural act for the civilization that carried it out.

One might also imagine that the surest way to become a civilization that routinely operates in Earth orbit is to promote the development of (commercially viable and growing) orbital passenger travel. NASA has never shown interest in doing this (Dan Goldin: "Space tourism is not my job!"), although it would be directly analogous to the role of its predecessor NACA, which developed the airfoil technology that made the jet age possible. I've maintained a list since about 1995 of specific technologies NASA would develop were this to become its charter. It would be an alternative mission for NASA – one which arguably would be more enabling of a long-term, sustainable, growth-oriented human breakout into space.

This interlinked argument – that big next steps are taken when civilizations are ready, that we are not ready yet for lunar settlement or Mars expeditions, that commercial development of Earth orbital passenger travel is a precursor linchpin, and that the government could accelerate such development through strategic investments not currently part of NASA's charter – is the view I believe has the most merit and the best prospects. In 1990 I presented a paper at the ISDC called "Commercial Resort Hotels in Low Earth Orbit." It was the seminal paper of that era on space hotels (Hilton's was the first, decades earlier) and catalyzed the creation of the "space business park" line of thinking by Chuck Lauer, Joe Hopkins, and me. The original draft (which I still have) began, "The next major step in space exploration will not be taken by a half-dozen government employees walking on another planet, but rather by hundreds of thousands of ordinary people flying routinely in Earth orbit." My organization director at Boeing made me change it to, "The next major step in space exploration will be taken by government employees walking on another planet," to toe the SEI party line of the time. But I still believe what I originally wrote. Reaching for too much too fast puts us far out on a thin limb waving in the wind. The Apollo limb snapped off because its socio-psychological and economic root was weak. The way to build a towering redwood is by slowly growing a stout trunk.

* * *

5.2 The Moon, then Mars

The raging battle among those converted-to-space in the late 1990s and early 2000s was whether to include the Moon on the way to Mars. The Mars-Direct crowd deemed the Moon to be, at best, a diversion from the real goal of colonizing Mars. "Been there, done that" was a common refrain against the return to the Moon.

The Moon-First-on-the-Way-to-Mars group supported the eventuality and possible dominance of a Martian civilization. However, there were clear benefits to colonizing a planetary body three days from Earth versus another that is about a year away. From any perspective except public relations, the clear and rational way for Man to go to space was via the Moon. Mars-Direct ignored critical technical, physiological and financial issues that were unresolved at that time, pretending that existing technology need not be tested extensively before being sent on a yearlong mission to Mars with human life depending on its reliability.

The risk of a Mars-Direct program was that, even in the best circumstances of no catastrophic failures, it would have become another Apollo – we come to Mars, plant the flag a few times, and then go back home for another thirty to fifty years of hibernation from manned space.

The Moon (and ISS) had to be, and was, our stepping-stone to Mars, the Solar System, and beyond. Today in 2169, our economic activities on the Moon finance many of our efforts to colonize the Solar System.

The following section is a discussion of some of these issues from the perspective of the early 21st century.

5.2.1 Framework for understanding the engineering of a manned trip to Mars (early 21st century)

Two documents will be referred to in this discussion in order to highlight that Mars-Direct is not just highly risky, but rather at the level of Russian roulette in its cavalier attitude towards a responsible approach to the building of the necessary experience for a successful mission.

The first document is titled "Engineering Stages of New Product Development," published by the National Society of Professional Engineers (NSPE). The second document is in two parts and is published by the Federal Aviation Administration. One is the Advisory Circular (AC 25.1309-1A) "System Design and Analysis" issued on 21 June 1988 and the other is the Federal Aviation Regulations (FAR) 25.1309. We will refer to these by their respective acronyms.

We refer to and quote from these documents to emphasize what is considered to be accepted practice in any engineering venture. While Apollo was a high-risk venture, it too followed accepted engineering practice on its way to the Moon. What is proposed by Mars-Direct is a stretch of the design envelope to a point where extrapolation cannot be substantiated in any way.

From NSPE, the six stages of an engineering project are:

1. Conceptual
2. Technical feasibility

3. Development
4. Commercial validation and production preparation
5. Full scale production
6. Product support.

"The objective of each stage of the engineering project is to establish the engineering information (technical, economic and risk assessment) necessary to make the decision to proceed or not." The key here is that, in addition to technical and economic decisions, it is imperative to be able to assess risk. In order to do that, relevant engineering experience with appropriate experimental and computational data are required.

To date, manned space flight has had its longest flight between the Earth and the Moon, a round trip of about seven days. Continuous manned presence in space has been on the Mir space station – with a maximum of about fourteen months. A round trip to Mars is approximately two years in duration, plus the time on the planet. We have no experience with manned systems operating in the space environment for such a long duration without access to repair. We do not have the necessary experience to determine how such complex spacecraft components and equipment will fare in such harsh environments. The Moon-First strategy is one that accounts for these unknowns and uncertainties by using the lunar base to perform long-duration tests.

FAR 25.1309 requires that "... compliance with the requirements [that the occurrence of any failure condition be extremely improbable] must be shown by analysis, and where necessary, by appropriate ground, flight, or simulator tests." How is this to be done for a Mars-Direct scenario? The assumption that by using existing technologies we can be confident of a reliable system is fallacious. Long-term reliability of components in one set of environmental conditions does not imply that complex systems of these components will be equally reliable even in the same set of environmental conditions, much less a whole new set of conditions. The aforementioned compliance requires a systematic procedure to gain confidence in space technologies and the ability of physiological systems (humans) to survive. This can most easily be done on the Moon, where there are the added benefits of being able to construct a different kind of spacecraft for travel to Mars than would otherwise be possible on Earth or in LEO.

Page 7 of AC 25-1309-1A discusses the need for an assessment to identify and classify failure conditions in both qualitative and quantitative terms. Furthermore, "... an analysis should show that the system and its installation can tolerate failures to the extent that major failure conditions are improbable and catastrophic failure conditions are extremely improbable." A Mars-Direct mission is composed of too many non-quantifiable risks.

During this nation's almost ten year series of missions that culminated with the landings of men on the Moon, an incremental approach was used, moving from Mercury to Gemini and finally to Apollo. We did not begin with Apollo – the type of approach advocated by those who see no need to first set up bases on the Moon but rather see Mars as the first port of call. Clearly, the base on the Moon is our Mercury and Gemini phase of an eventual trip to Mars.

Examples of potential failures (for aircraft but equally valid for spacecraft) "would include rapid release of energy from concentrated sources such as uncontained failures of rotating parts or pressure vessels, pressure differentials, non-catastrophic structural failures, loss of environmental conditioning, disconnection of more than one subsystem or component by overtemperature protection devices, contamination by fluids, damage from localized fires, loss of power, excessive voltage, physical or environmental interactions among parts, use of incorrect, faulty or bogus parts, human or machine errors, and unforeseeable adverse operational conditions, environmental conditions, or events external to the system ... ." How do we quantify these possibilities without having a history of operations in similar conditions? Which of the above list of possibilities was the cause of the destruction of Mars Observer, which apparently was destroyed upon entering Mars orbit? It was launched on 25 September 1992, but contact was lost shortly before it was to enter orbit around Mars on 22 August 1993. We do not know.

On page 9 of the AC there is an interesting statement about the use of software assessments to demonstrate compliance with risk guidelines. "In general, the means of compliance described in this AC are not directly applicable to software assessments because it is not feasible to assess the number or kinds of software errors, if any, that may remain after the completion of system design, development, and test." On the first launch of the Ariane V, on 4 June 1996, control was lost on take-off due to the use of old software. Simple errors can lead to catastrophic results.

If we cannot justify Mars-Direct on engineering and safety grounds, can we do so for economic reasons? Will entrepreneurs fund a journey to Mars for resource recovery that they seem unwilling to do for trips to the asteroids? Are there processes that can be designed for the Martian environment that would not work on the Moon? Is there something so valuable on Mars that the costs of a one to two year round trip pale by comparison? The answer to all of these questions is "no." In fact, due to the distance between the Earth and Mars, there is no current hope of an economic break-even point.

Finally, from a science perspective, will Mars provide us with so much more scientific return than will the Moon that we need to risk many lives in the process? Can the case be made that the "return-on-investment" is so much higher in a Martian expeditionary trip than one to the Moon? The case has not been made. Science depends on reasoned progress in an incremental process. There is so much yet to learn from a permanent base on the Moon. It is the ideal next step in space.

In summary, therefore, Mars-Direct cannot be defended on any rational engineering grounds. It cannot be justified purely on economic grounds. It cannot be defended from a scientific perspective. It would circumvent the difficult work that lies on the path to the Moon, and the much more difficult work needed to send people to Mars.

The only rational way to Mars is by way of the Moon.

5.2.2 An historical interview with Brand Griffin (October 2008)

Can you give us a one or two paragraph bio?

Brand Griffin is recognized as an innovator in space systems analysis and design and is currently working at Gray Research on NASA programs for a return to the Moon and a human mission to Mars. Before joining Gray, he was Vice President at Genesis, Inc. where he was in charge of the Huntsville, Alabama office managing engineering operations for contracts on the Space Station Propulsion Module, External Carriers, and Cargo Transport Container.

Prior to this, he was Boeing's lead configurator on the Space Station Program, Habitation Module Manager, and during the Space Exploration Initiative was responsible for New Technology and piloted lunar rovers. He is credited with developing many innovative spacecraft and space suit concepts including open-cockpit lunar hoppers, an advanced space suit, wheeled landing for pressurized rovers, and the first concept for a horizontal lunar lander. His design for the next generation space suit was on display in the Smithsonian's National Air and Space Museum for ten years.

Brand Griffin was one of the original editors for the college textbook, *Human Spaceflight: Mission Analysis and Design* and was the lead author for the "Extravehicular Mobility" chapter. He has over 20 technical publications and numerous articles in books and periodicals. His work has been featured on covers of *Aviation Week and Space Technology* and was contracted by TIME-LIFE for their *Voyage Through the Universe* series.

It looks as though the U.S. will be without heavy launch capability beginning in 2010 when the Shuttle is retired. It seems almost unbelievable that we have gotten into this situation. What do you think will happen?

A gap in U.S. human launch capability may be bothersome, but not tragic. The Shuttle, albeit a great engineering accomplishment, is of another century and should gracefully retire. Now, it stands in the way of a more efficient launch system, one that will enable human exploration beyond low Earth orbit.

Are you optimistic that President Bush's timeline for the return to the Moon will approximately be kept?

Can the schedule be kept? Yes! I don't believe it will.

Why is settling the Moon so important for civilization?

It's doubtful that we will settle the Moon, but much of space is about making and achieving bold plans. Because we've lost the generational experience that got us to the Moon, we need to rebuild it. It won't be easy, but that is why we need to press on.

What do you see as the major hurdles for our return to the Moon for permanent manned settlements? Are they technical, financial, physiological, psychological?

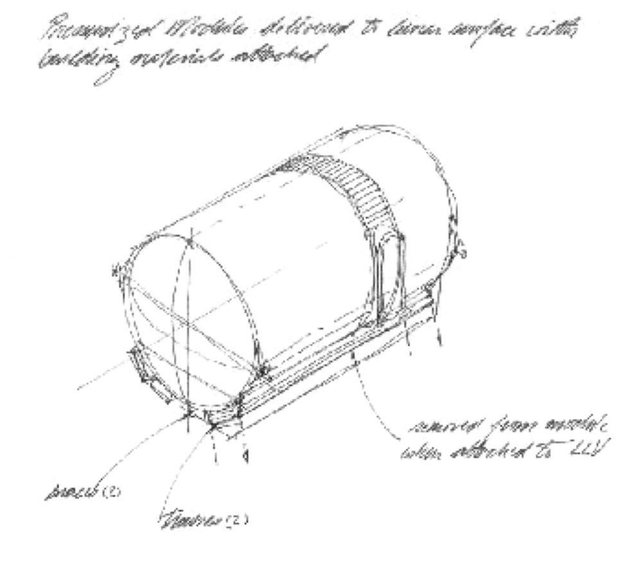

Fig. 5.2 Pressurized module delivered to the lunar surface with building materials attached. (Courtesy Brand Griffin)

None of the above. We need an aggressive schedule with decisive leaders. No new technologies are required for a return to the Moon. Our consensus style of management will not produce daring victories or glorious defeats.

How do we answer critics who say space is too expensive and that there are numerous problems on Earth to take care of first?

First, space is about Earth. It represents the intellectual capital of the future. Give it up and some other country will take your place. Regarding the expense, space is small potatoes. Get rid of the whole program and it won't affect the U.S. budget.

When you envision the future, where do you see us in 50 years? In 100 years? How do you see this evolving from the present?

In 50 years we will have a modest presence on the Moon and in 100 years only a few humans will have visited Mars. Unfortunately, this is a gloomy forecast. Space exploration is not about more money and greater cooperation; it's about a passion to do the difficult and work overtime.

What do you see as NASA's role?

For me, NASA's Apollo Program was such a great inspiration; visionary, bold, aggressive, exposed and successful. Was every person on the NASA team the best at what they did? No. Did they do the best they could? Yes. NASA is a different creature now.

Can the private sector do this with minimal government participation?

Minimal participation, not interference. Government needs to be involved because an ongoing space program should have sustained oversight. We have an incredible transcontinental road system because of a relationship between public and private interests. Space could be the same way.

You have sketched some wonderful renderings of space-related engineering concepts – how did you become interested in both engineering and art?

I've always loved to draw but have never really taken it seriously. Space has been my passion and drawing is the medium of ideas. Drawing helps create many solutions to the problem rather than falling in love with your first concept. This process blossomed when Paul Hudson and I teamed up to portray some of these ideas. I'd develop the spacecraft and Paul would put them into reality with his paintings. He is second to none and was a true partner in visualizing some very intriguing concepts. We inspired each other for 10 years.

Do you think Americans will be the first to send Man back the Moon in the 21st century, or will it be the Chinese?

Excellent question. Going to the Moon is about a sustained vision. With the possibility of a new administration every four years, it's hard to keep the flame alive. Even more, an "American" return to the Moon will most likely be an "International" mission which means more time. National pride is a key factor for going to the Moon and it appears the Chinese can not only do it by themselves, but want the bragging rights as well.

* * *

How did the early pioneer lunar engineers go about developing concepts for structures that were to house people on a permanent basis? The need to bring a large amount of mass to the Moon has always been the bottleneck to large scale space exploration and settlement. Costs of about $10,000 per pound to low Earth orbit made any foray into space a very expensive proposition. Even so, the early steps in the return to the Moon were accomplished by utilizing massive rockets and Apollo-like logistics and trajectories with lunar landers. And it is likely that everything that had to be brought to the Moon would have been, eventually. But it was the engineering and construction of Earth and later lunar space elevators that put the settlements on a fast track. Not only did launch costs fall by several orders of magnitude but the rate at which quantities could be moved from the Earth to the Moon increased by several orders of magnitude. We saved decades in our efforts to settle the Moon and beyond.

A review of some of the technical reports written during the later part of the 20th century and the beginning of the 21st century helps us trace how settlement concepts evolved. Of course, looking at the lunar cities of today makes us forget how primitive and dangerous were the initial settlements. Those were truly like the wagon trains of the Western expansion of the 1800s.

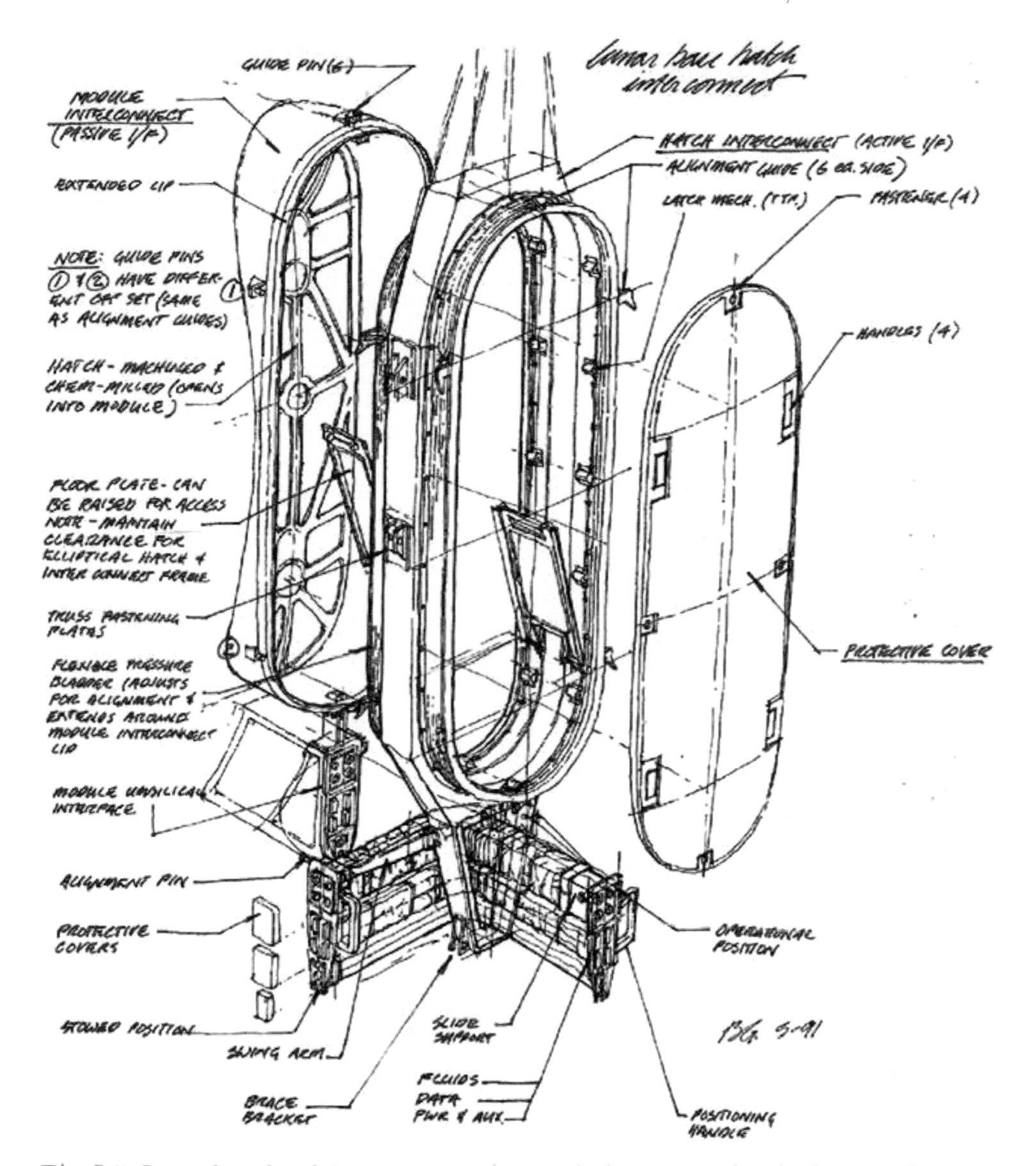

Fig. 5.3 Lunar base hatch interconnect, along with the structural and other attachments. (Courtesy Brand Griffin)

Concepts for lunar base structures have been proposed since long before the dawn of the space age. Significant studies of lunar base concepts have been made since the days of the Apollo program when it appeared likely that the Moon would become a second home to humans. For an early example of the gearing up of R&D

efforts, see the Army Corps of Engineers study.[6] During the decade between the mid-1980s to the mid-1990s these studies had intensified, both within NASA and outside the Government in industry and academe.

There were numerous studies of what the lunar habitation structure would look like and what engineering issues needed to be solved. Fundamental studies considered the structural mechanics[7] of the base.[8] Specialized studies of lunar structures have been collected.[9,10] And the study of the lunar regolith from the engineering perspective had also been already started.[11] Numerous other studies discussed science on the Moon, the economics of lunar development, and human physiology in space and on planetary bodies. An equally large literature existed about related policy issues.

Unfortunately, by the mid-1990s, the political climate turned against a return to the Moon, and the community began to look at Mars as the "appropriate" destination, essentially skipping the Moon. The debate between "Moon-First" and "Mars-Direct" continued until U.S. President G.W. Bush refocused NASA for a manned and permanent return to the Moon in 2004.

The emphasis of the discussion below is on structures for human habitation, a technically challenging fraction of the total number of structures likely to comprise the lunar facility. The test for any proposed lunar base structure is how it meets certain basic as well as special requirements. Designs for the lunar surface must satisfy numerous constraints that are different from those for terrestrial structures.

A number of structural types had been proposed for lunar base structures. These included concrete, metal frame, pneumatic, and hybrid structures. In addition, due to the hazardous environment, some preferred subsurface architectures and the use of natural features such as lava tubes for the sites of these structures.

The first step for the structural engineer is always to define the environment for the structure. All else flows from this knowledge.

---

[6]Department of the Army, "Special Study of the Research and Development Effort Required to Provide a U.S. Lunar Construction Capability," Office of the Chief of Engineers, 1963.

[7]Mechanics is a branch of engineering and physics that studies how materials and structures deform under forcing, how structures such as aircraft fly and interact with the atmosphere, and how structures such as ships and submarines move in the water and interact with the water waves and currents.

[8]H. Benaroya, Editor, "Applied Mechanics of a Lunar Base," *Applied Mechanics Reviews*, 1993, Vol. 46, No. 6, pp. 265–358.

[9]H. Benaroya, Editor, "Lunar Structures," *Journal of the British Interplanetary Society*, Vol. 48, No. 1, 1995.

[10]W. Mendell, Editor, *Lunar Bases and Space Activities of the 21st Century*, Proceedings of the Lunar and Planetary Institute, Houston, 1985.

[11]M. Ettouney, and H. Benaroya, "Regolith Mechanics, Dynamics, and Foundations," *Journal of Aerospace Engineering*, Vol. 5, No. 2, 1992, pp. 214–229.

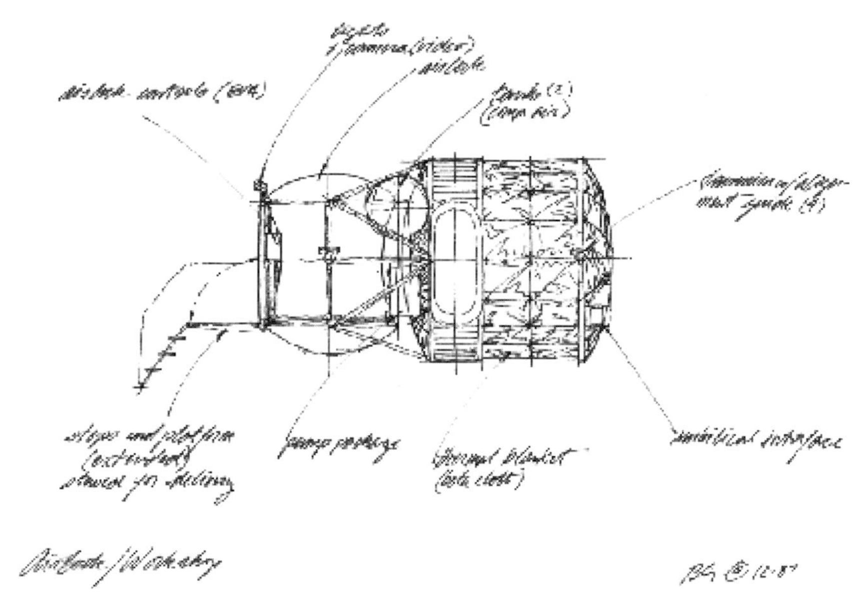

Fig. 5.4 Airlock and attached workshop. Steps are shown in an extended position on the left. (Courtesy Brand Griffin)

5.3 The Environment

This discussion is partly based on two early works[12,13] and reflects the current thinking of the time. The lunar physical environment is dramatically different than that of the Earth. A summary of key characteristics is presented in Table 5.1.

Gravity: At the lunar surface gravitational acceleration is about $1/6\ g$, where $g = 9.8\ \mathrm{m/s^2}$ on Earth. This means that the same structure will have six times the weight-bearing capacity on the Moon as it would on the Earth. Conversely, to support a certain loading condition, one-sixth the load bearing strength is required on the Moon as on the Earth. Therefore, the concepts of dead loads and live loads[14] within the lunar gravitational environment had to be reconsidered. Mass based rather than weight-based criteria were developed for lunar structural design codes because mass is invariant whereas weight depends on the gravitational acceleration, as per the equation

$$weight = mass \times gravitational\ acceleration.$$

[12] B. Sherwood and L. Toups, "Technical Issues for Lunar Base Structures," *Journal of Aerospace Engineering*, Vol. 5, No. 2, p. 175–186, April 1992.

[13] A.M. Jablonski and K.A. Ogden, "Technical Requirements for Lunar Structures," *Journal of Aerospace Engineering*, Vol. 21, No. 2, pp. 72–90, April 2008.

[14] A dead load is a weight that is permanently in the structure, for example, floor weight or machine weight. A live load is a weight that can move. It does not have to be alive, it just has to be able to be at different locations at different times. For example, cars moving on a bridge are live loads.

Table 5.1 Comparison of Earth and lunar physical parameters. Seismic energy does not account for seismic activity due to meteoroid impacts. The magnetic vector field is in units of Ampere/meter. For the Moon there is a small paleofield, that is, a very small ancient magnetic field.

Property	**Moon**	**Earth**
Surface area [km^2]	37.9×10^6	510.1×10^6
Radius [m]	1,738	6,371
Gravity at Equator [m/s^2]	1.62	9.81
Escape velocity at Equator [km/s]	2.38	11.2
Surface temperature range [°C, °F]	$\begin{bmatrix} -173 \text{ to } 127 \\ -279 \text{ to } 261 \end{bmatrix}$	$\begin{bmatrix} -89 \text{ to } 58 \\ -128 \text{ to } 136 \end{bmatrix}$
Seismic energy [J/year]	$\simeq 10^9$ to 10^{13}	10^{17} to 10^{18}
Magnetic vector field [A/m]	0	24 to 56 A/m
Surface atmospheric pressure [kPa, psi]	0	101.3, 14.7
Day length [Earth Days]	29.5	1
Communication delay [s]	2.6	0

Lower gravity on the Moon means that much longer spanned structures are possible, but it also means that gravity plays a much smaller role is anchoring the structure.

Temperature extremes: The diurnal cycle on the Moon is 29.53 Earth days, almost evenly split between daylight and nighttime. The lack of atmosphere has many implications for lunar dwellers, such as a lack of shielding against radiation and micrometeoroids, but also that daylight is in extreme contrast, which has implications for outposts at the poles. The temperature transition from daylight to nighttime is rapid (about $5°C$/hr). At the Apollo landing site the range of temperatures was from 111°C to –171°C, resulting in major thermal expansion/contraction and thermal cycling challenges to surface structures. If a structure is to be directly exposed to these extreme temperatures, it must be made of highly elastic materials. Material fatigue due to thermal cycling is generally a problem that needs to be ameliorated. Even those structures that are shielded are susceptible to material fatigue and brittle fracture.

Radiation: The lack of atmosphere and negligible magnetic field raises challenges for the design of structures with shielding against various forms of radiation from deep space. There are high energy galactic cosmic rays (GCR) composed of heavy nuclei, protons and alpha particles. And there are the products of solar flares, or solar particle events (SPE), which are a flow of high energy protons that result from solar eruptions. Such radiation has serious implications for human and plant survivability, as well as possible effects on materials.

Micrometeoroids are a serious threat in space and on the lunar surface due to their speed, concurrent with the lack of atmospheric shielding. These speeds are on the order of 10 km/s (22,369 mi/h). On Earth most of these dust-size particles burn up in the atmosphere. The larger rocks that hit the ground can cause great damage. But these are rare. Therefore, a lunar surface structure needs to be shielded from this sort of attack. For certain types of structures, such as depots and telescopes, shielding is only practical when they are not in use. New failure modes due to high-velocity micrometeorite impacts were analyzed and probabilities of impact were developed.

Regolith dust: The lunar surface has a layer of fine particles referred to as regolith that has formed over billions of years, primarily due to continued impact by meteoroids and larger rocks. This regular action against the surface of the Moon has homogenized, pulverized and compacted the local lunar rocks. As a practical matter, the lunar regolith is a serious environmental hazard to human and machine. A photoelectric change in the conductivity of the dust particles causes them to levitate and adhere to surfaces. The dust is also abrasive, toxic if breathed, and poses serious challenges for the utility of construction equipment, air locks, and all exposed surfaces.

Pressure differentials: The lunar structure is in fact a life-supporting closed environment under internal air pressurization. It is a pressurized enclosed volume with an internal pressure of between 10 and 15 psi. The enclosed structure must contain this pressure and must be designed to be "fail-safe" against catastrophic and other decompression caused by accidental and natural impacts. In some situations, for our early lunar surface settlements, a layer of regolith was placed atop the structure for shielding, but the added weight only partially (in the range of 10% to 20%) balanced the forces on the structure caused by internal pressurization.

Hard vacuum outgassing for exposed steels and other effects of high vacuum on steel, alloys, and advanced materials occur on the Moon. This precludes the use of certain materials that may not be chemically or molecularly stable under such conditions.

Regolith mechanics: Some published reports[15] on the properties of regolith indicated that an accurate evaluation of regolith stresses and strains would require a greater use of nonlinear stress-analysis methods than such calculations for the Earth environment. This is because regolith interlocking stress levels are very small, about 0.73 psi (5 kPa), and crushing stress levels may be as small as 4.35 to 10.1 psi (30 to 70 kPa). These numbers are considerably smaller than their equivalents on Earth, even after accounting for the differences in gravitational accelerations. The nonlinearity of regolith stress-strain behavior is therefore taken into account in lunar engineering projects. The dominance of nonlinear effects in regolith, as compared with Earth soil materials, is not necessarily an unwanted engineering effect. Since interlocking occurs for very small stress levels the regolith will become stiffer and relatively smaller strains will develop, resulting in better system behavior at these stress levels. On the other hand, lower crushing stress levels cause problems in larger projects.

Seismicity was not viewed as significant to the analysis and design of lunar structures as compared to Earth structures, but it could not be ignored. The energy released in lunar seismic activity is about 10^9 smaller than in earthquakes. On

[15]S. Johnson, "Extraterrestrial Facilities Engineering." *Encyclopedia of Physical Science and Technology; 1989 Year Book*, Academic Press, 1989.

the Richter scale, based on Apollo data, it was known that magnitudes of 1 to 2 were usual, perhaps a few that might have reached magnitudes of 4. On Earth a significant earthquake can reach magnitudes of 6 to 8, with 8 being devastating. As we know, since the scale is logarithmic, each increase in magnitude of one unit on the scale is a tenfold increase in energy released. A magnitude 7 quake releases ten times as much energy as a magnitude 6 quake.

Lunar dust and moonquakes are discussed in a bit more detail next.

Lunar dust

The lunar dust is quite problematic, which has been known since the Apollo days. One of the Surveyor sightings of such was a glow above the western horizon about one hour after sunset.[16] It was observed that positively charged 5–10 micron particles levitate due to local electric fields. These can move up to 1 m above the lunar surface. After sunset, light scatters from these particles, resulting in the glow. These are called lunar horizon glow particles. Particles of sub-micron size can move to much higher altitudes on ballistic trajectories at speeds of about 1 m/s.

Such mobile particles can impact lunar surface activities, which also add to the quantity of charged suspended particles. That the regolith particles are jagged and cling add to the design challenges for spacesuits, pressure doors and seals, machine joints, just about anything that the particles can come into contact with.

We have developed autonomous technologies that collect particles and dampen their mobility.

Moonquakes

While the Moon is a body without a molten core like Earth's, it is still susceptible to seismic activity. The source for these energy releases is tidal – the Moon is pulled on by Earth's gravity field causing strains and stresses. Repetitive cycles of such loading leads to relatively minor quakes. Data mining of the Apollo lunar seismic data set[17] had revealed that there had been 7245 deep moonquakes from 77 regions within the Moon. This analysis invigorated further research of the structure of the Moon, but also had implications for the designers of habitable structures.

The largest recorded moonquake in the Apollo data was 2–3 on the Richter scale, and usually the magnitudes were in the range of 1–2.[18] "The deep moonquakes showed some periodic regularity depending on the point when the Moon's orbit is closest to Earth. They occurred at these centers at opposite phases of the tidal pull, so that the most active periods are 14 days apart. ... Much of the data collected in the Apollo seismic experiments were due to meteoroid impact, demonstrating

[16] J.E. Colwell, C.J. Grund, and D.T. Britt, "Mechanical and Electrostatic Behavior of Lunar Dust," 2143.pdf, NLSI Lunar Science Conference, 2008.

[17] Y. Nakamura, "New identification of deep moonquakes in the Apollo lunar seismic data," *Physics of the Earth and Planetary Interiors*, Vol. 139, 2003, pp. 197–205.

[18] Jablonski and Ogden, cited earlier.

that meteoroids will be a much more significant concern than moonquakes with an original source from the Moon interior."

Moonquakes, while never a serious issue, are a non-issue in the present. However, as we began to build underground, we realized that seismic effects could be potentially more serious. So we incorporated a variety of vibration isolation devices into these buried structures, much like the ones used in tall buildings and bridges in seismic region on Earth.

* * *

In the last century and a half on the Moon we have built numerous structures. In the next several sections an historical summary of the early structural concepts is provided.

5.4 Early structural concepts

Today, we are essentially self-sufficient on the Moon. This is because we have developed a sophisticated infrastructure that allows us to create almost everything we need from *in situ* resources. Of course, this was not always the case. When we returned to the Moon in the mid-21st century, we had to bring everything we needed for survival from Earth. The first lunar habitat was brought to the Moon in one piece from the Earth and placed on the lunar surface by astronauts and robots.

Those first structures were known as Class I structures. Once we were on the Moon for a while and were creating an infrastructure, we went on to Class II structures, where the components were brought from Earth but we were able to put the components together on the lunar surface. And finally, almost thirty years after the first return, we began to be able to use local materials to build our structures and synthesize many of our needs. Those were called Class III structures.

5.4.1 Constraints on the design of lunar structures

The design of any structure must satisfy four overall objectives. These are to meet functional requirements, resist loads safely, and meet constructability and economical requirements. In evaluating these objectives, the designer must provide for the special constraints imposed by the harsh lunar environment and its remoteness. The design constraints include some general aspects, but also specific aspects to where the structure was to be erected.

While NASA was building its new rockets, it became an assumption that the first return to the Moon would be to the South Pole, in particular to one of the cold depressions known as Shackleton crater. The primary reason for this was the belief that due to permanent darkness there was the possibility that water ice might exist there. The Pentagon's *Clementine* lunar orbiter mission of 1994 gave rise to such belief due to interpreted radar data. Clementine was a joint project between NASA and the Strategic Defense Initiative Organization. NASA's *Lunar Prospector* mission of 1998–1999 recorded an enhanced signal of hydrogen in some of these craters, leading to some to conclude that the source was water ice.

When in the summer of 2008 the Japanese sent their orbiter SELENE (KAGUYA) to the South Pole, their data and images showed no obvious brightness that would have indicated an area of water ice. "The inside of Shackleton Crater at the lunar south pole is permanently shadowed; it has been inferred to hold water-ice deposits. The Terrain Camera (TC), a 10 m resolution stereo camera onboard, succeeded in imaging the inside of the crater, which was faintly lit by sunlight scattered from the upper inner wall near the rim. The estimated temperature of the crater floor based on the crater shape model derived from the TC data is less than approximately 90 °K, cold enough to hold water-ice. However, the derived albedo indicates that exposed relatively pure water-ice deposits are lacking on the floor at the TC's spatial resolution. Water-ice may be disseminated and mixed with soil at a few percent, or may not exist at all."[19] As mentioned in the last chapter, the LCROSS mission of 2009 demonstrated the existence of water-ice deposits on the Moon.

Design constraints for the first lunar structures were based on the fact that there existed absolutely no infrastructure awaiting the initial visitors from Earth. The design engineers had to take the following into account:

Environment: The structure had to resist the effects of the environment in which it would be built. The lunar environment, compared with the Earth environment, has a number of unique characteristics that have significant effects on basing concepts and structures. These included a one-sixth Earth gravity, hard vacuum, continuous solar and cosmic radiation, extremes in temperature and radiation, meteorites (direct strikes and seismic vibration from nearby strikes), abrasive and adhesive dust, and a 28 day diurnal cycle. Approximately 10 ft of regolith overburden was considered sufficient to shield humans from the dangers of radiation, and some of the meteoritic activity. Today in 2169 new shielding materials reduce that required thickness considerably.

Constructability: The remoteness of the lunar site, in conjunction with the high costs associated with transportation from Earth, played a key role in lunar base construction. Construction had to be easy with a minimum of on-site support equipment, manpower, and project risk. Construction components needed to be practical, simple, reliable, durable, maintainable, and multipurpose in order to minimize local fabrication. Since the lunar base was to be an evolving, growing facility, it was necessary that it be easily expandable. It was recognized that local materials needed to be used as much as possible.

Functionality encompassed, among other things, concerns regarding lunar base operations – items related to the basic purposes for the base, sunlight ingress, easy and safe access to the lunar surface for occupants and equipment, and the facility's effective and efficient maintenance. Also of concern here were issues related to the base as a living, working, recreational, and psychological haven for

[19] J. Haruyama, M. Ohtake, T. Matsunaga, T. Morota, C. Honda, Y. Yokota, C.M. Pieters, S. Hara, K. Hioki, K. Saiki, H. Miyamoto, A. Iwasaki, M. Abe, Y. Ogawa, H. Takeda, M. Shirao, A. Yamaji, J.-L. Josset, "Lack of Exposed Ice Inside Lunar South Pole Shackleton Crater," originally published in *Science Express* on 23 October 2008, and in *Science,* Vol. 322, No. 5903, 7 November 2008, pp. 938–939.

occupants. "Shirt-sleeve" activity required internal pressurization. An important issue facing planners was how to reconcile structural issues that conflicted with issues of architectural function.

Reliability: Factors of safety, originally developed to account for uncertainties in the Earth design and construction process, were adjusted for the lunar environment. At first it was unclear whether the adjustment had to be up or down – it depended on one's perspective and tolerance for risk. It ended up that early structures had a high factor while subsequent versions and generations were designed based on lower factors as knowledge and infrastructure evolved. Human safety and the minimization of risk to "acceptable" levels were always at the top of the list of considerations for any engineering project. The Moon offered new challenges to the engineering designer. Minimization of risk implies, in particular, structural redundancy, and when all else fails, easy escape for the inhabitants. The key word is "acceptable." It is a subjective consideration, deeply rooted in economic considerations. What is an acceptable level of safety and reliability for a lunar site, one that must be considered highly hazardous? Such questions go beyond engineering considerations and must include policy considerations: Could we have afforded to fail without bringing the return to the Moon to a grinding halt?

Operations are the processes of running of the facility. Such processes are related to science, exploration, use of the lunar resources, manufacturing, agriculture, power generation and storage, use of the facility as a test bed for technology development and new operations procedures, resource utilization, and a spaceport for landings and launchings.

Maintenance includes the following: inspectability for structural degradation, repairability – in particular, mitigation of dust effects on seals and movable parts, of ultraviolet degradation of coatings and organics, in counteraction of detrimental effects of outgassing in the vacuum, of amelioration of radiation damage to computers and software, and of micrometeoric impact damage. Maintenance is a crucial component of the cost analysis of the structural life cycle. In the present – 2169 – technology developments such as self-healing materials and structures have automated many of the maintenance tasks.

ISRU: This was an extremely important aspect in the long-term view of extraterrestrial habitation. But feasibility had to wait until a minimal presence had been established on the Moon. Initial lunar structures were transported from the Earth as complete units.

The design process

The "design process" warrants a few general comments. Engineering design is the process of creating the procedure by which something is built or manufactured. Materials are selected with the needed strength and geometry and put together in a way that effectively meets the needs of the people for whom it was designed.

The design process involves an iteration between conceptualization and preliminary designs that are critically evaluated, eventually leading to the final design.

Various concepts are generally considered as possible designs. In order to assess these concepts and then compare how, and to what degree, they satisfy the constraints of the system, the concepts must be evaluated. This is normally performed via a detailed analysis.

Important aspects of a design process are the creation of a detailed design and prototyping. For a structure in the lunar environment, such building and realistic testing cannot be performed on the Earth or even in orbit. It was not possible to experimentally assess the effects of suspended (due to one-sixth g) lunar regolith fines on lunar machinery in the early 21st century. The Apollo experience was extrapolated, but only to a boundary beyond which new information was necessary.

Another crucial aspect of a design involves an evaluation of the total life cycle, that is, taking a system from conception through retirement and disposition – the recycling of the system and its components. Many factors affecting system life could not have been predicted in the early 21st century due to the nature of the lunar environment and the inability to realistically assess the system before it was to be built and utilized. Such knowledge was gained after our return to the Moon.

Finally, concurrent engineering[20] became a byword for lunar structural analysis, design, and erection. Concurrent engineering simultaneously considers system design, manufacturing, and construction, moving major items in the cycle to as early a stage as possible in order to anticipate potential problems. Given the extreme nature of the environment for the structure, concurrency implied flexibility of design and construction. Parallelism in the design space guaranteed that at each juncture of the construction, alternate possibilities existed, permitting the continuation of the construction even in the face of completely unanticipated difficulties.

Concepts

Numerous studies were done at the turn of the 21st century on what a basic habitable lunar structure needed to be for human survival on the Moon. Some documents from that era that had been lost were recently recovered. Our vantage point in the 22nd century could have provided a straighter path to the Moon, but decisions are always made in the shadow of uncertainty.

Lunar structural concepts can be categorized in a variety of ways. One way is the following: rigid structures, inflatable structures, erectable structures, and structures from *in situ* materials. The *in situ* structures were always the ultimate goal – as in the old American West, we need to be able to live off the land in order to be considered a self-sufficient settlement. But in addition to the food and tools, a trip to the Moon required that we bring our own air and water, and be able to shield ourselves from the very dangerous elements. And this was not possible until an infrastructure was built.

There was and is no one concept for a lunar structure. Different requirements led to different designs. Different concepts were contrasted and compared using de-

[20] H. Benaroya, L. Bernold, "Engineering of Lunar Bases: Review," *Acta Astronautica*, Vol. 62, 2008, pp. 277–299.

cision science and operations research tools.[21] One approach compared concepts[22] using a points system for an extraterrestrial building system, including pneumatic, framed/rigid foam, prefabricated, and hybrid (inflatable/rigid) concepts.

A very early lunar structural design study[23] presented the available information with the goal of furthering the development of criteria for the design of permanent lunar structures. In this work, the lunar environment was detailed, regolith in foundation design was discussed, and excavation concepts were reviewed. An historical review of the evolution of concepts for lunar bases up through 1990 is interesting to read.[24,25] Surface and subsurface concepts for lunar bases were considered as options.[26] America's future on the Moon was predicted to be in support of scientific research, the recovery of lunar resources for use in building a space infrastructure, and the attainment of self-sufficiency in the lunar environment as a first step in planetary settlement. The complexities and costs of constructing a particular settlement structure depended on its mission and location.

Parametric models have been used to quantify the lunar base.[27] Such a model is based on hundreds of equations that are functions of thousands of parameters. Example parameters include crew size, the location of the lunar base, and all of the environmental conditions.

All structures are analyzed and designed using equations, but what purpose can there be to create a mathematical model of the whole base and all its processes? We have seen the kinds of results obtained from a parametric economic model of lunar activities in Section 3.3. From the parametric model of a lunar base it is possible to see the larger picture of lunar base operations. We can examine how the various aspects of those operations interact. A lunar base is essentially a small city. There are transportation issues, energy needs, economic and scientific activities, and they all depend on each other. A parametric model embodies the

---

[21] H. Benaroya and M. Ettouney, "Design Codes for Lunar Structures," *SPACE 92, Engineering, Construction, and Operations in Space*, pp. 1-12, 1992.

H. Benaroya and M. Ettouney, "Design and Construction Considerations for a Lunar Outpost," *Journal of Aerospace Engineering*, Vol. 5, No. 3, 1992, pp. 261–273.

[22] P.J. Richter and R.M. Drake, "A Preliminary Evaluation of Extraterrestrial Building Systems," *SPACE 90 Engineering, Construction, and Operations in Space*, 1990, pp. 409–418.

[23] S.W. Johnson, *Criteria for the Design of Structures for a Permanent Lunar Base* (Ph.D. Dissertation), Univ. of Illinois, Urbana, 1964.

[24] S.W. Johnson and R.S. Leonard, "Evolution of Concepts for Lunar Bases," *Lunar Bases and Space Activities of the 21st Century*, Proceedings of the Lunar and Planetary Institute, Houston, 1985, pp. 47–56.

[25] S.W. Johnson and J.P. Wetzel, "Science and Engineering for Space: Technologies from SPACE 88," *SPACE 90, Engineering, Construction, and Operations in Space*, Proceedings of the ASCE, 1990.

[26] W.D. Hypes and R.L. Wright, "A Survey of Surface Structures and Subsurface Developments for Lunar Bases," *SPACE 90 Engineering, Construction, and Operations in Space*, Proceedings of the ASCE, 1990, pp. 468–479.

[27] P. Eckart, "Lunar Base Parametric Model," *Journal of Aerospace Engineering*, Vol. 10, No. 2, April 1997.

mathematical relationships and interactions of all the components of the system with each other and with the environment.

Since the cost of bringing material from Earth to the Moon determines its economical viability, mass is used as the units of choice. Quoting from the earlier-cited paper,[25]

> "... it is possible to manipulate hundreds of input parameters to determine their impact on the overall system mass and the interdependencies among the different systems. From these results, recommendations can be derived concerning the selection of a system like the power supply system, thermal control system, or life support system as a function of input parameters (like crew size or lunar base location). The following ... is a list of lunar base element system models:

- lunar surface environment model
- crew metabolic load model
- shielding model
- communications system model
- EVA/airlock operation model
- lunar surface transportation systems model
- life support system model
- low/high-temperature thermal control system model
- *in situ* oxygen production model
- power supply system model."

All of these models are characterized by equations that are then linked to the equations of the other models. These are known as a *system* of equations. Such systems can be represented by "block diagrams," with a representative diagram shown in Figure 5.5.

In the block diagram, the *plant* represents a structure, a machine or a (chemical) process. The *input* is a particular environmental force. The *output* is how the structure responds to that environmental force. The *sensor* measures the response

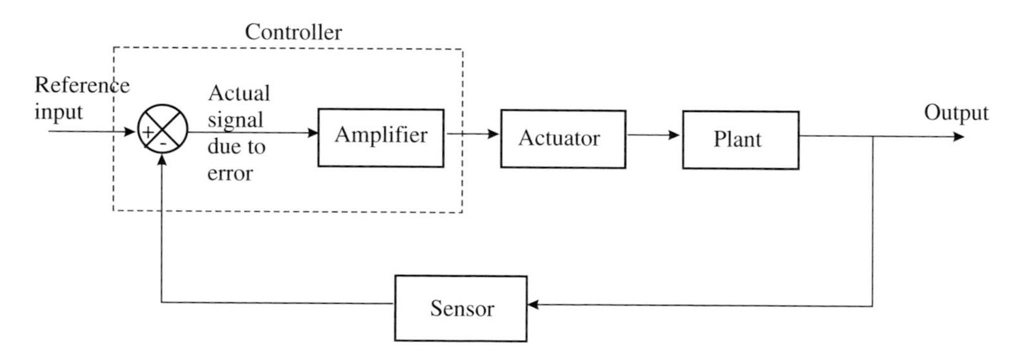

Fig. 5.5 This block diagram relates the system output to the system input. Along the path are an actuator that applies forces to the plant (structure) and a sensor that measures the output and adjusts the actuator such that the desired output is attained.

and determines if it is acceptable. Any deviation or *error* results in a force (or other input) to be generated of a magnitude that is related to the error. The force is generated by an *actuator* and is applied to the plant. The new modified or controlled plant output is again measured and compared with the desired response. This occurs for each critical component of the structure and the larger system, with *feedback* used to keep the structure and system operating safely. Such an approach is used to design *self-healing* systems.

Rigid structures

During the late 20th century, there was still hope that a manned return to the Moon could proceed rapidly given the go-ahead of the President of the United States. A quick return could have utilized proven technologies and the National Space Transportation System (NSTS) for the early development of a lunar outpost.[28] Transfer vehicles and surface systems would have been developed so that the payload bay of the Shuttle could have been used for transport. The lunar outpost structure would have been easily broken into parts, separating radiation protection from module support allowing easy access, installation, and removal of the elements attached to the Shuttle trusses.

Preliminary designs of permanently manned lunar surface research facilities were also a focus[29] with criteria for the base design to include scientific objectives as well as the transportation requirements to establish and support its continued operations.

Figure 5.6 shows us how relatively simple the first lunar settlements were planned to be – of course, they only look simple from where we sit today. At the time of their conception – over 200 years ago – there were many challenges to place even such simple structures on the Moon. We see how regolith for shielding has been placed atop the cylinder on the right side of the figure.

A concept for a subsequent and more ambitious settlement can be seen in Figures 5.8 and 5.9, where cylindrical structures are placed on their sides and attached to a truss framework. We can see the regolith shielding being placed on the framework. Eventually the complete network of cylinders would be covered.

A related concept was[30] the use of the liquid oxygen tank portions of the Space Shuttle external tank assembly for a basic lunar habitat. The modifications of the tank, to take place in low Earth orbit, included the installation of living quarters, instrumentation, air locks, life support systems, and environmental control systems. The habitat would then have been transported to the Moon for a soft landing. Originally, the idea was to use the Space Shuttle external tanks left in

[28] B.N. Griffin, "An Infrastructure for Early Lunar Development," *SPACE 90 Engineering, Construction, and Operations in Space*, pp. 389–398, 1990.

[29] S.J. Hoffman and J.C. Niehoff, "Preliminary Design of a Permanently Manned Lunar Surface Research Base," *Lunar Bases and Space Activities of the 21st Century*, Proceedings of the Lunar and Planetary Institute, Houston, pp. 69–76, 1985.

[30] C.B. King, A.J. Butterfield, W.D. Hyper, and J.E. Nealy, "A Concept for Using the External Tank from a NSTS for a Lunar Habitat," Proceedings, *9th Biennial SSI/Princeton Conference on Space Manufacturing*, Princeton, May 1989, pp. 47–56.

Fig. 5.6 A 1963 Boeing Concept. The LESA (Lunar Exploration System for Apollo) initial concept was designed to accomodate six people for six months. It would hold 46,000 pounds of payload, a 10 kW nuclear reactor, a 3765 lb rover, and equipment to move regolith for shielding use. (Courtesy Lunar and Planetary Institute)

Fig. 5.7 An old sketch of how it was once envisioned that regolith would be placed atop the cylinder habitat.

orbit, thus saving launch costs. But this never happened. But a variant of this idea was implemented once space elevators were operational. Large tanks were towed via elevator to a very high orbit and fitted for their final destinations.

Inflatables

The pillow-shaped structure shown in Figure 5.10 was evaluated[31] as a possible concept for a permanent lunar base. The proposed base consists of quilted inflatable pressurized tensile structures using fiber composites. Shielding is provided by an overburden of regolith, with accommodation for sunlight ingress.

[31]W.Z. Sadeh and M.E. Criswell, "Inflatable Structures – A Concept for Lunar and Martian Structures," AIAA 93-0995, AIAA/AHS/ASEE Aerospace Design Conference, Irvine, 1993.

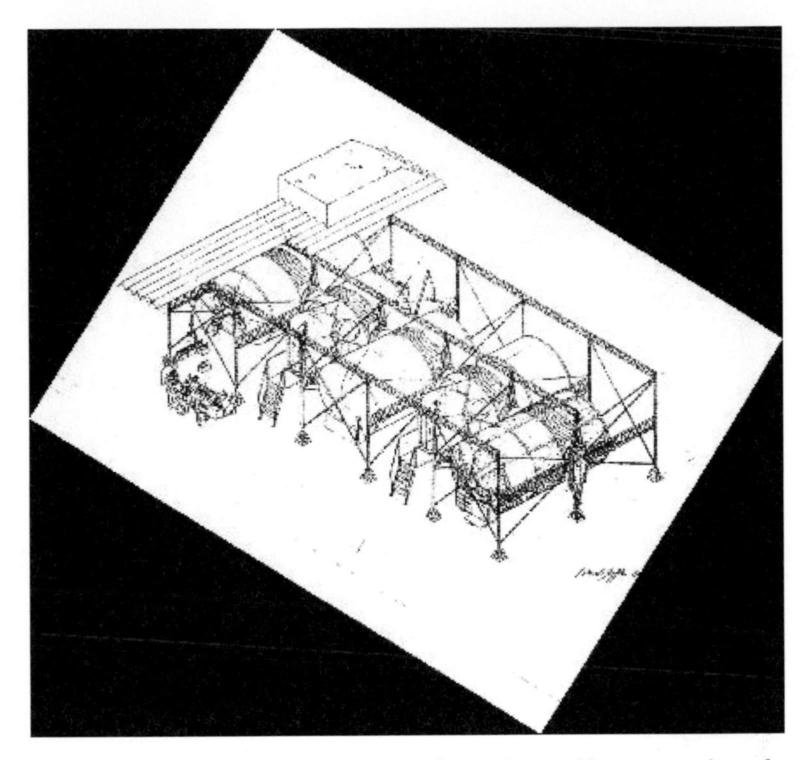

Fig. 5.8 This sketch shows an isometric view from above of how a number of structural cylinders, on their sides, are connected by way of a framework. Regolith shielding would be placed on the framework. (Courtesy Brand Griffin)

Inflatable structural concepts for a lunar base were proposed[32] as a means to simplify and speed up the process of creating a habitable lunar facility while lessening transport costs. The inflatable structure was suggested as a generic test bed structure for a variety of application needs for the Moon.[33] Design criteria were developed since few structural engineers had experience with inflatable structures.[34] Because of the possibility of deflation during construction due to accident or meteorite strike, temporary columns were used to support the inflated structure – these were removed after the structure was made rigid.

Inflatable concepts were also very popular because they could be rolled up or folded into a small volume and easily transported to the surface of the Moon. There, the "balloon" structure would be inflated on top of its final location – the surface

[32]W.J. Broad, "Lab Offers to Develop an Inflatable Space Base," *The New York Times*, 14 November 1989.

[33]W.Z. Sadeh and M.E. Criswell, "A Generic Inflatable Structure for a Lunar/Martian Base," *SPACE 94 Engineering, Construction, and Operations in Space*, 1994, pp. 1146–1156.

[34]M.E. Criswell, W.Z. Sadeh, and J.E. Abarbanel, "Design and Performance Criteria for Inflatable Structures in Space," *SPACE 96, Engineering, Construction, and Operations in Space*, 1996, pp. 1045–1051.

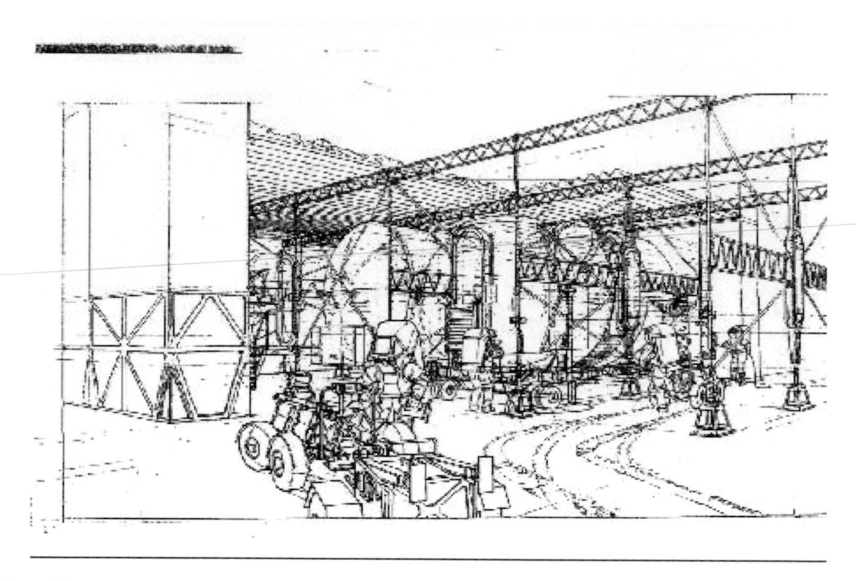

Fig. 5.9 Here we see a ground level view of cylindrical habitats being placed in a structural framework as well as the initial placement of a cover upon which regolith shielding would be placed. (Courtesy Brand Griffin)

Fig. 5.10 A pillow-shaped inflatable structure. The large grid of inflatable structures is partially covered by regolith shielding, with light being directed inside via mirrors. (From Sadeh and Criswell [29]; reprinted with permission of the American Institute of Aeronautics and Astronautics, Inc.)

would have been smoothed out and otherwise made ready for the structure – and once fully inflated, various ways would have been used to make the structure rigid. This could be by injecting foams that would harden with time within interstices and within the space between the layers of the inflatable structure. Once the inflated structure became rigid, a structural support framework could be erected inside and the structure pressurized.

The deployment of the inflatable structure required the design of a foundation and reliability concerns were significant for such structures.[35] While the concept of folding a deflated membrane is simple, the practicalities are major. If there is too much pressure on one fold for too long a period of time will that region be weakened? If the folded structure shifts is abrasion a problem? Such concerns led to studies to minimize the risks of damage during the packaging of the structure for shipment and during its inflation.

An example of a more elaborate pressurized membrane structure[36] for a permanent lunar base is shown in Figure 5.11. It is constructed of a double-skin membrane filled with structural foam. A pressurized torus-shaped substructure provides edge support. Shielding is provided by an overburden of regolith. Briefly, the construction procedure requires shaping the ground and spreading the uninflated structure upon it, after which the torus-shaped substructure is pressurized. Structural foam that hardens quickly is then injected into the inflatable components, and the internal compartment is pressurized. The bottoms of both inflated structures are filled with compacted soil to provide stability and a flat interior floor surface. Backfilling is a difficult operation to carry out through an airlock.

Habitat Design Workshops were popular in the early 21st century and such groups were very creative in the concepts[37] they produced for lunar and Martian surface structures.[38] The workshop brought together over thirty people of varied backgrounds for a week for a concurrent design approach to the design of surface habitable structures.

It was recognized early on that it was necessary to bring the whole spectrum of professions into the design of a lunar habitat for living. As such, in addition to the engineering aspects of design, there were physiological concerns as well as other human factors that needed to be incorporated into the design.

In the workshop design imaged here, the base consists of the lunar lander modules with inflatable ovoid domes – the individual modules would be connected so that the base can be expanded.

Given the number of inflatable structures that were sent to the Moon, an assembly facility had been used to build modular components into the complete structure.

---

[35] P.S. Nowak, M.E. Criswell, and W.Z. Sadeh, "Inflatable Structures for a Lunar Base," *SPACE 90 Engineering, Construction, and Operations in Space*, 1990, pp. 510–519.

P.S. Nowak, W.Z. Sadeh, and M.E. Criswell, "An Analysis of an Inflatable Module for Planetary Surfaces," *SPACE 92 Engineering, Construction, and Operations in Space*, 1992, pp. 78–87.

[36] P.Y. Chow and T.Y. Lin, "Structures for the Moon," *SPACE 88 Engineering, Construction, and Operations in Space*, 1988, pp. 362–374.

P.Y. Chow and T.Y. Lin, "Structural Engineer's Concept of Lunar Structures," *Journal of Aerospace Engineering*, Vol. 2, No. 1, January 1989.

[37] Project: Moon 1 - Fram Studentteam: A. Fischer, J. Jorgensen, J. Tizard, H. Västinsaldo, A. Wielders, S.Zanini © ESA Habitat Design Workshop 2005 (M. Aguzzi, S. Häuplik-Meusburger, E. Laan, K. Özdemir, D. Robinson, G. Sterenborg et al.).

[38] D.K.R. Robinson, G. Sterenborg, M. Aguzzi, S. Häuplik, K. Özdemir, E. Laan, R.S. Drummond, D. Haslam, J. Hendrikse, and S. Lorenz, "Concepts for 1st Generation Hybrid and Inflatable Habitats with In-Situ Resource Utilisation for the Moon, Mars, and Phobos: Results of the Habitat Design Workshop 2005," IAC-05-D4.1.03.

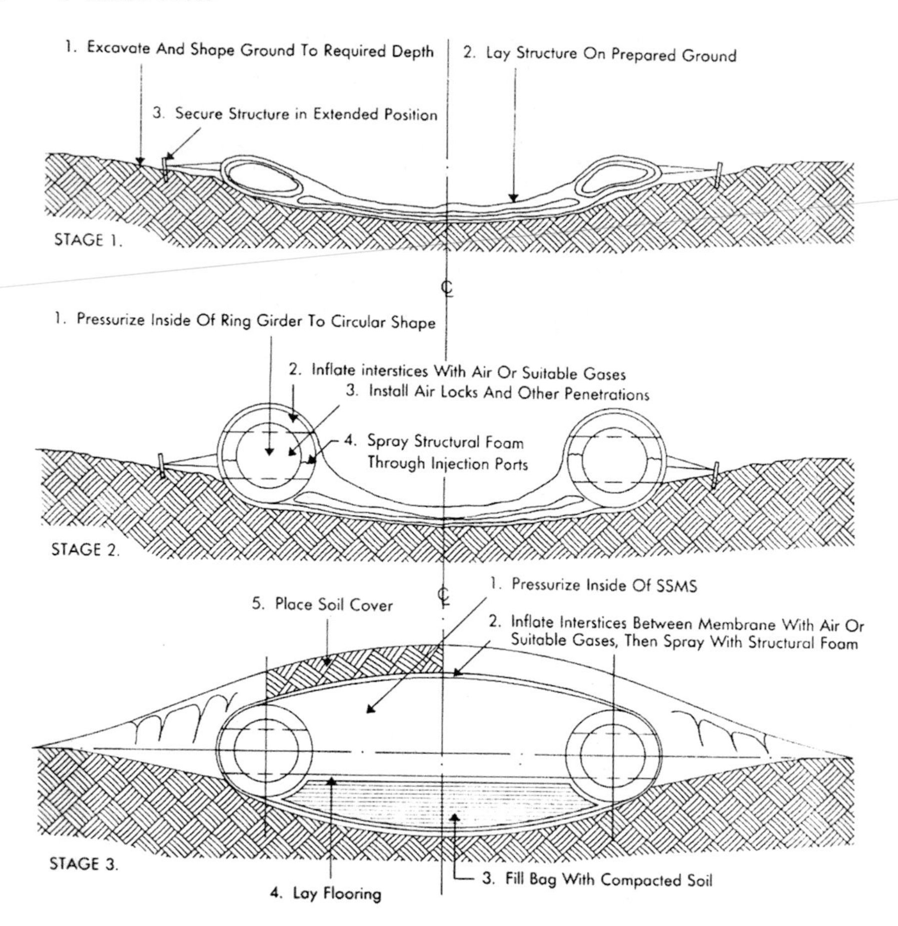

Fig. 5.11 This inflatable membrane structure concept is shown from the time it is placed on a prepared site on the lunar surface, inflated, and then rigidized in its final form. This concept was also patented by Phil Chow and T.Y. Lin International; U.S. Patent 5058330. (Courtesy T.Y. Lin International)

Inflatable structures required mechanical equipment to initiate deployment and inflation.[39] Air for inflation was initially transported to the Moon in liquid form. As our infrastructure evolved we created the air we needed from regolith resource recovery.

Approximately one-third of our permanent structures on the lunar surface are of this type. They are mostly older structures that were brought here by standard transportation methods while costs were still very large. After the mid 21st century, beginning in 2046, when the suite of space elevators came online, first around the Earth and then around the Moon, cost and mass became less of a factor and lunar

[39] J.M. Hines, C.E. Miller, and R.M. Drake, "Mechanical Equipment Requirements for Inflatable Lunar Structures," *Journal of Aerospace Engineering*, Vol. 5, No. 2, April, 1992, pp. 248–256.

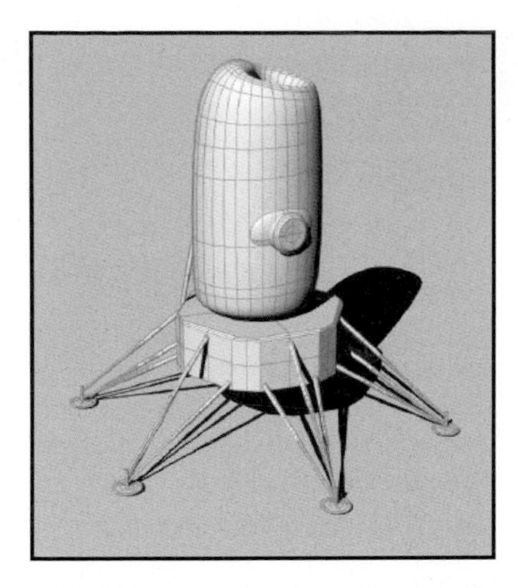

Fig. 5.12 Landing of structure. (Courtesy of group listed in Figure 5.15)

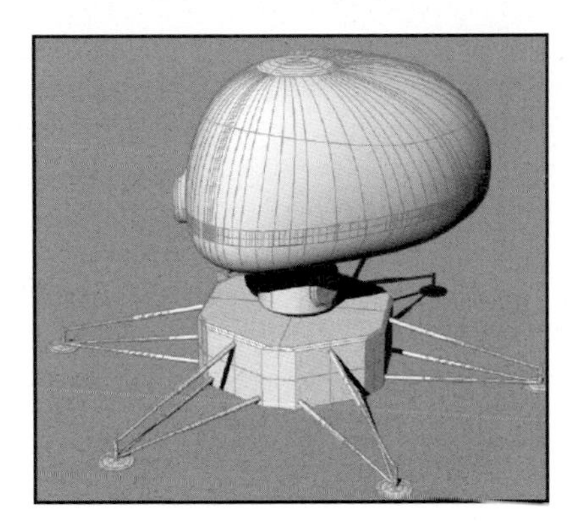

Fig. 5.13 Module rotates into place. (Courtesy of group listed in Figure 5.15)

Fig. 5.14 Group of five modules – top view. See Figure 5.15 for side view. (Courtesy of group listed in Figure 5.15)

Fig. 5.15 Ovoid dome colony. This image and the prior three images are courtesy of the members of the Habitat Design Workshop, 2005. Project: Moon 1 - Fram Studentteam: A. Fischer, J. Jorgensen, J. Tizard, H. Västinsaldo, A. Wielders, S.Zanini © ESA Habitat Design Workshop 2005 (M. Aguzzi, R. Drummond, S. Häuplik-Meusburger, J. Hendrikse, J. van der Horst, S. Lorenz, E. Laan, K. Özdemir, D. Robinson, G. Sterenborg and P. Messino / ESA.) (Courtesy of all listed)

structural concepts began to have more flair. But by then we also were starting to have a lunar infrastructure worth bragging about, and we were well into our automated ISRU construction.

Erectables

Most structures on Earth are erectable – they are made of components such as beams and columns with pipes and conduits – and these structures are put together using bolts and welds by a construction crew. We have the most experience with such structures, resulting in a very large database and very high reliability. It is therefore no surprise that many concepts for lunar structures were based on the erectable concepts perfected for the Earth environment.

On Earth, such structures almost always rely on a foundation to hold the structure in place as it is pounded by wind forces, or earthquake loading if the structure is in a seismic zone, or wave forces for ocean structures. For example, the Petronas Twin Tower in Kuala Lumpur, completed in 1998, has a height of 1483 ft (452 m) and its foundation is 394 ft (120 m) deep. Generally, the foundation must go into bedrock. If bedrock is too deep, then other types of foundations are more effective. On the Moon, the regolith – except for the dust – is very dense. Over the millions of years of meteorite bombardment the crumbled pieces were continually sifted and voids in the ground filled, resulting in an extremely dense medium.

Therefore, early structures built on the Moon had minimal, if any, foundation. They were designed to sit on a prepared smooth surface. Sometimes shallow craters were used to cup the structure.

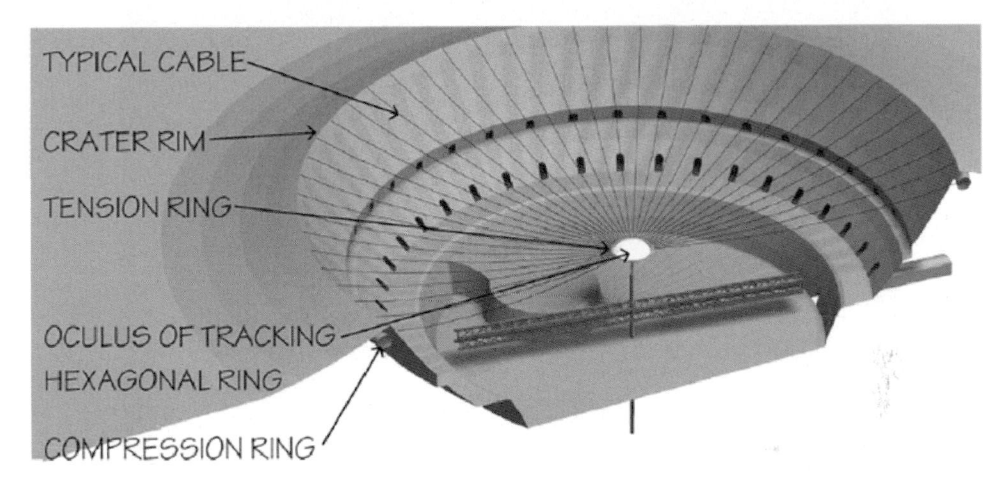

Fig. 5.16 A base inside an adapted crater. The crater is cleared of debris, leveled and reinforced. Cables are placed in the pattern shown. An impermeable cover is place on top and on the surface of the crater and the interior is pressurized. (Courtesy (the late) Alice Eichold)

The unique concept illustrated in Figure 5.16 utilized a "cleaned-up" crater upon which cables were laid in a circular-symmetric pattern.[40] These would be covered by a high-strength impermeable material capable of withstanding internal pressurization. Once complete, the large interior volume could be used for a variety of uses with no spacesuits required. This concept, known today as an Eichold structure, was implemented in the early 22nd century in the crater *Jules Verne* (35.0S - 147.0E - 143.0; Figure 5.17), a large lunar crater on the far side of the Moon, seen after modification in Figure 5.18. It is located to the west-southwest of the *Mare Ingenii*, one of the few lunar mares on the far side.

Most of the interior floor of this crater has been flooded with basaltic lava, leaving a dark, low-albedo surface that is relatively level and flat – thus making it attractive for the cable concept. It is somewhat unusual for a crater feature on the far side to be flooded with lava, as the crust is generally thicker than on the near

[40] A. Eichold, "Conceptual Design of a Crater Lunar Base," International Conference on Environmental Systems, Monterey, CA, July 1996.

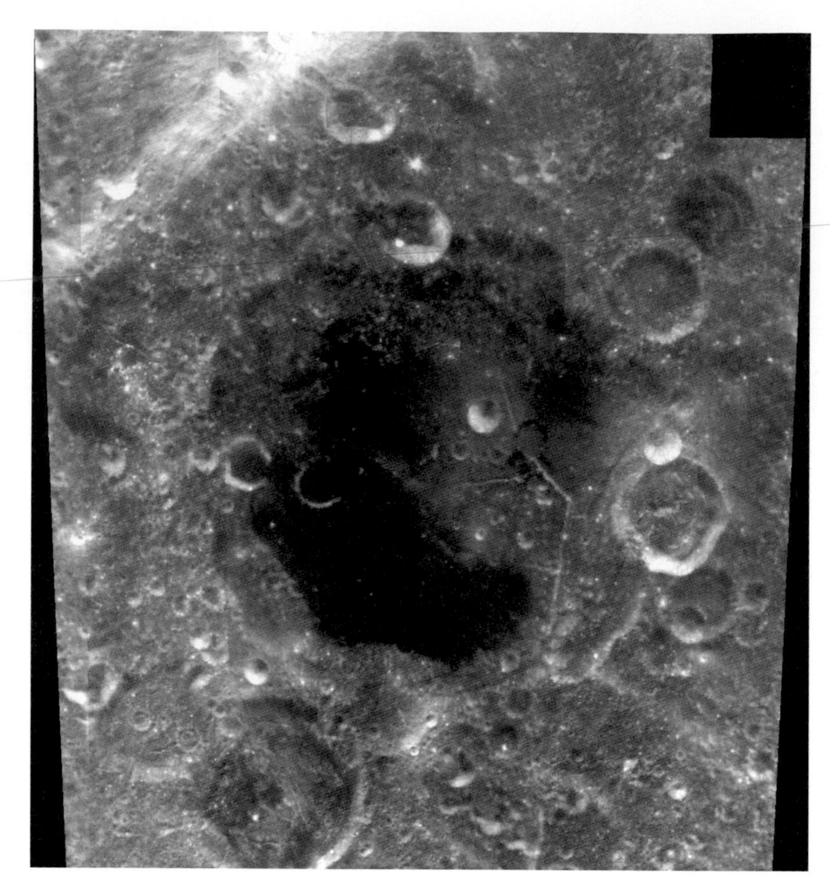

Fig. 5.17 Crater Jules Verne as photographed by the Clementine spacecraft on 1994, long before its conversion to a pressurized facility. It has a diameter of 143 km. (Courtesy NASA)

side. The outer rim of Jules Verne is worn and eroded, with several craters laying across the edge.

Various other concepts were floated,[41] including modular,[42] membrane,[43] and tensegrity structures.[44]

---

[41]H.M. Kelso, J. Hopkins, R. Morris, and M. Thomas, "Design of a Second Generation Lunar Base," *SPACE 88 Engineering, Construction, and Operations in Space*, 1988, pp. 389–399.

[42]M.E. Schroeder, P.J. Richter, and J. Day, "Design Techniques for Rectangular Lunar Modules," *SPACE 94 Engineering, Construction, and Operations in Space*, 1994, pp. 176–185.

[43]M.E. Schroeder, P.J. Richter, "A Membrane Structure for a Lunar Assembly Building," *SPACE 94 Engineering, Construction, and Operations in Space*, 1994, pp. 186–195.

[44]H. Benaroya, "Tensile-Integrity Structures for the Moon," Applied Mechanics of a Lunar Base, *Applied Mechanics Reviews*, Vol. 46, No. 6, 1993, pp. 326–335.

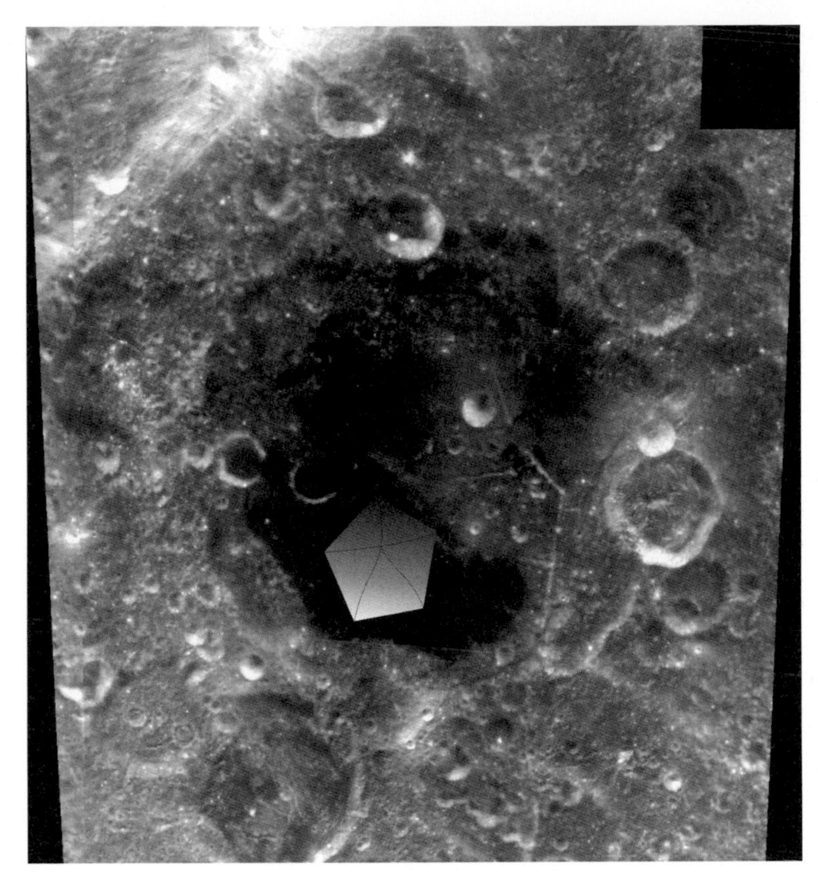

Fig. 5.18 The Jules Verne crater facility for astronomical studies. A portion of the crater has been converted into a pressurized Eichold structure, after the ideas of Alice Eichold.

5.5 Concrete and Lunar Materials

ISRU – *in situ* resource utilization – is the foundation for our survival beyond Earth. We have created an infrastructure that allows us, in 2169, to essentially fabricate everything we need for survival. The first items fabricated were building materials, in particular, a sulfur-based concrete.[45]

"Sulfur-based concrete, using sulfur as the cement, is commonly used in harsh terrestrial environments. It has properties different from Portland cement, in that

[45] V. Gracia, I. Casanova, "Sulfur Concrete: A Viable Alternative for Lunar Construction," *SPACE 98 Engineering, Construction, and Operations in Space*, 1998, pp. 585–591.

W.N. Agosto, J.H. Wickman, and E. James, "Lunar Cement/Concrete for Orbital Structures," *SPACE 88 Engineering, Construction, and Operations in Space*, 1988, pp. 157–168.

R.S. Leonard, S.W. Johnson, "Sulfur-Based Construction Materials for Lunar Construction," *SPACE 88 Engineering, Construction, and Operations in Space*, 1988, pp. 1295–1307.

it doesn't have capillary porosity, and this makes it impermeable to fluids. The concrete would be composed of mineral aggregates cemented together with highly polymerized native sulfur, thereby making a durable concrete. The typical percentages of ingredients for the terrestrial sulfur concrete are around 80% of aggregates, 12% of sulfur, and 8% of fly ashes. It is ... used in some civil construction [on Earth] because of its properties, especially in chemically aggressive environments and in the presence of salts. It is capable of being operational within 24 hours from the time of casting and also has the possibility of being cast at temperatures well below 0 °C."[46]

When lunar structural concepts were envisioned, roads between the different structures needed to be considered. The regolith dust needed to be contained so that it did not get tracked into machines and habitats. An approach conceived of in the early 21st century and further developed over the decades is the microwave sintering of the lunar soil. A robotic paving machine passes over the path that is to be sintered and it sends microwave energy into the regolith, melting and solidifying it to a prescribed depth, depending on the purpose of the road. Roads for heavy loads are solidified to one meter thickness.

A side benefit of the process is the release and capture of most of the solar-wind particles within the soil – hydrogen, helium, carbon, and nitrogen. "Sintering of pre-formed blocks of soil can be used to form solid bricks for various construction purposes, for example, igloos!. An impact crater might be selected and smoothed out to a parabolic shape. Subsequent microwave treatment of the surface could produce an antenna dish, complete with a smooth glass surface. Or, this dish might be cut into sections for movement and reassembly to another location. ... Processing of premolded soil can produce strong structural components. Melting of the mare low-viscosity soil [used] for blowing glass wool or pulling of glass fibers."[47]

Fused regolith structures have been suggested[48] where the proposed structures were small and many, and reside on the surface. Prime advantages offered for

---

H. Namba, T. Yoshida, S. Matsumoto, K. Sugihara, and Y. Kai, "Concrete Habitable Structure on the Moon," *SPACE 88 Engineering, Construction, and Operations in Space*, 1988, pp. 178–189.

H. Namba, N. Ishikawa, H. Kanamori, and T. Okada, "Concrete Production Method for Construction of Lunar Bases," *SPACE 88 Engineering, Construction, and Operations in Space*, 1988, pp. 169–177.

D. Strenski, S. Yankee, R. Holasek, B. Pletka, and A. Hellawell, "Brick Design for the Lunar Surface," *SPACE 90 Engineering, Construction, and Operations in Space*, 1990, pp. 458–467.

[46]M. Pinni, "Lunar Concrete," in *Lunar Settlements*, ed. H. Benaroya, CRC Press, 2010.

[47]L.A. Taylor and T.T. Meek, "Microwave Sintering of Lunar Soil: Properties, Theory, and Practice," *Journal of Aerospace Engineering*, Vol. 18, No. 3, July 2005.

[48]E.W. Cliffton, "A Fused Regolith Structure," *SPACE 90 Engineering, Construction, and Operations in Space*, 1990, pp. 541–550.

R.S. Crockett, B.D. Fabes, T. Nakamura, and C.L. Senior, "Construction of Large Lunar Structures by Fusion Welding of Sintered Regolith," *SPACE 94 Engineering, Construction, and Operations in Space*, 1994, pp. 1116–1127.

planning numerous smaller structures were safety and reliability. The sun's energy was to be used to fuse regolith into components.

Various related concepts were also studied over the years: a precast, prestressed concrete lunar base with a floating foundation in order to minimize differential settlement,[49] utilizing unprocessed or minimally processed lunar materials for base structures as well as for shielding,[50] the use of indigenous materials for the design of a tied-arch structure,[51] construction using layered embankments of regolith and filmy materials (geotextiles) via robotic construction,[52] and fabric-confined soil structures.[53]

5.6 Lava tubes

Lava tubes were assumed to exist on the Moon and Mars long before they were actually located and a few accessed for our settlements.[54] The advantage of utilizing caves for habitable bases is that the underground location offers natural protection against some of the most severe environmental problems, namely, radiation and micrometeorites. Early speculations were that such caves could be shored up structurally, sealed with a polymeric material, and then pressurized. Once pressurized, structures could be built inside the cave – these structures do not have to be pressurized and are therefore of much simpler design.

The first lava tubes were discovered from data gathered by the JAXA Kaguya mission.[55] A Japanese team analyzed this data and found holes in the lunar surface, called "skylights." One was found in the Marius Hills region of the Earth-facing

[49]T.D. Lin, "Concrete for Lunar Base Construction," *Concrete International* (ACI), Vol. 9, No. 7, 1987.

T.D. Lin, J.A. Senseney, L.D. Arp, and C. Lindbergh, "Concrete Lunar Base Investigation," *Journal of Aerospace Engineering*, Vol. 2, No. 1, January 1989.

[50]E.N. Khalili, "Lunar Structures Generated and Shielded with On-Site Materials," *Journal of Aerospace Engineering*, Vol. 2, No. 3, July 1989.

[51]J.A. Happel, "The Design of Lunar Structures Using Indigenous Construction Materials," A thesis submitted to the Faculty of the Graduate School of the University of Colorado in partial fulfillment of the Master of Science in Civil Engineering, 1992.

J.A. Happel, "Prototype Lunar Base Construction Using Indigenous Materials," *SPACE 92 Engineering, Construction, and Operations in Space*, 1992, pp. 112–122.

[52]M. Okumura, Y. Ohashi, T. Ueno, S. Motoyui, and K. Murakawa, "Lunar Base Construction Using the Reinforced Earth Method with Geotextiles," *SPACE 94 Engineering, Construction, and Operations in Space*, 1994, pp. 1106–1115.

[53]R.A. Harrison, "Cylindrical Fabric-Confined Soil Structures," *SPACE 92 Engineering, Construction, and Operations in Space*, 1992, pp. 123–134.

[54]R.R. Britt, "Lunar Caves: The Ultimate Cool, Dry Place," posted on Space.com, 1pm ET, 21 March 2000.

[55]The Japanese SELENE (Selenological and Engineering Explorer) mission, also nicknamed Kaguya, was launched on 14 September 2007, and, at the end of its mission, intentionally crashed into the lunar surface on 10 June 2009.

Fig. 5.19 A sketch of a structure buried in a sealed volume within a lava tube under the lunar surface. The volume is pressurized and can accomodate numerous unpressurized structures. (Courtesy Ana Benaroya)

side of the Moon. The skylight was approximated to be 60 m wide, and its discovery cause quite a stir among those who had been proponents of lava tube bases.[56]

Ideas regarding the utility of constructing the first outposts under the lunar surface have been long-proposed. Preliminary assessments[57] recommended subselene development for the most effective evolutionary potential for settlement.

The initial difficulty of utilizing this concept was the lack of an infrastructure for digging. Preliminary expeditions were minimally robotic and depended on the astronauts to take on the role of construction workers. Eventually though, most of our structures were buried with only a tip peaking through the surface, the "iceberg" model – 95% below ground and 5% above. The Russian concepts along this line, shown in Chapter 11, are more like our current habitats. Structures are mostly buried even though we have advanced shielding materials because over very long periods of time radiation dosage can still be a problem. These shielding materials are very useful however for months-long expeditions.

We excavated regions that were allocated for purposes of development and dug several hundred meters below the surface. We built our structures, and then buried

[56] I. O'Neill, "Living in Lunar Lava Tubes," Discovery Channel, posted online 27 October 2009.

[57] A.W. Daga, M.A. Daga, W.R. Wendell, "A Preliminary Assessment of the Potential of Lava Tube-Situated Lunar Base Architecture," *SPACE 90 Engineering, Construction, and Operations in Space*, 1990, pp. 568–577.

them allowing for access tubes. The 5% of the structures that remained above-ground had views of the stars and the surface – for scientific purposes, links to surface transportation, and tourist and sports activities.

5.7 Soil mechanics

Soil mechanics[58] is the engineering and scientific study of the behavior of soils under all environmental forces. Soil and regolith are very complex media. They are uneven in geometry and strength with properties changing when compressed or, the case of soil, when pores become filled with water. And yet, all of our land structures on Earth sit in soil and engineers design the foundations that hold up those structures. Regolith is not soil. The term soil implies a process of biological decomposition.

The foundation engineer needs a precise mathematical model of the soil at the site where the structure is to be erected. Soil properties vary over small distances. This is why two adjacent structures respond very differently to the same earthquake. Slight differences in soil properties lead to earthquake energy being transmitted with very different magnitudes and directions to the structures above. One structure will survive and the other nearby will collapse.

Assuming we have access to the soil where we will build a structure, we can test for its properties – density, stiffness, damping, void ratios – and determine its bearing capacity, thus allowing us to design a foundation that will work for the structure that will be placed atop it. One of the difficulties we had as we geared up to return to the Moon in the early-to-mid 21st century was that we did not have direct access to the site where we placed the first lunar bases. We did not have direct data. We guessed at the properties.

An interesting example of this difficulty takes us way back to the Apollo era. There was little knowledge of the regolith properties where Apollo 11 was to land and, therefore, much uncertainty about how deep the lander would sink into the regolith. Figure 5.20 shows one of the lander's legs with the large saucer at the base designed to distribute the lander weight.

Clearly and fortunately, the lander did not sink very much. To gather additional information about the regolith, a picture was taken of an astronaut footprint – see Figure 5.21 – so that the depth of penetration could be used to determine its bearing strength. Other such tests were performed during Apollo.

Early studies compared Earth soil mechanics to lunar regolith mechanics and discussed how the analysis and design procedures would differ.[59] Many other studies were performed using actual and simulated lunar regolith.

---

[58]Soil and regolith need to be distinguished. Soil, such as that found on Earth, includes the elements of life. Regolith does not.

[59]M. Ettouney, H. Benaroya, "Regolith Mechanics, Dynamics, and Foundations," *Journal of Aerospace Engineering*, Vol. 5, No. 2, April 1992.

Fig. 5.20 Astronaut Edwin E. Aldrin, Jr. walks on the surface of the Moon near a leg of the Lunar Module during the Apollo 11 extravehicular activity. Astronaut Neil A. Armstrong, Apollo 11 commander, took this photograph with a 70mm lunar surface camera. The astronauts' footprints are clearly visible in the foreground. (AS11-40-5902, 20 July 1969. Courtesy NASA)

5.8 Materials issues

The severe lunar environment not only caused concerns about human survivability but also about material degradation. Issues included thermal stability, fatigue resistance, ultraviolet radiation effects, solar flare protons, galactic cosmic rays, high charge and high energy particles, vacuum stability, and interior humidity gradients.[60] Materials at particular risk were composites – those fabricated by merging

[60]D.W. Radford, W.Z. Sadeh, and B.C. Cheng, "Material Issues for Lunar/Martian Structures," IAF-91-302, 42nd Congress of the International Astronautical Federation, 5-11 October 1991, Montreal.

Fig. 5.21 A close-up view of an astronaut's bootprint in the lunar soil, photographed with a 70mm lunar surface camera during the Apollo 11 extravehicular activity (EVA) on the Moon. The first footprints on the Moon will be there for a million years. There is no wind to blow them away. (AS11-40-5878, 20 July 1969. Courtesy NASA)

a suite of materials, each of which addressed part of the spectrum of strength or durability needed in a modern structure.

Thermal gradients and cyclic thermal inputs due to various operations and changes in the lunar environment can lead to fatigue. Cyclic loads of all types on a structure lead to a weakening of the material due to the breaking of bonds at the atomic level. This is a fact in the design of all structures, whether for the Moon or for Earth. Engineers had quite a bit of experience designing composites for extreme environments. Some of this experience was transferred to the design of lunar structures.

Radiation effects on composites such as polymeric matrix materials led to rapid aging, resulting in embrittlement and the development of microcracking. Radiation shields were utilized and other composites were developed that were less susceptible to these issues.

Vacuum stability, sometimes called outgassing, is the process in which some of the elements in a material emerge from the compound. This can lead to problems in sensitive equipment such as optical and other instrumentation. Also, if the outgassing occurs into the living quarters an added burden is placed on the life support systems. Outgassing of hazardous elements was a critical consideration.

Humidity gradients within the structure can result in damage of certain composite materials. Cyclic variations in humidity levels can redistribute internal stresses.

Some of these problems are a characteristic of the interaction of the material and the lunar environment. We found that burying the bulk of the structure enabled us to avoid many of these problems. As we developed the lunar ISRU infrastructure, we found that by using regolith-derived components we avoided many of the other issues.

5.9 Radiation shielding concepts

While regolith shielding and the burial of habitable structures provided the foremost protection against radiation and micrometeorites, other concepts were developed and utilized. Radiation shielding concepts included those summarized below. The figures referred to in this section are from an early but advanced space concept study.[61]

Shielding concepts were either passive or active. Passive concepts were those that anticipated and were designed for a particular environment. Once placed on site, the passive shielding was not easily altered. Regolith shielding placed atop a structure is a passive design solution. Active concepts were more robust and adapted to the changing environment. But active concept designs, relying on sensors to detect environmental changes and actuators to alter the design attributes, were more complex and expensive, and needed to be maintained. Both active and passive concepts were eventually utilized.

Various categories of radiation are shielded. *Solar particle events* originate in the Sun and are correlated to solar activity. These pose little hazard except during solar storms at which time living beings on the Moon require extra protection. *Galactic cosmic radiation* is due to the stars and short-term effects are not hazardous, but long-term exposure can increase the risks of cancer. A third type of radiation are X-rays that result from high-energy electron collisions with metal conductors or passive radiation shields. In this case the shielding can cause more biological damage than the original particles being shielded.

[61] C.R. Buhler and L. Wichmann, "Analysis of a Lunar Base Electrostatic Radiation Shield Concept," NIAC CP 04-01, 2005.

Galactic cosmic radiation is very difficult to shield.[62] A 1 GeV proton[63] has a range of about 2 m in regolith and secondary particles released due to the primary particle collisions penetrate deeper. The high-energy particles from these collisions within spacecraft materials and lunar regolith produce secondary radiation that is more dangerous than the primary radiation. There is also the added difficulty that when solar activity is at a minimum galactic cosmic radiation peaks.

Much of the early difficulties for the settlers were due to our lack of precise models for the behavior of the solar cycles. We could not predict when there would be more solar activity and thus more danger to our astronauts.[64] Later, as we settled the Moon, solar and space physics evolved rapidly, providing us with improved characterization and understanding of radiation sources in space. It became possible to plan missions on the lunar surface in a precise way with a high probability of safety. Specialists in space "weather" prediction became as popular on the Moon as their earlier counterparts on Earth, the meteorologists. The new predictive tools reduced uncertainties and significantly increased the number of days allowable for human crews to be without the shielding of a shelter. We can now predict solar activity months in advance, thanks to an improved understanding of the physics but also with data from the large network of satellites placed in orbit around the sun for this purpose in the last century. With such lead times we can also adjust schedules for interplanetary travel so that they do not significantly overlap with dangerous solar activity.

Space radiation also affects our structures and tools. Solar cells are degraded by radiation. Spacecraft operations and communications are sensitive to radiation levels.

5.9.1 Active and passive shielding

Charged particles are deflected by active shielding using electric or magnetics fields. The charged particle can be deflected by a force that is proportional to the charge, its velocity, and the electric and/or magnetic field strength utilized. An electrostatic shield utilizes only a static electric field. A magnetic shield utilizes a static magnetic field. And a plasma shield utilizes both electric and magnetic fields. Dynamic fields – time-dependent fields – can also be utilized.

Early shielding devices had to deal with a number of design constraints. Engineers had to contend with potential electrical breakdown of structural materials if the field strength was too high. In order for the shielding devices to perform, effectively 100 MV voltages were needed, but this strength was not initially available. The electrodes needed to have the mechanical strength able to withstand the electrical forces between them. Since the initial settlements were constructed

[62] T. Straume, "Ionizing Radiation Hazards on the Moon," 2167.pdf, NLSI Lunar Science Conference, 2008.

[63] GeV is a gigaelectronvolt is a unit of energy.

[64] *Space Radiation Hazards and the Vision for Space Exploration: Report of a Workshop*, Ad Hoc Committee on the Solar System Radiation Environment and NASA's Vision for Space Exploration, National Research Council, 2006.

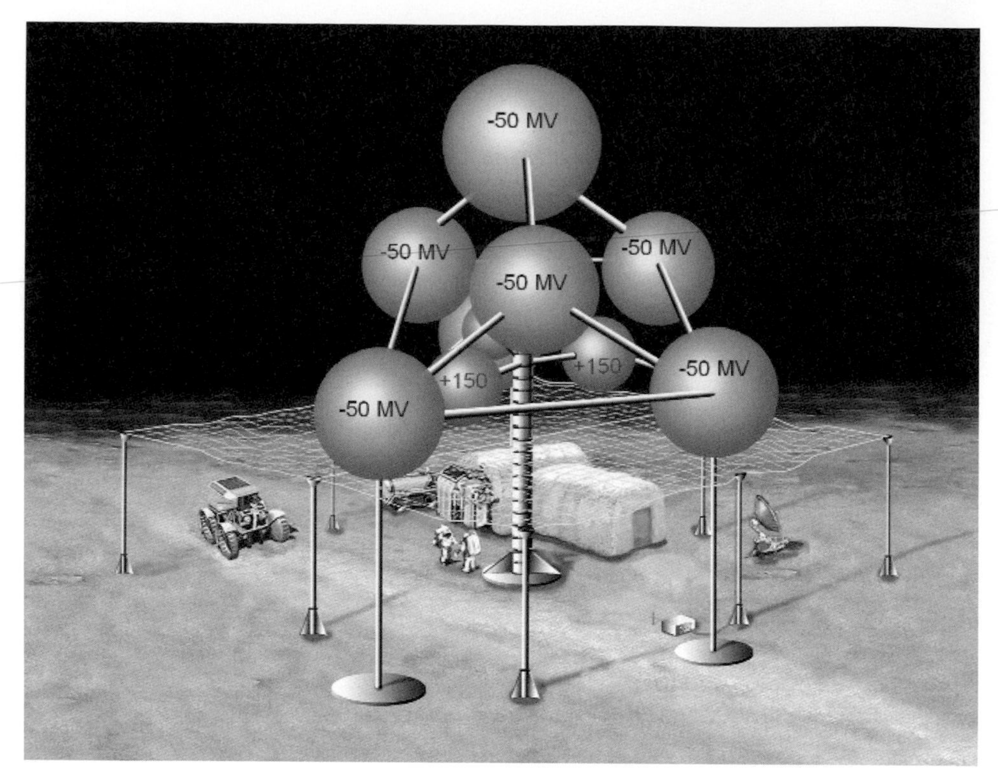

Fig. 5.22 Artist's concept of a "sphere tree" consisting of positively charged inner spheres and negatively charged outer spheres. The screen net is connected to ground potential. (Courtesy NASA)

using items transported from Earth, the size and weight of the structures were limited. The electrode designs also needed to avoid attracting regolith particles, and needed to avoid generating X-rays when impacted by high speed electrons. There were many conflicting challenges that the engineers needed to resolve.

The sphere tree design shown in Figure 5.22 utilizes weak, negatively charged spheres distributed along the shield's outer regions to deflect electrons with a negative potential of less than −40 MV at approximately 75 m while strong, positively charged generators that are clustered at the center deflect high-energy protons. The spheres are placed high above the surface structure so that the electric and magnetic fields they generate do not draw electrons from the structure or surrounding area.

Figure 5.23 depicts a tree of positive-ion-repelling spheres. A magnetic field is created by the solenoids, deflecting the electrons. This concept also prevents the electrons from colliding with the material of the electrostatic spheres and the solenoids, thus preventing the lethal X-rays discussed above.

Figure 5.24 shows an early settlement where a passive shield wall is used to deflect low-angle radiation. In each of these design case studies, the deflection properties are different. Ultimately, as our settlements grew, we used hybrid concepts

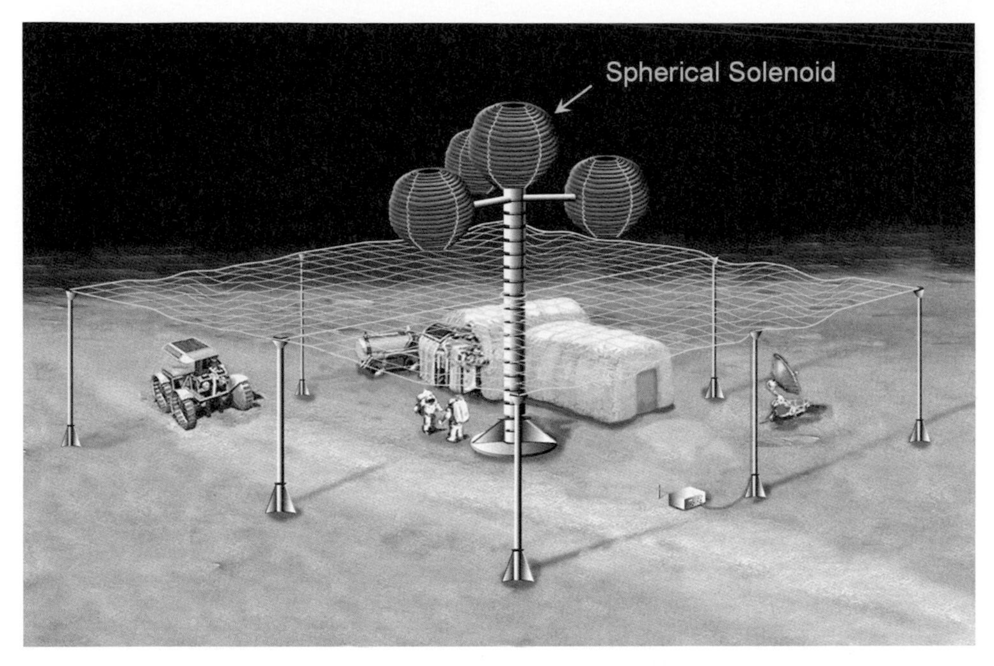

Fig. 5.23 Electrostatic shield concept using only positively charged spheres, wrapped with a current-carrying wire. Electrons are repelled by the magnetic field generated in the spherical wire loops. (Courtesy NASA)

that included all of these shielding strategies, as well as more recently-developed technologies.

The Apollo astronauts were basically lucky that during their spaceflights and lunar sojourns they were not subjected to major radiation events. Had there been a major solar flare, the astronauts would have had no safe haven from the spike in radiation. Being on a planetary body like the Moon, however, offers natural protection against cosmic radiation since the planetary body provides shielding from half the sky. When the astronauts are flying between planets, the radiation is striking from all directions and, when compared to being on a planetary body, the radiation is twice as potent. The short trip to the Moon had this advantage as compared to the much longer trip to Mars.

5.10 Risk and reliability

A key issue of concern that affected everything we did in space – and everything we do in space today – is the assessment of risks. All of our designs are explicitly a result of an assessment of risks. The key concerns of a reliability-based design have been known for a long time, and have been applied to lunar settlements.[65]

[65]H. Benaroya, "Reliability of Structures for the Moon," *Structural Safety*, Vol. 15, 1994, pp. 67–84.

Fig. 5.24 Assuming that the lunar surface under the protected habitat volume is grounded, the low-angle radiation from the horizon is not sufficiently impeded by the shield's electrostatic field. Thus, a passive shield wall is needed in this concept in order to mitigate that low-angle radiation. (Courtesy NASA)

In particular: What failure rate is acceptable? What factors of safety and levels of redundancy are necessary to assure this failure rate?

In the early days, every loss of life brought space exploration to a halt. While all such losses are tragic, they resulted in a disproportionate disruption to progress. Because space was a relatively new environment for people to inhabit, the novelty and high risk led to over-caution.

On Earth, there were losses in the development of air flight vehicles. Test pilots were regularly lost in the line of work. But rather than stop all test flights for years, as was done in the case of the Space Shuttle, lessons were learned and efforts were multiplied, assuring that these losses were not in vain.

The difficulty with space was that it was (and is) such a hostile environment for which we had almost no data upon which to base risk estimates. Risk is another word for probability, the study of events that are only predictable on average. For example, tossing a fair die results in a number from one to six – which number is unknown. A single toss yields, say, a two. Our knowledge of the structure of the die – knowing that it has six possible sides with all equally likely to come up – leads us to the estimate that each number can come up with a fraction or probability of

1/6. Similarly a coin toss leads us to the probability of heads or tails being equal to 1/2 because we know that a coin has two sides.

But what if we did not know the structure of the die or the coin? What if we did not know that a die has six sides or a coin two sides? And what if all the sides did not have the same probability of showing up? If our die had 27 sides and each side had a different probability of showing up, it would take us a long time to figure out the probabilities of each possible outcome. This is the difficulty assessing risk for space activities.

We did not have enough information or data in the early 21st century to estimate the risks. We did not know how many sides the die had and we did not know what was the likelihood of each side coming up. And even with that level of uncertainty we had to design structures for the lunar surface and rockets to land on Mars with precursor infrastructure. Even with technology that was more mainstream, like the Shuttle, we lost two with many lives. Such an event was a shock for the space program because each loss indicated a significant lack of knowledge and understanding of the Shuttle. We thought the Shuttle was a 6-sided die, but we found out it was really a 10-sided die after losing many astronauts. There seems no way around this dilemma. We either accept the losses or we stay in our caves.

So back to our original questions. We put ourselves in the minds of the engineers who designed the first lunar bases. What failure rate was acceptable to them? Since it is generally accepted that one cannot economically (or in reality) design for zero risk, the next logical consideration is the level of acceptable risk. One way to begin to answer such a question is to study the sources of natural risks to a system in its intended environment. In particular, we examine all the natural phenomena and determine the risk exposure of the system to each phenomenon. Some, such as meteorites of a certain size, can destroy a facility, but occur infrequently and therefore need not necessarily be designed against. Each of these risks defines a time limit (in the probabilistic sense) to structural or system life; these may be independent or correlated. Thus, the probability of occurrence of a catastrophic meteorite hit is a small risk, perhaps the smallest encountered risk. It therefore may be viewed as the base risk against which other risks can be weighed. Other natural risks may be ascertained as best as possible – compared to the base risk – and then considered within the overall reliability analysis.

Next, the probability of man-made risks need to be assessed. Examples are: the probability of an explosion of a liquid oxygen tank, the likelihood of projectiles piercing a critical structural component due to an accident, the likelihood of thermal cycle fatigue, and the probability of various human errors. These can be more easily estimated and compared to the above base risk. All of these "component" risk factors need to be assessed. Using engineering judgment and calculations, weighed somewhat by political considerations, acceptable risk can then be decided upon. To this day we recall the financial and psychological costs to the space program due to the Shuttle Challenger disaster. We needed to develop methodologies for assessing risk.

Suppose that we define $R_m = \Pr\{\text{meteorite}\}$ to be the probability that a destructive meteorite will strike a site on the Moon during a year's time. Also $R_f = \Pr\{\text{thermal fatigue}\}$ is defined as the probability that a certain number of

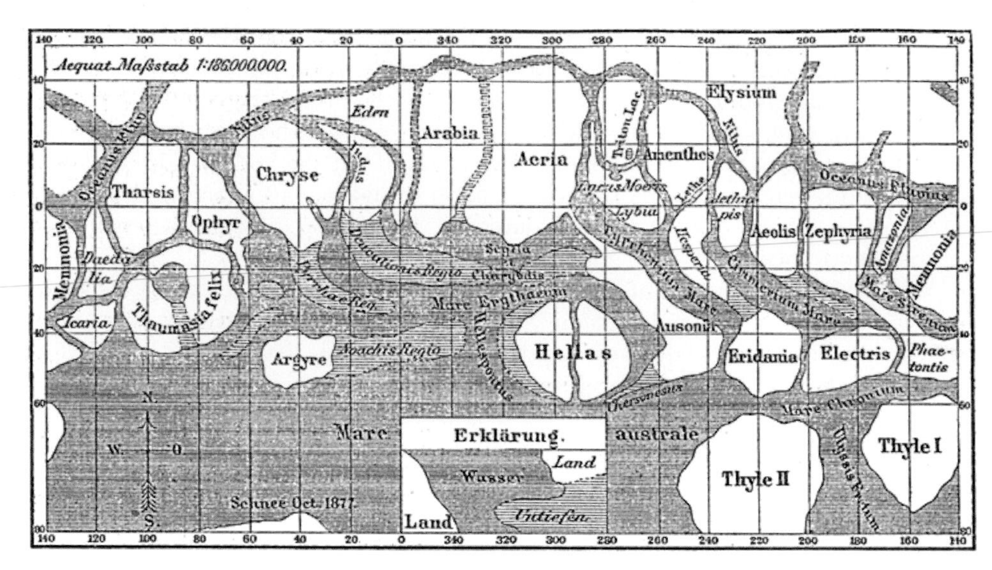

Fig. 5.25 Historical map of planet Mars from Giovanni Schiaparelli. Meyers Konversations-Lexikon (German Encyclopaedia), 1888.

thermal cycles in one year will result in material failure. Each such risk probability can be estimated independently, any correlation established, and then we can define a minimum prescribed design risk as $\min\{R_m, R_f\}$. This will be a measure of the smallest acceptable risk. This risk may be too small to be economically acceptable, but it is a starting point for an analysis. When we further consider that structures are designed to be compartmentalized and modular, accessible and repairable, then it begins to appear possible to increase the value of the acceptable design risk, R_{min}, to be used in the preliminary designs.

As engineers, we try to design as much warning as possible into the structure so that the inhabitants will have time to exit in the event of an impending failure. Therefore, we cannot accept a first-excursion failure. We design for progressive failure. First-excursion failure occurs the first time a design is exceeded. There is no warning. Designers attempt to build into their structures a robustness so that an impending failure gives warning before occurring. For example, the ceiling bends in a large way letting occupants of the room know of the impending problem and giving them enough time to leave before the collapse.

What factors of safety[66] and levels of redundancy are necessary to assure this failure rate? Given an agreed upon acceptable level of risk, it becomes necessary

[66] A factor of safety is an engineering parameter that is an indicator of the level of knowledge of the system that is being designed. If everything is exactly known and there are no uncertainties regarding the structure or its intended environment, then the factor is 1. If there is some uncertainty, then the engineer will design a stronger structure so that there is some overdesign in the structure. If the structure is 30% stronger to account for the uncertainties, then the factor of safety is 1.3. Generally the range of safety factors for well-known structures and systems is between 1.1–1.3. If there is too much uncertainty, then testing is necessary – thus, flight tests.

Color Plates

Plate 1 Earth's Moon, just 3 days away, is a good place to test hardware and operations for a human mission to Mars. A simulated mission, including the landing of an adapted Mars excursion vehicle, could test many relevant Mars systems and technologies. (Artwork done for NASA by Pat Rawlings, of SAIC. S95-01563, February 1995. Courtesy NASA) See Fig. 1.6 on page 11.

Plate 2 A large Arecibo-like radio telescope on the Moon uses a crater for structural support. In the background are 2 steerable radio telescopes. (Artwork done for NASA by Pat Rawlings of SAIC. S95-01561, February 1995. Courtesy NASA) See Fig. 1.12 on page 27.

Plate 3 Astronaut Edwin E. Aldrin, Jr., lunar module pilot of the first lunar landing mission, poses for a photograph beside the deployed United States flag during an Apollo 11 Extravehicular Activity (EVA) on the lunar surface. The Lunar Module (LM) is on the left, and the footprints of the astronauts are clearly visible in the soil of the Moon. (Astronaut Neil A. Armstrong, commander, took this picture with a 70mm Hasselblad lunar surface camera. AS11-40-5875, 20 July 1969. Courtesy NASA) See Fig. 1.14 on page 34.

Plate 4 Shimizu space hotel. This illustration shows a large facility that is accessed by a space plane, shown docked at the lower part of the station. The hotel is within the centrally-located inverse pyramid. (Courtesy Shimizu Corporation) See Fig. 2.11 on page 54.

Plate 5 A lunar mining operation. (Drawing by Pat Rawlings for NASA. Courtesy NASA) See Fig. 3.9 on page 121.

Plate 6 At the Lunar Marriott, people are flying at the left of the image and there is surface activity at the lower left. (Image created by Paul DiMare for Popular Mechanics – popularmechanics.com. Courtesy Popular Mechanics) See Fig. 4.5 on page 148.

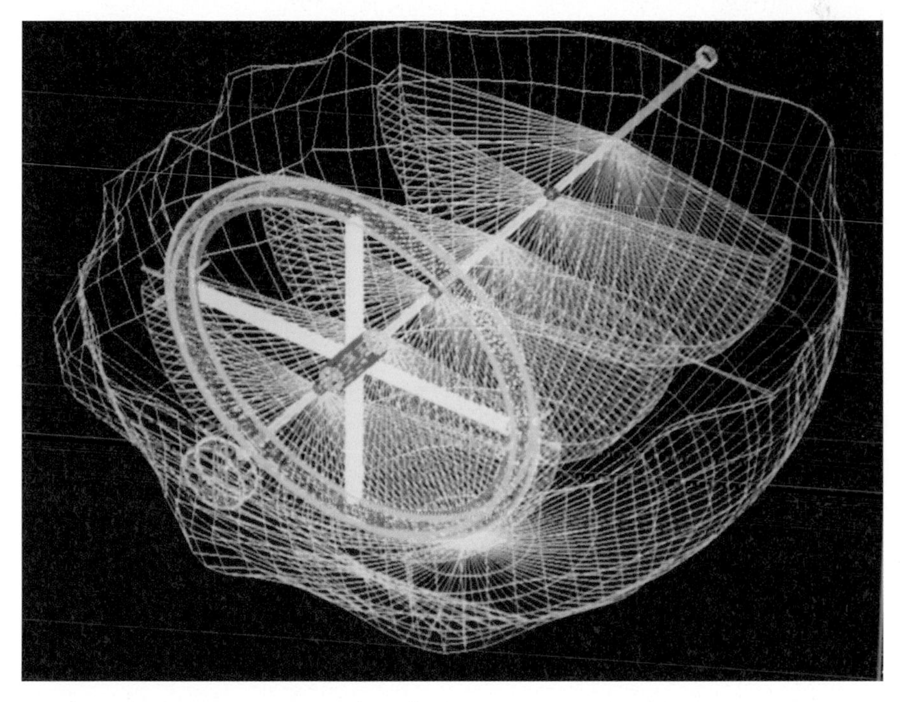

Plate 7 An early digital design (1993) of the comet's interior, including the central tunnel, the caves and the first habitat. (Courtesy Tom Taylor and Werner Grandl) See Fig. 7.3 on page 252.

Plate 8 This painting shows an asteroid mining mission to an Earth-approaching asteroid. Asteroids contain many of the major elements that provide the basis for industry and life on Earth, the Moon and Mars. A NASA-sponsored study on space manufacturing held at Ames Research Center in the summer of 1977 provided much of the technical basis for the painting. "Asteroid-1" is the central long structure and the propulsion unit is the long tubular structure enveloped by stiffening yard arms and guy wires. Solar cells running the length of the propulsion system convert the sunlight into electricity that is used to power the propulsion system. During the mission these solar arrays would be oriented toward the Sun to gather maximum power. In the left foreground is an asteroid mining unit, doing actual mining work. An orbital construction platform in permanent orbit provides power, supplies depot and work volume within which work proceeds. (Artist concept by Denise Watt. S78-27139, June 1977. Courtesy NASA) See Fig. 7.6 on page 255.

Plate 9 A space elevator being propelled up the ribbon by an ocean platform-based laser. Present-day climbers are being propelled by very small fusion reactors. These also power the shielding devices on board each elevator. (Courtesy Space Elevator Visualization Group and Alan Chan) See Fig. 9.12 on page 304.

Plate 10 Space elevator being propelled further up the ribbon by an ocean platform-based laser. Here the elevator has passed through an orbital station. (Courtesy Space Elevator Visualization Group and Alan Chan) See Fig. 9.13 on page 306.

Plate 11 A closeup of the "Luna Ring" concept for lunar solar power generation. (Courtesy Shimizu Corporation) See Fig. 10.13 on page 339.

Plate 12 Phobos propellant processing facility in the shadow of Mars. (Courtesy Phillip Richter, Fluor Daniel and Rockwell International) See Fig. 10.15 on page 341.

Plate 13 Manned Lunar Base 2050: Energia-Sternberg Project, residential zone in crater, general view. (Courtesy Vladislav Shevchenko) See Fig. 11.8 on page 353.

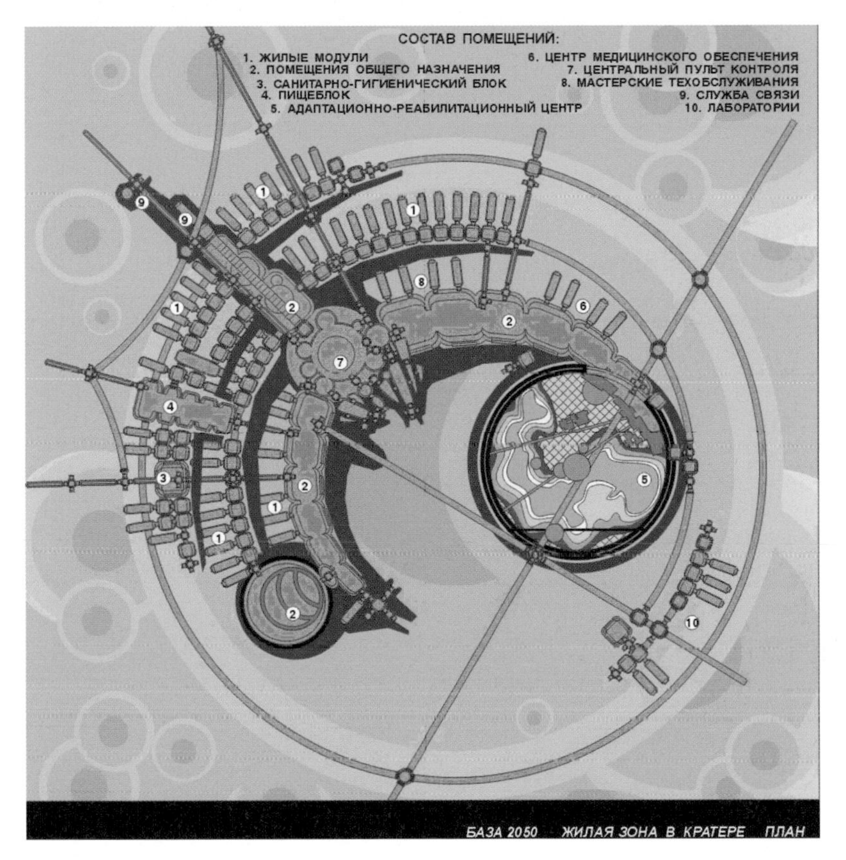

Plate 14 Manned base in crater: plan view. **Legend:** 1. inhabited modules 2. general purposes 3. clean facilities 4. kitchens 5. adaptation and rehabilitation 6. medical facilities 7. control center 8. maintenance 9. communications 10. labs. (Courtesy Vladislav Shevchenko) See Fig. 11.9 on page 354.

Plate 15 An advanced lunar settlement which pictures concepts that have been under study for many years, including a payload shooting off the end of a rail-launch system, an astronomical observatory in the upper left and a nuclear power plant in the upper right. (Courtesy JAXA) See Fig. 11.14 on page 360.

Plate 16 A Mars settlement with much activity. (Courtesy JAXA) See Fig. 12.1 on page 366.

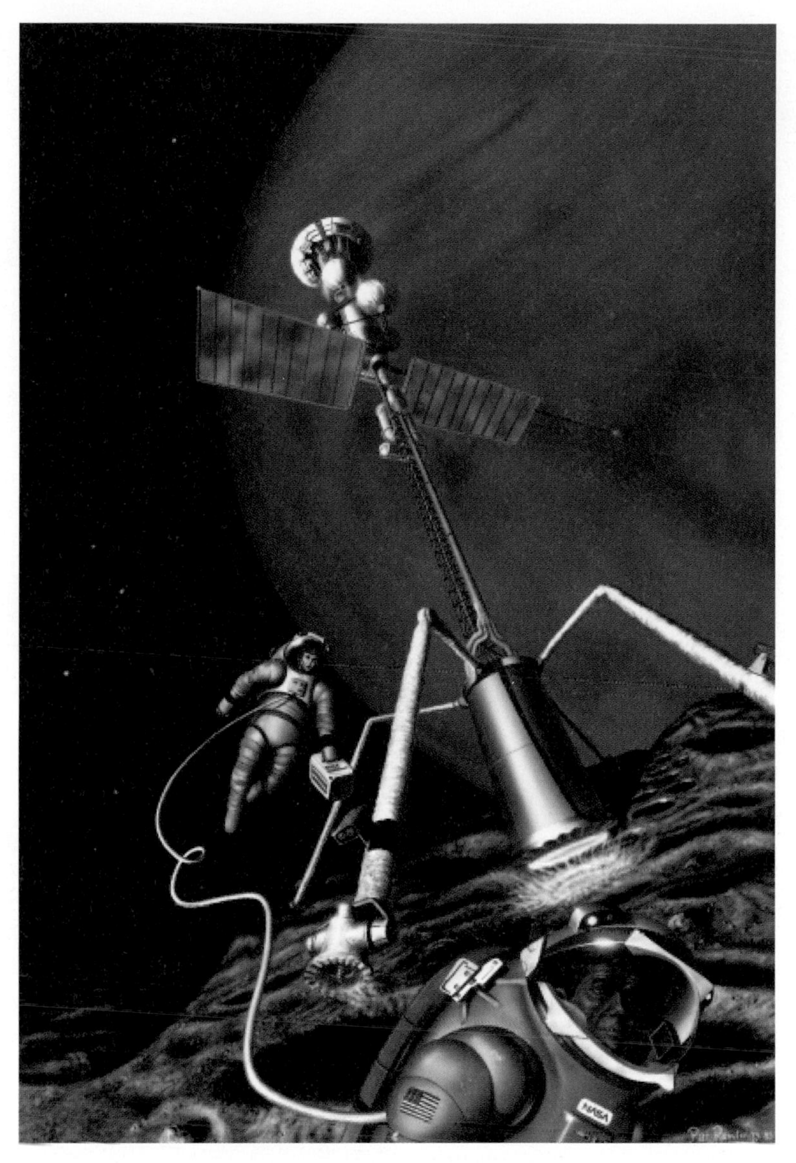

Plate 17 On Phobos, the innermost moon of Mars and likely location for extraterrestrial resources, a mobile propellant-production plant lumbers across the irregular surface. Using a nuclear reactor the large tower melts into the surface, generating steam that is converted into liquid hydrogen and liquid oxygen. (Artwork by Pat Rawlings of Eagle Engineering. S86-25375, 1986. Courtesy NASA) See Fig. 12.2 on page 369.

Plate 18 The Surface Stereo Imager on NASA's Phoenix Mars Lander took this image (original in false color) on 21 October 2008, during the 145th Martian day, or sol, since landing. The white areas seen in these trenches are part of an ice layer beneath the soil. The trench on the upper left, called "Upper Cupboard," is about 60 cm (24 in) long and 3 cm (1 in) deep. The trench in the middle, called "Ice Man," is about 30 cm (12 in) long and 3 cm (1 in) deep. The trench on the right, called "La Mancha," is about 31 cm (12 in) and 5 cm (2 in) deep. (Courtesy NASA) See Fig. 12.5 on page 373.

Plate 19 Artistic conceptualization of robotic ISRU construction of Mars habitats. These robots use ballistic particle manufacturing procedures. (Courtesy Ana Benaroya) See Fig. 12.10 on page 376.

Plate 20 *Earthrise* is the name given to a photograph of the Earth taken by astronaut William Anders in 1968 during the Apollo 8 mission. (Courtesy NASA) See Fig. 13.1 on page 385.

Plate 21 A 16 m diameter inflatable habitat is depicted and could accommodate the needs of a dozen astronauts living and working on the surface of the Moon. Depicted are astronauts exercising, a base operations center, a pressurized lunar rover, a small clean room, a fully equipped life sciences lab, a lunar lander, selenological work, hydroponic gardens, a wardroom, private crew quarters, dust-removing devices for lunar surface work and an airlock. (This artist concept reflects the evaluation and study at Johnson Space Center by the Man Systems Division and Johnson Engineering personnel. S89-20084, July 1989. Courtesy NASA) See Fig. 13.3 on page 387.

as a practical matter to establish a design philosophy. For example, what factors of safety did the engineers of the early 21st century build into their "lunar design codes?" Since the lunar site provided designers with the most uncertainties of any engineering project, with few opportunities to obtain experience or data until the last century, one philosophy demanded higher-than-Earth factors of safety. However, it is possible to approach this question from another design perspective. Consider the site to be inherently high-risk and – just as we accept high risks for test pilots – accept a high-risk approach to a lunar outpost design concept. Both approaches can be justified.

In addition, the above risks can be compared to statistical data from occupational and normal population groups. Such data allowed us to back out mission requirements such as the allowable (acceptable) individual death by illness during the mission, allowable death by injury, and allowable death from all causes, including spacecraft failure. Based on such an analysis it was possible to set up reliability objectives for each mission scenario, and to make sure that the mission safety objectives were met.[67]

Redundancy is a separate question. Once a basis has been set for acceptable risk and safety factors, the designer must be ingenuous in the conceptual design, optimizing the design so that overall risk is as close as possible to the acceptable level. Risk should also be distributed throughout the lunar habitat site in accordance with the criticality of the various parts to the overall mission. This is a difficult problem requiring the study of competing structural and system concepts.

How did logistics interplay with considerations of risk and reliability? The link is quite close. Generally, one has two options when a component or system fails: replace or repair. During the early days of the return to the Moon, inventories on the Moon were not large enough to always be able to replace components. Thus, some failures were repaired rather than replaced. Others led to replacement. All potential failures that were viewed to be high risk had to be accounted for in any design.

In the early days, reliability and safety were linked to the maximum amount of payload that could have been brought to the lunar facility from Earth in the minimum amount of time. This minimum-time-to-maximum-payload defined the absolute necessary self-sufficiency time for the lunar inhabitants. During this time, local replacements and/or repairs were mandatory in order to recover from and survive significant failures. Logistic requirements, therefore, became important at an early stage of the design development cycle.

As we developed our ISRU infrastructure on the Moon, we could in essence manufacture any component needed for the safe operation of the settlements. We designed redundancy into our designs as well as ease of repair and reconditioning. Commonality of parts has always been a strategic goal and therefore a design constraint. This is not new; this is how lunar and Martian systems have been designed for over a hundred years.

---

[67] G. Horneck and B. Comet, "General Human Health Issues for Moon and Mars Missions: Results from the HUMEX Study," *Advances in Space Research,* Vol. 37, 2006, pp. 100–108.

Our design approach has the following philosophy. A large-scale lunar outpost, if designed for low risk to inhabitants, would be a complex and expensive undertaking, primarily because humans are very delicate. Instead, our lunar settlements are designed to higher risk tolerances with significant cost savings. But to ensure an overall high level of human safety, a number of smaller but much safer facilities are placed throughout the settlements at easily accessible locations. These smaller facilities are designed to support the population for a significant amount of time – the time needed for rescue missions to arrive from the other settlements. The added cost of the smaller facilities is much less than the cost would be to bring all the lunar settlements to those same high standards. Our fleet of pressurized rovers is a part of this safety net strategy. With time, we expect to be able to evolve beyond the current frontier standards for safety. We are quite comfortable with our safety, and the strategy just outlined has worked on several occasions with no loss of life.

Also part of our safety strategy is the use of "smart" structures for the critical parts of our settlements. Such structures are completely sensored and monitored so that there is warning of impending failures and problems. There are self-repairing capabilities as well – air leakages are sealed automatically – and excess regolith dust on machines is tracked and cleaned.

5.11 Building systems

These lists organize structures, their infrastructure, and their applications. They show how many interrelated systems are integrated and considered in the design of a structure and a settlement.

5.11.1 Types of applications

Habitats

- people (living/working)
- agriculture
- airlocks (ingress/egress)
- temporary storm shelters for emergencies and radiation
- open volumes

Storage facilities/shelters

- cryogenic (fuels/science)
- hazardous materials
- general supplies
- surface equipment storage
- servicing and maintenance
- temporary protective structures

Supporting infrastructure

- foundations/roadbeds/launchpads
- communication towers and antennas
- waste management/life support
- power generation, conditioning and distribution
- mobile systems
- industrial processing facilities
- conduits/pipes

5.11.2 Application requirements

Habitats

- pressure containment
- atmosphere composition/control
- thermal control (active/passive)
- acoustic control
- radiation protection
- meteoroid protection
- integrated/natural lighting
- local waste management/recycling
- airlocks with scrub areas
- emergency systems
- psychological/social factors

Storage facilities/shelters

- refrigeration/insulation/cryogenic systems
- pressurization/atmospheric control
- thermal control (active/passive)
- radiation protection
- meteoroid protection
- hazardous material containment
- maintenance equipment/tools

Supporting infrastructure

- all of the above
- regenerative life support (physical/chemical/biological)
- industrial waste management

5.11.3 Types of structures

Habitats

- landed self-contained structures
- rigid modules (prefabricated/*in situ*)
- inflatable modules/membranes (prefabricated/*in situ*)
- tunneling/coring
- exploited caverns

Storage facilities/shelters

- open tensile (tents/awning)
- "tinker toy"
- modules (rigid/inflatable)
- trenches/underground
- ceramic/masonry (arches/tubes)
- mobile
- shells

Supporting infrastructure

- all of the above
- slabs (melts/compaction/additives)
- trusses/frames

5.11.4 Material considerations

Habitats

- shelf life/life cycle
- resistance to space environment (uv/thermal/radiation/abrasion/vacuum)
- resistance to fatigue (acoustic and machine vibration/pressurization/thermal)
- resistance to acute stresses (launch loads/pressurization/impact)
- resistance to penetration (meteoroids/mechanical impacts)
- biological/chemical inertness
- repairability (process/materials)

Operational suitability/economy

- availability (lunar/planetary sources)
- ease of production/use (labor/equipment/power/automation/robotics)
- versatility (materials and related processes/equipment)
- radiation/thermal shielding characteristics
- meteoroid/debris shielding characteristics
- acoustic properties
- launch weight/compactability (Earth sources)
- transmission of visible light
- pressurization leak resistance (permeability/bonding)
- thermal and electrical properties (conductivity/specific heat)

Safety

- process operations (chemical/heat)
- flammability/smoke/explosive potential
- outgassing
- toxicity

5.11.5 Structures and technology drivers

Mission/application influences

- mission objectives and size
- specific site-related conditions (resources/terrain features)
- site preparation requirements (excavation/infrastructure)
- available equipment/tools (construction/maintenance)
- surface transportation/infrastructure
- crew size/specialization
- available power
- priority given to use of lunar material/material processing
- evolutionary growth/reconfiguration requirements
- resupply versus reuse strategies

General planning/design considerations

- automation/robotics
- EVA time for assembly
- ease and safety of assembly (handling/connections)
- optimization of teleoperated/automated systems
- influences of reduced gravity (anchorage/excavation/traction)
- quality control and validation
- reliability/risk analysis
- optimization of *in situ* materials utilization
- maintenance procedures/requirements
- cost/availability of materials
- flexibility for reconfiguration/expansion
- utility interfaces (lines/structures)
- emergency procedures/equipment
- logistics (delivery of equipment/materials)
- evolutionary system upgrades/changeouts
- tribology

5.11.6 Requirement definition/evaluations

Requirement/option studies

- identify site implications (Lunar soil/geologic models)
- identify mission-driven requirements (function/purpose/staging of structures)
- identify conceptual options (site preparation/construction)
- identify evaluation criteria (costs/equipment/labor)
- identify architectural program (human environmental needs)

Evaluation studies

- technology development requirements
- cost/benefit models (early/long-term)
- system design optimization/analysis

* * *

5.12 Quotes

- “We shape our buildings; thereafter they shape us.” Winston Churchill
- “Buildings, too, are children of Earth and Sun.” Frank Lloyd Wright
- “Architecture aims at Eternity.” Christopher Wren
- “Architecture is a visual art, and the buildings speak for themselves.” Julia Morgan
- “I see Earth! It is so beautiful!” Yuri Gagarin
- “The greatest advances of civilization, whether in architecture or painting, in science and literature, in industry or agriculture, have never come from centralized government.” Milton Friedman
- “I must study politics and war, that my sons may have the liberty to study mathematics and philosophy, natural history and naval architecture, in order to give their children a right to study painting, poetry, music, architecture, tapestry, and porcelain.” John Adams

6 An early design and its construction

"I told them how excited I would be to go into space and how thrilled I was when Alan Shepard made his historic flight, and when ... men had landed safely on the Moon, and how jealous I was of those men."

Christa McAuliffe

The first structures placed on the Moon for human habitation were prefabricated cylindrical modules. These were placed on the lunar surface and shielded using regolith. That is how we lived on the Moon for decades. Even though quarters were cramped, the first inhabitants could hardly believe their senses that they were on the Moon for an extended stay. Nothing else mattered.

In 2029, five years after the 2024 manned return to the Moon, a facility that looked like an igloo was erected that became a habitat for permanent settlement. It was somewhat larger than the cylinders. But the real improvements were the friendlier interiors. With time the initial igloo was expanded and within twenty years we had a small city on the Moon. At the same time, other settlements were planted.

6.0.1 An historical interview with Neil Armstrong (March 2009)

Can you please provide us with a brief bio?

I am the retired Chairman of the EDO Corporation, an electronics and aerospace manufacturer. As a Naval Aviator, I flew 78 combat missions from an aircraft carrier. Subsequently, I served 17 years with the NACA and NASA as an engineer, test pilot, astronaut and administrator.

As a research pilot at NASA's Flight Research Center at Edwards, California, I was a project pilot on many pioneering high speed aircraft, including most of the early supersonic jets, the rocket powered X-1, the variable sweep wing X-5, the vertical take off and landing X-14, and the well known X-15, which I flew to over 60 kilometers altitude and 6400 kilometers per hour.

I transferred to astronaut status in 1962 and was the commander of the Gemini 8 flight in 1966 when I performed the first successful docking of two vehicles in space. As spacecraft commander for Apollo 11, I, with colleagues Mike Collins and Buzz Aldrin, completed the first landing mission to the Moon.

I have been the holder of 13 world records in aviation and space.

I received my engineering degrees from Purdue University and the University of Southern California. During the years 1971-1979, I was Professor of Aerospace Engineering at the University of Cincinnati. I am a member of the U.S. National Academy of Engineering and the Royal Academy of the Kingdom of Morocco.

Given all of the years you spent preparing for your mission to the Moon, and then the achievement of that mission over a matter of a week's time, was your transition from that peak experience back to "everyday life" a difficult one? Did you have a plan?

Fortunately, I was employed by a great organization, NASA. I looked forward to going to work and there was much to do. There was never such a thing as 'everyday life' at the Johnson Space Flight Center.

As a follow-up, after returning from the lunar surface the astronauts needed to move from their extraordinary experiences to more ordinary daily lives. Each did this in his own way. Did you find the transition within your expectations and to your satisfaction?

My work at NASA was primarily engineering. Astronauts spend very little time in flight compared to the work they do on the ground. While flying in space is an extraordinary experience, there is great satisfaction in the work that is done on the ground in planning space flight techniques and testing equipment. Crew members really get an opportunity to contribute in major ways to the productivity and success of the flights that they and others will experience.

When I left active flight crew duties, I went on to other engineering duties. I was assigned to engineering management responsibilities. That also was very satisfying although, I must admit, it was sometimes not as much fun as basic engineering work. In my experience, engineering managers love to get into situations where they can revert to their basic engineering instincts and use their engineering skills.

So my entire career was involved in working as an engineer, teaching engineering, or managing engineering projects and companies. I was extremely fortunate and don't regret any part of it.

Do you keep in touch with the other Apollo astronauts?

Yes, our paths seem to cross frequently. I had an opportunity to attend the Apollo 7 40th anniversary in Dallas and the Apollo 8 40th anniversary in San Diego where a number of the Apollo crew members, flight controllers and other Apollo contributors were present.

Is settling the Moon important for civilization?

We haven't spent any money in space. I look forward to the day when we can. All our investments in space exploration are investments in technology development and all the money is spent here on Earth.

How do we answer critics who say that space is too expensive and that there are numerous problems on Earth to take care of first?

Space exploration is expensive. Not nearly as expensive as many of our government enterprises: Defense is 30 times more expensive, Intelligence is 3 times more expensive, Health and Human Services is 38 times more expensive. NASA requires less than 1% of the U.S. national budget. NASA's responsibility is to develop options for future generations. If they do it properly, our grandchildren and great-grandchildren will benefit in many ways and it will have proven to be a very excellent investment.

What do you see as the major hurdles for our return to the Moon with permanent manned settlements? Are they technical, financial, physiological, psychological?

There are no impassable barriers to returning to the Moon. It only requires that our society agree that it is a worthy investment.

When you envision the future, where do you see us in 50 years? In 100 years? How do you see this evolving from the present?

I suspect that inventions and discoveries that we have not even dreamed of will occur and change anything we might predict. They will change our expectation of what is possible in myriad ways.

If there were a city on the Moon, would you think about living there?

By the time we return to the Moon and live in a cabin, I will be an upper tier octogenarian. The time for me to be moving to a new city may well have passed (unless one of those unexpected developments comes along :)).

* * *

In this chapter we summarize the design of the "igloo" lunar structural concept, some of the considerations that led to its size and configuration, and how it finally looked. It also went by the name the "Rutgers Lunar Base" since it was initially conceived at Rutgers University. The design discussion that follows provides the reader with a glimpse of the engineering approach not only to structural design but also to problem solving in general.

6.1 An early design (2004)

The discussion that follows is about an early design based on research and design efforts that culminated in a published paper in 2006.[1] It can be viewed as a modified

[1] F. Ruess, "Structural Analysis of a Lunar Base." Master's Thesis, University at Stuttgart/Rutgers University, 2004.

F. Ruess, J. Schaenzlin, and H. Benaroya, "Structural Design of a Lunar Habitat," *Journal of Aerospace Engineering*, Vol. 19, No. 3, July 2006, pp. 133–157.

version of the arch structure proposed in 1992[2] and is written from the perspective of the period of time during which the design was conceived.

6.1.1 Preliminary considerations

Determining the dimensions of a lunar base habitat is a very complex task. Numerous factors such as crew size, mission duration and function of the base as an industrial or scientific outpost influence the necessary habitat size. Hence, the approach taken is based on the parameter: habitable volume per person. Habitable volume excludes volume occupied by equipment or stowage.

As demonstrated by Gemini,[3] relatively short duration missions can be endured by a person restrained to a chair most of the time. The habitable volume per crew member in Gemini was 0.57 m^3. In 2008, the NASA Man Systems Integration Standards[4] (NASA STD 3000) recommended a minimum habitable volume of about 20 m^3 in order that crew performance could be maintained for mission durations of four months or longer. Despite this, a design volume (living and working areas) of 120 m^3 per person (approximately 25 m^2 of floor space) for a lunar habitat has been recommended, based on research of long-term habitation and confined spaces. This value is about equivalent to the volume available per crew member on board the International Space Station.

An arch is selected as the cross-section of the structure – thus the igloo. The arch profile has been used since the dawn of construction, in bridges, tunnels, hangars and even cars. The next step is to decide on the shape and rise of the arch. A single floor layout is preferred to avoid additional structural mass for internal flooring and to simultaneously reduce the size of the main structural members. A rise of 5 m was chosen for the arch. Figure 6.1 shows how the space within the arch will be divided into the different functional areas.

The optimum floor height is then selected based on the minimum habitable volume. Proposed floor heights for lunar habitats range from 2.44 m to 4.0 m. People moving in low gravity require more vertical space than on Earth. They lift off the floor higher while walking and especially when trying to run. Therefore, a floor height of 4.0 m is selected and used in this design – designs are iterative, meaning that a first design concept is evaluated, modified, and then re-designed with improvements.

However, floor height is not equal to clear height. Support systems like lighting and ventilation will use between 0.5 m and 1.0 m of this space. This leaves in most cases about 3.5 m for the actual habitable volume. With these numbers fixed, we calculate 34.4 m^2 floor area per person. The total floor area depends not only on crew size but also on the amount of equipment and stowage space that is needed. A summary for different crew sizes is given in Table 6.1.

---

[2]M. Ettouney, H. Benaroya, and N. Agassi, "Cable Structures and the Lunar Environment," *Journal of Aerospace Engineering*, Vol. 5, No. 3, 1992, pp. 297–310.

[3]The Gemini missions were for two men. They flew between 23 March 1965 (5 hours) and 15 November 1966 (4 days).

[4]The NASA-STD-3000 was created to provide a single, comprehensive document defining all the generic requirements for space facilities and related equipment that directly interface with crew members. Revision B July 1995.

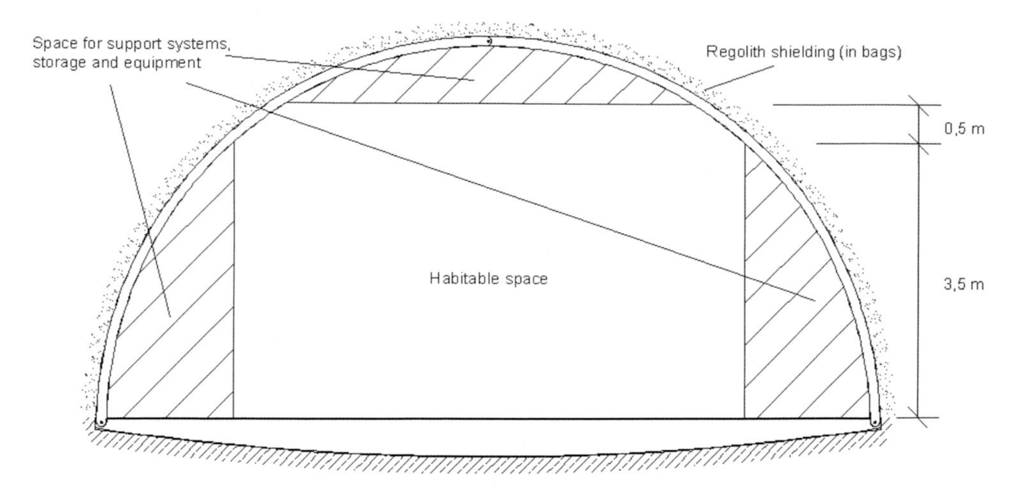

Fig. 6.1 Structure cross-section and habitable space. The two arches are hinged at the top and connected to the tie at the base. (Courtesy Florian Ruess)

Table 6.1 Total needed floor area with respect to crew size.

Crew size	6	8	10	12
Habitable area [m^2]	206	275	343	412
+20% for equipment and stowage [m^2]	41	55	69	82
Total area (rounded up) [m^2]	250	320	415	500

Having determined the total floor area, we can begin to size the structure. Depending on the chosen structural system, we need to find the most efficient span of the main structure and the spacing between primary structural elements. The necessary clear floor height can govern the span for some concepts, such as arches. The layout of the habitat is also very important at this point.

On Earth, where gravitational loads usually govern our design, parabolic arch shapes are most efficient. An arch can be designed to be only in compression with little or no bending moment introduced into the structure. Bending moments should be avoided wherever possible. It is a very inefficient way to transfer the loads to the foundations. Simple tension or compression is much more efficient and is one of the design goals.

On the Moon, however, the governing load is not gravitational. Parabolic and circular arches under internal pressure can be compared in Figures 6.2 and 6.3. A comparison clearly shows that the circular arch is the more suitable structure because no bending moments are introduced. The arrows represent pressure and the shaded areas represent bending moments in the structure in response to the pressure. The calculations that were performed to find these results were based on an in-plane two-dimensional analysis, found to be sufficient since no major three-

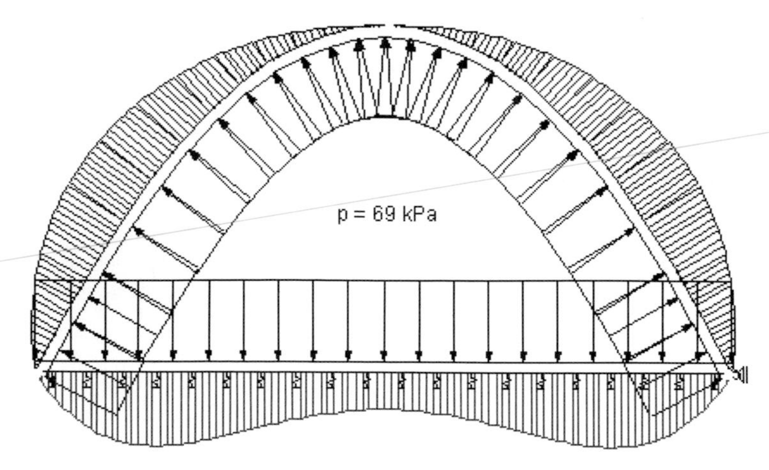

Fig. 6.2 A comparable parabolic arch for the lunar surface, shown with internal, external and foundation force distributions. The shaded regions depict the bending moment in the structure due to the internal forces. We see moments on all parts of the structure. (Courtesy Florian Ruess)

dimensional effects are expected because the structure runs continuously in the third direction.

A decision is needed next as to whether hinges or a rigid connection should be used to connect the arch to the base. Hinges are easier to construct than rigid connections. They also enable the designer to easily divide the structure into different segments, reducing the transportation dimensions. The maximum allowable number of hinges is three. With four or more hinges, the structure turns into a mechanism and is not statically stable any more. The two- and three-hinge arrangements are better than a one-hinge arrangement because they do not introduce bending moments in the arch.

The bending moment in the tie is a result of the interaction between the soil and the structure – how the soil reacts to the pressure from the structure's weight. It depends on the ratio of foundation-to-soil stiffness. The final bending moment in the tie can only be determined iteratively because every change in tie stiffness results in a change in bending moment that in turn may require a different tie cross section. Thus, the final bending moment distribution will only be available after the structural design is finished.

The three-hinge layout shown in Figure 6.4 is chosen for ease of construction. The three-hinge arch is a statically determinate structure. Therefore, temperature loading during construction will not introduce stresses in the members, only deflections. This is another advantage of the three-hinge concept. Figure 6.5 shows the foundation reaction to the structural weight.

6.1.2 Internal and external forces

The mass of a body never changes. However, its weight depends on the gravitational acceleration. On the Moon, the designer has to be careful when applying

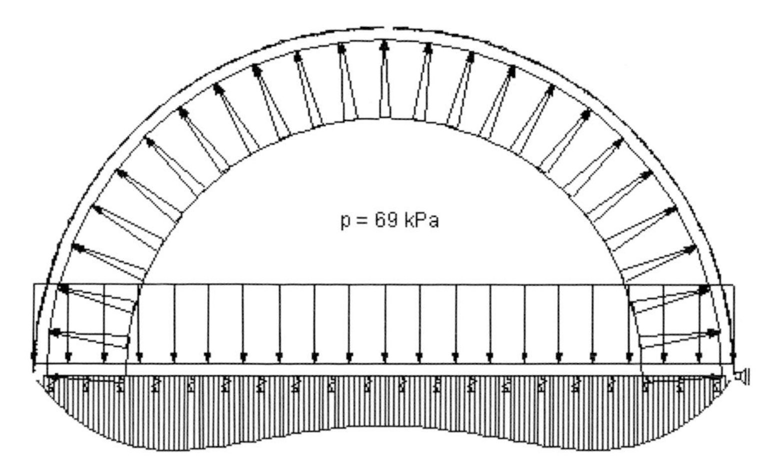

Fig. 6.3 Circular arch lunar surface structure shown with internal and foundation force distributions. We only see a bending moment in the tie; the arches show no bending moment. (Courtesy Florian Ruess)

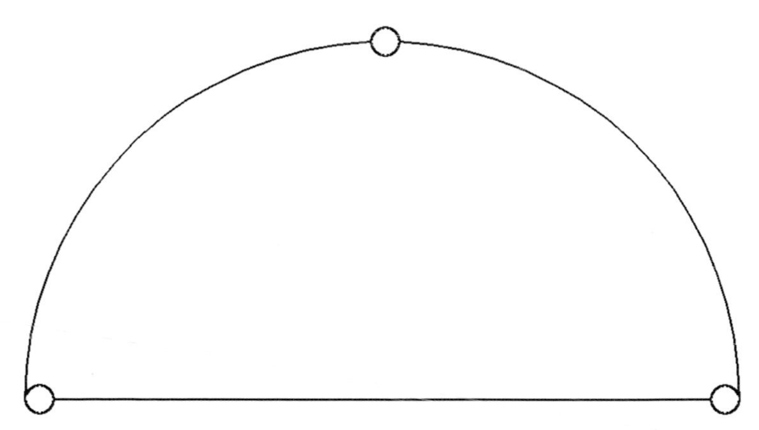

Fig. 6.4 A schematic of the cross section of the three-hinge structure. (Courtesy Florian Ruess)

gravitational loads, which are usually given as a weight in units of kiloNewton or kiloNewton-meters. For lunar analysis, the weight has to be calculated using the lunar gravitational acceleration. The resulting loads will be only about 1/6 of those on Earth.

Load combinations

For the static analysis of the structure, five main load cases were identified, in addition to the structure's self-weight, and are illustrated in Figures 6.6 to 6.10:

1. Internal pressure $p = 69$ kPa (Figure 6.6)
2. Regolith covering the whole structure $q = 8.3$ kPa (Figure 6.7)

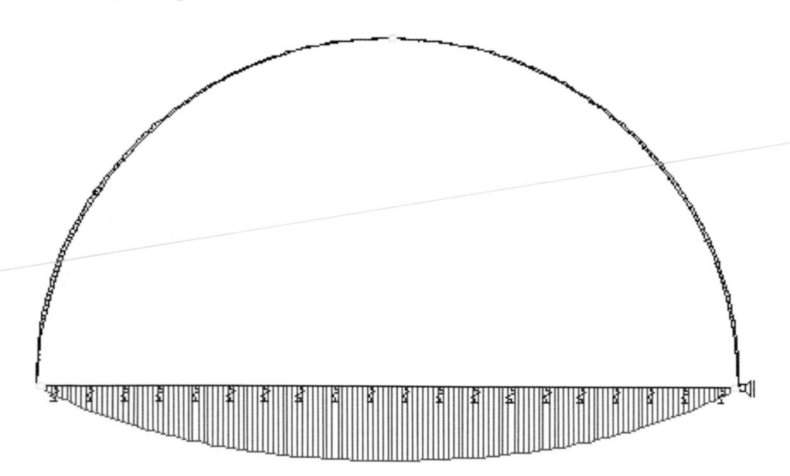

Fig. 6.5 A schematic of the three-hinge structure with foundation reaction forces. (Courtesy Florian Ruess)

3. Regolith covering one-half of the structure $q = 8.3$ kPa (Figure 6.8)
4. Floor service loads $q = 1$ kPa (Figure 6.9)
5. Installation loads attached to the roof $q = 0.25$ kPa (Figure 6.10)

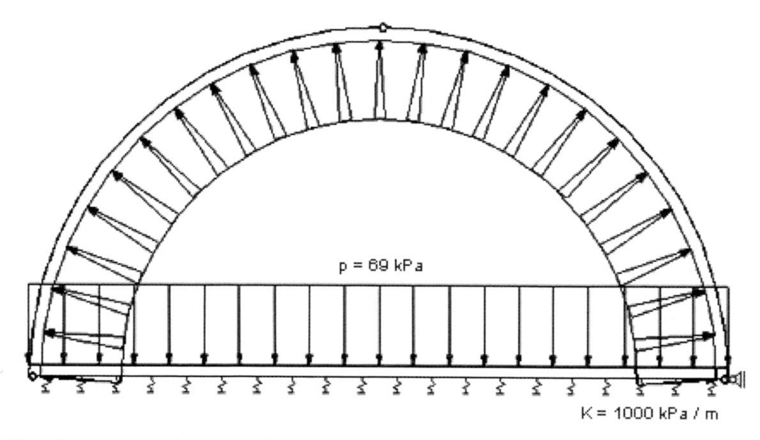

Fig. 6.6 Load combination one: Internal pressurization loads. (Courtesy Florian Ruess)

All of the figures show small springs along the foundation/tie with the stiffness property $K = 1000$ kPa/m. This value represents the stiffness of the regolith in resistance to the weight of the structure.[5]

The loads for the regolith cover assume that the regolith is bagged and can therefore be placed uniformly on the structure. If instead of bagging, loose soil is

[5]A kilopascal, kPa, is a unit of pressure. One kPa equals 0.145 lb/in^2. One psi equals 6.895 kPa.

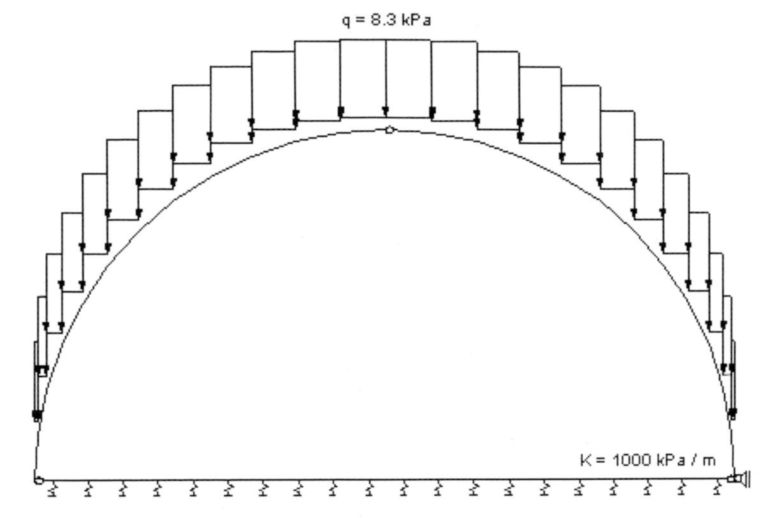

Fig. 6.7 Load combination two: Regolith loading. (Courtesy Florian Ruess)

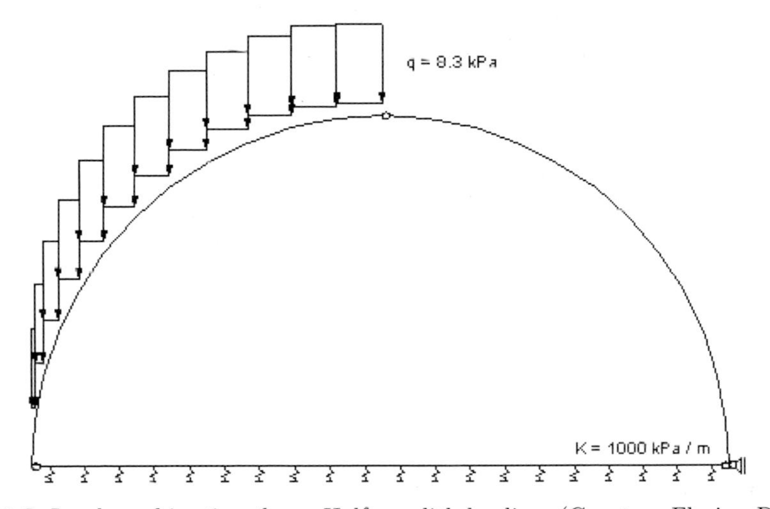

Fig. 6.8 Load combination three: Half-regolith loading. (Courtesy Florian Ruess)

simply heaped upon the top of the structure, the resulting load will be trapezoidal and not uniform.

6.1.3 Structural considerations

To minimize structural mass, it is crucial to optimize the cross-section. An infinite number of cross section types is possible. Four likely types of structural members were examined in this study, as shown in Figures 6.11 and 6.12. The cross-sectional area and its shape are chosen to best resist shear forces and bending moments in the structure.

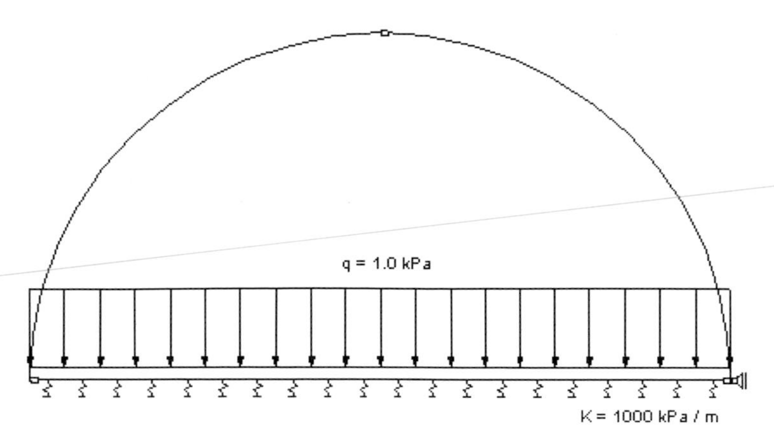

Fig. 6.9 Load combination four: Interior loads. (Courtesy Florian Ruess)

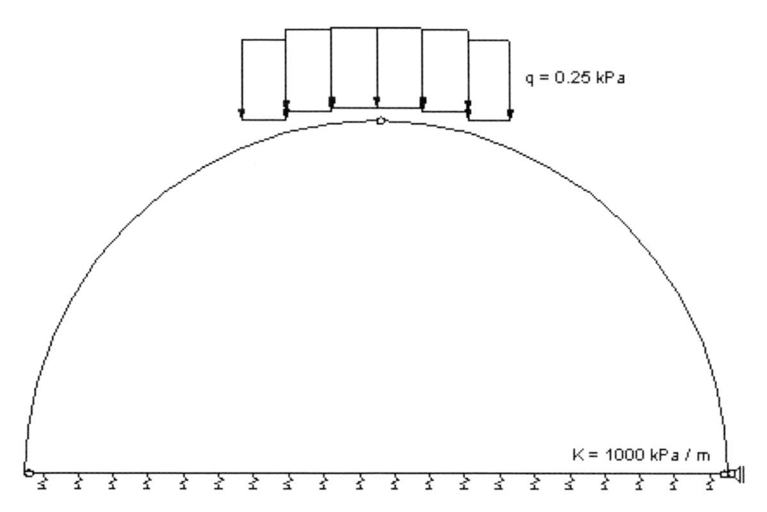

Fig. 6.10 Load combination five: Additional loads during construction. (Courtesy Florian Ruess)

Table 6.2 SI and English/US units for key physical parameters.

Parameter	SI	English/US
Force	N (kg-m/s^2)	0.2248 lb$_f$
Mass	kg	2.2046 lb$_m$ = 0.0685 slug
Length	m	3.28 ft
Acceleration	m/s^2	3.28 ft/s^2
Spring constant	N/m	6.8543×10^{-2} lb/ft
Pressure	Pa (N/m^2)	1.45×10^{-4} psi (lb/in^2)
Damping constant	N-s/m	6.8543×10^{-2} lb s/ft
Mass moment of inertia	kg-m^2	0.7375 lb-ft-s^2
Angle	rad	$360/2\pi$ deg

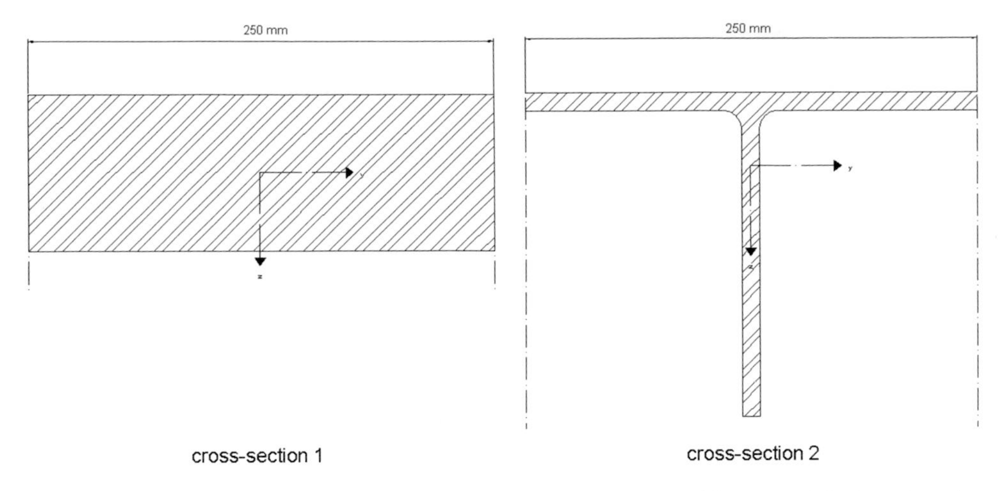

Fig. 6.11 Cross-sections 1 and 2: Rectangular and T-beam. (Courtesy Florian Ruess)

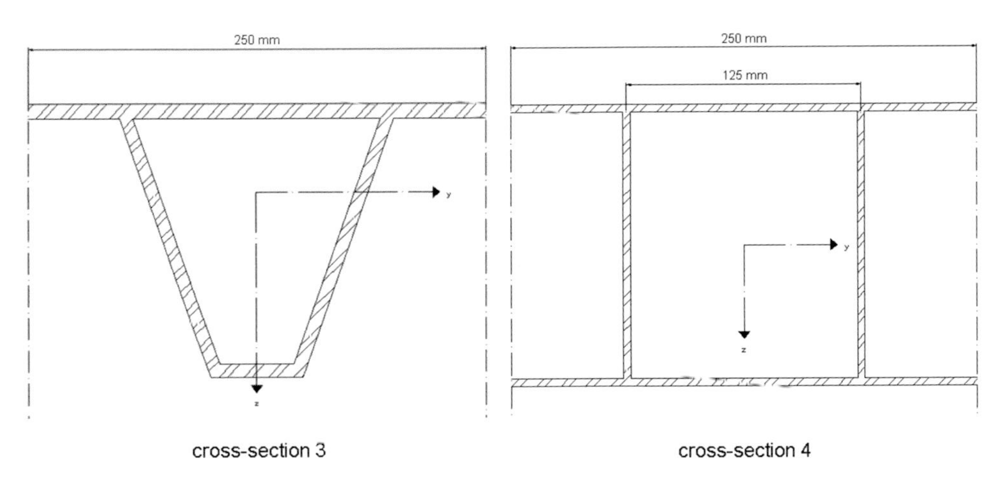

Fig. 6.12 Cross-sections 3 and 4: V-Box, Double I-beam. (Courtesy Florian Ruess)

Determining the stresses in the structural members for the different load combinations can be done by applying the principle of linear superposition.[6] Load Combination One was found to give the highest stresses for the tie, and Load Combination Three gives the maximum stresses for the arch segments. The regular operation mode for the lunar structure will be with all loads acting.

In the early designs, since it was not yet possible to use local lunar materials for the structures, lightweight but strong elements were brought from Earth. In this design, the construction material was assumed to be high-strength aluminum with the members prefabricated on Earth and transported to the Moon. Composite

[6]The principle of linear superposition is used to simplify an analysis by considering each force separately and then combining all of the respective displacements to obtain the total structural displacement.

materials that could further save mass were not used mainly because of a lack of experience at the time with those materials. So, all calculations for this design are based on a high-strength aluminum alloy with a yield strength of 500 N/mm^2. The depth of the legs in cross-sections 2 to 4 was limited to a maximum of 15 cm for the arch segments in order to limit transportation volume.

The design of the end walls is discussed in a subsequent subsection. However, one of the results is already needed at this point. Internal air pressure on the end walls introduces tensile forces into the arch structure perpendicular to the already calculated forces. These need to be taken into account when designing the arch and tie members. The forces are 130 kN/m for the arch segments, and 170 kN/m for the tie, respectively. The sandwich type cross-section number four is the most efficient and is used for both arch and tie. It is optimal regarding strength and weight.

The arch member design is relatively easy. It is a structure governed by the interaction of bending moment and tension force. The design of the tie/foundation member is more complicated. It is governed by the bending moment, which depends on the ratio of foundation-to-soil stiffness. A stiffness increase in the foundation yields a higher bending moment, since stiffness attracts forces. So designing the foundation plate is an iterative process.

To find out the influence of a difference in the factor of safety on the design, two different global safety factors are considered and compared: safety factor 4.0 and safety factor 5.0. The range for the factor of safety on Earth is, for example, in the German design code DIN,[7] approximately between 1.7 and 3.5. Since there are numerous uncertainties on the Moon we require higher safety factors – a factor of safety of 4.0 or higher seems reasonable.

Two main conclusions result from the structural analysis. First, the arch segments can have a uniform cross-section. It is possible, but not necessary, to adjust the arch cross-section to the distribution of internal forces since these are almost uniform. Second, in order to get an efficient cross-section for the tie, it has to be adjusted to the distribution of internal forces. The bending moment has the shape of a parabola, so it was decided to give the tie a similar shape. The section height is affine to the square root of the bending moment, that is, the depth of the tie cross-section varies, with the smallest depth at the ends and the largest in the middle. Figure 6.13 shows the principal shape of the tie/floor/foundation.

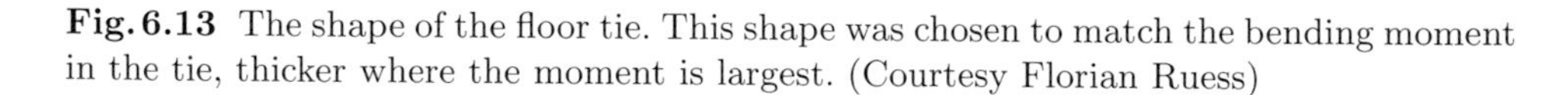

Fig. 6.13 The shape of the floor tie. This shape was chosen to match the bending moment in the tie, thicker where the moment is largest. (Courtesy Florian Ruess)

Deflections under all loads, with the structural members designed using a safety factor of 5.0, can be viewed in Figure 6.14. The allowable deflection is 50 mm. If the members are designed using a global safety factor of 4.0, the cross-sections

[7] Deutsches Institut für Normung e.V. (DIN; in English, the German Institute for Standardization).

become lighter and this results in slightly larger deflections, with a maximum of 68 mm. Since the expected deflections only slightly exceed the allowable deflections, we adjusted the design to reduce them.

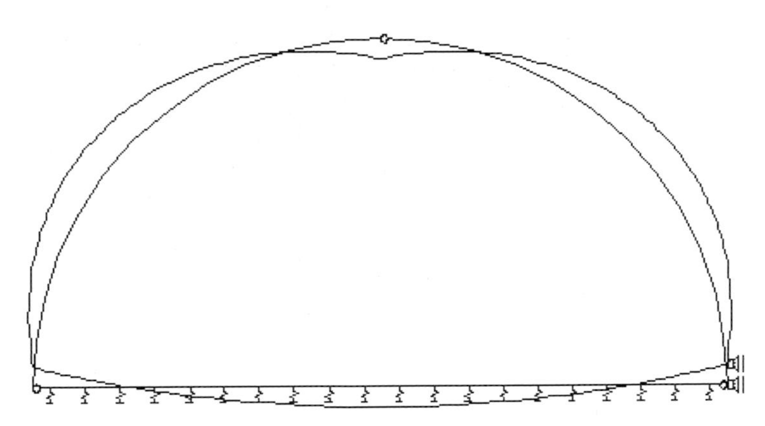

Fig. 6.14 Exaggerated static deflection of the structure. We see the effects of the internal pressurization. (Courtesy Florian Ruess)

Reducing deflections can be done in many different ways:

Strengthen the members: This approach results in extra mass that is not needed from a strength point of view. Since low mass is crucial for an economic design, this is not a desirable approach.

Strengthen the soil: Sintering the regolith before erecting the structure will result in a higher modulus of subgrade reaction, and therefore lower deflections of the tie. It does not affect the arch deflections. Calculations show that the modulus of subgrade reaction would have to be increased about tenfold to obtain the desired deflections. This is very likely not achievable by sintering the regolith. More research data is needed for this topic.

Cambering: All of the loads can be assumed permanent. Therefore, cambering of the members is a solution. The members can simply be manufactured in a shape opposite to the calculated deflections. Manufacturing has to be done very carefully and exactly to enable construction onsite.

End walls

The end walls are also designed with sandwich-type cross-section four. This cross-section type is simply the most efficient. The most efficient shape for the end walls is the quarter sphere where end moments are reduced to a minimum. However, for various reasons, such as the need for low transportation volume, easy implementation of air locks, and expandability of the structure, the flat end-plate is chosen. A schematic is shown in Figure 6.15, where the different shading represents different stresses.

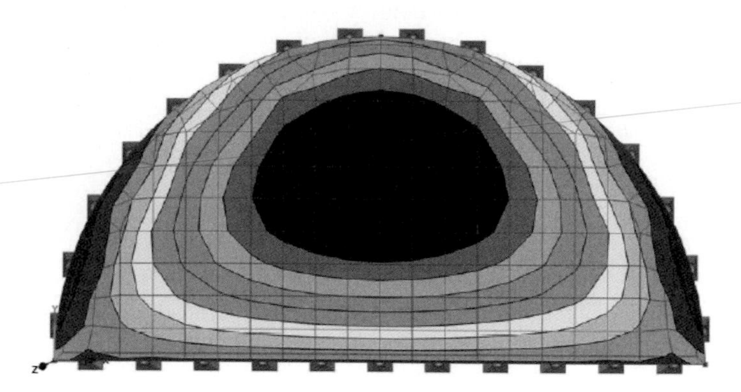

Fig. 6.15 A computational model of one of the end walls subjected to internal pressure. The different shades of grey represent different stresses, with the highest stresses at the center and at the edges. (Courtesy Florian Ruess)

Vibration

A vibration analysis is done for the design using a safety factor of five. In order to simulate vibrating machinery, the dynamic response of the structure for a cam mechanism with an amplitude of approximately 4.6 cm and a frequency of 172 Hz (Hertz or cycles per second) was calculated. No significant deflections or additional internal forces were found.

Resonance is the primary concern in structures that vibrate. This is the phenomenon when the vibration frequencies of structural components and forces overlap. When this happens, the vibration amplitudes can become very large, much larger than when there is no such overlap, and structural failure can occur.

Construction sequence

The habitat dimensions in the longitudinal direction depend mainly on the crew size. The transportable length for structural members is limited by the transportation system, so it is necessary to divide the structural members into different segments for transportation.[8] A reasonable segment length is 2.5 m. Ten segments put together would then provide enough space for a crew of six astronauts. When the segments are connected it is necessary to ensure that all the loads can be transferred between segments and that the pressure is sealed. Welding the full section

[8] At the time of this design study during the early 21st century, the transportation system in use was the Space Shuttle. The structural members mentioned here had to fit in the Shuttle cargo bay. The bay, 18.3 m long and 4.6 m wide (60 ft by 15 ft), had payload attachment points along its full length and was adaptable enough to accommodate as many as five unmanned spacecraft of various sizes and shapes in one mission. The cargo bay was equipped also with a 15 m (50 ft) robot arm for lifting and releasing, or grasping and retrieving satellites in space.

is the best way to seal the structure. Welding in space was performed during the Russian Soyuz T 12 mission, for example.

Erection of the structure can be done step by step. It is a linear process. First, the floor is constructed. The 2.5 m long panels are laid out on the site and joined using tongue and groove joints. Cables are used to pull all of the panels together. Welding all the panels provides sealing and guarantees that all forces can be transferred.

The arch panels are erected one after the other, by placing one arch segment on a temporary construction scaffolding and bolting it to the floor at the bottom. The second arch segment is placed in the same way. Finally, the two arch segments are bolted together. The temporary scaffolding is lowered and transferred to the next section where the placement of the next arch segments can begin. The panels then have to be welded together.

All welding is best done at night to minimize temperature-induced deformations. If construction is performed during the lunar day, unprotected members are exposed to the sun and temperatures on the structural surface climb up to 150 °C. However, the other side of the structure, which is shaded, will be at a temperature of −100 °C. This causes the structure to deflect approximately 0.5 m.

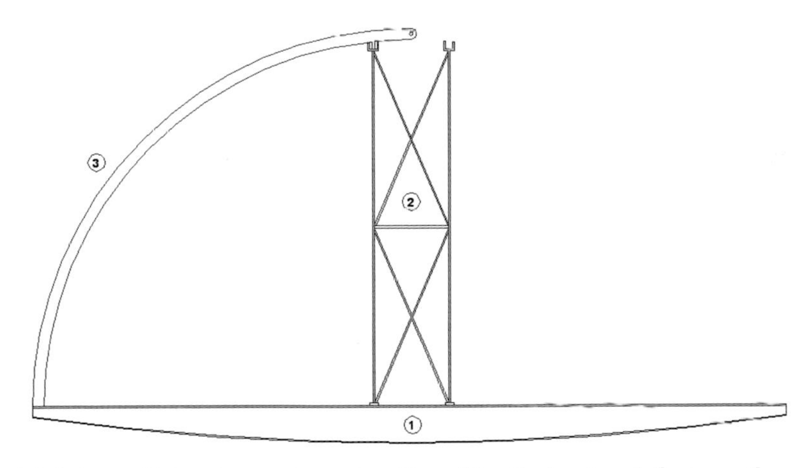

Fig. 6.16 Intermediate phase of construction. The tie is placed first on the prepared lunar surface, after which tomporary scaffolding is built. Then we see the first of two arch panels being placed and connected to the tie. Next, the second arch panel will be placed and connected to the surrounding structure. Finally the end panels are placed and all are welded together. (Courtesy Florian Ruess)

Figure 6.16 shows the structure after one arch panel is put in place. When the desired structural length is achieved, the end walls are slid in at the ends of the arch. The connections between the arch and the walls are implemented next. The structure is then sealed with membrane fabric strips glued to the structure along the structural connections where necessary. After completing all necessary seals, the habitat can be pressurized and finally the regolith cover is put into place.

The floor segments have a mass of 2,945 kg, so they weigh approximately 5 kN; the arch segments have a mass of 605 kg, and weigh about 1 kN each. Equipment is needed to move the parts from their landing site to the construction site and into place there, but no heavy equipment is needed to move 5 kN. A light crane is a good solution to put the segments in place on-site.

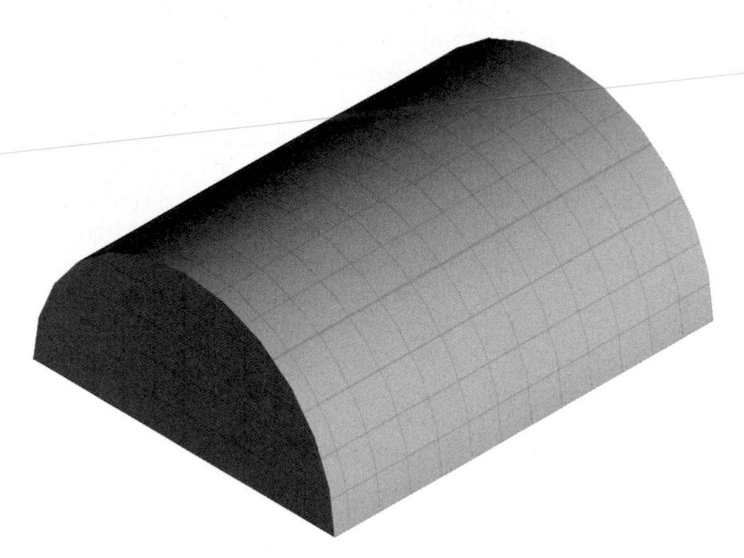

Fig. 6.17 Habitat module for three. (Courtesy Florian Ruess)

A solid rendering of a habitat module for a crew of three is illustrated in Figure 6.17. The lines on its surface are to enhance curvature and three-dimensional appearance.

Habitat layout

The question of possible lunar habitat layouts is a very complex one and deeply rooted in architectural and operations considerations. However, providing different modules for research, habitation, manufacturing and storage, for example, seems like a reasonable approach. These modules can be arranged in a multitude of scenarios. The five basic configurations are:[9]

Linear: The linear configuration is the simplest of all configurations. It is the repetition of modules with one primary hallway. The internal distances from one end to the other are maximized. The spatial characteristics are primarily public-type spaces that are noisy and conducive to space sharing with the hallway. One of the main problems with the linear system is safety.

Courtyard: When a linear configuration closes on itself, its basic characteristics change. The area coverage becomes greater and the enclosure minimizes. The courtyard is a unique identifiable space. There still exists one primary hallway; however, it now forms a closed loop. An additional attribute is the courtyard area itself.

[9]K.H. Reynolds, "Preliminary Design Study of Lunar Housing Configurations," NASA Conference Publication 3166, pp. 255–259, 1988.

Fig. 6.18 Three configurations for expanded habitat. One of the key criterion for a successful lunar structural design is that it be expandable in a relatively easy way. This graphic depicts three possible such expansions of the basic structural concept. (Courtesy Florian Ruess)

Radial: The radial configuration is a centralized space with linear extensions in more than two directions. The central area provides a major functional space with secondary areas radiating. These secondary spaces can be private and quiet areas, or hallways to additional functional zones. One advantage is the easy access to the central zone.

Branching: A linear growth system that expands with secondary paths from the main linear one characterizes a branching configuration.

Cluster: Cluster configurations have no dominant circulation patterns. Generally, a large area is created.

The proposed habitat module favors linear expansion over branching. Thus, linear and radial layouts are more suitable than courtyard, branching, or cluster layouts. A radial arrangement is most promising with respect to emergency egress, accessibility of facilities and the lunar surface, as well as for including an easily accessible safe haven to protect from solar flares. The central functional zone can be designed in different ways. Renderings of a radial lunar habitat configuration with different possible central zone layouts are shown in Figure 6.18.

Details

It is assumed that all parts described above – the arch segments, the floor members, and the end walls – can be transported to the Moon in one piece. However, it is also possible to further divide them into smaller parts if necessary, for example, dividing the end walls or the floor segments into two parts. The connections for this scenario are not designed here, but it is generally possible and should not impose great problems. A general solution is welding them back together on the Moon. Therefore, the only connections actually designed here are the hinged connections in the arch structure as well as the connection of the end wall to the arch, which is also idealized to be a hinged connection.

Connections

Two variants were considered for the connections between arches. Only the chosen design is shown here. If it were possible to design the hinges to be airtight, the structure would be cheaper and likely faster to erect. The result of these thoughts can be seen in Figures 6.19 and 6.20, and were conceived by Joerg Schaenzlin.[10]

End wall connections

The arch and the end walls may need some coaxing to come together. The wall panel rests on the floor, but there will be a gap between the wall and the arch segments due to construction tolerances. Aluminum plates extend from the wall and have to be welded to the floor and the arch. Welding has the advantage of also sealing the connection, but additional membrane fabric strips can be applied if desired.

Additional bracing is not necessary, but for a conservative design it could be a good idea to make sure the wall support reaction tensions are transferred to the full cross-section of both arch and floor by including an inclined plate in between the webs where the weld is made. A detailed drawing of the connections without additional bracing is shown in Figure 6.21.

---

[10]F. Ruess, J. Schaenzlin, and H. Benaroya, "Structural Design of a Lunar Habitat," *Journal of Aerospace Engineering*, Vol. 19, No. 3, 2006, pp. 133–157.

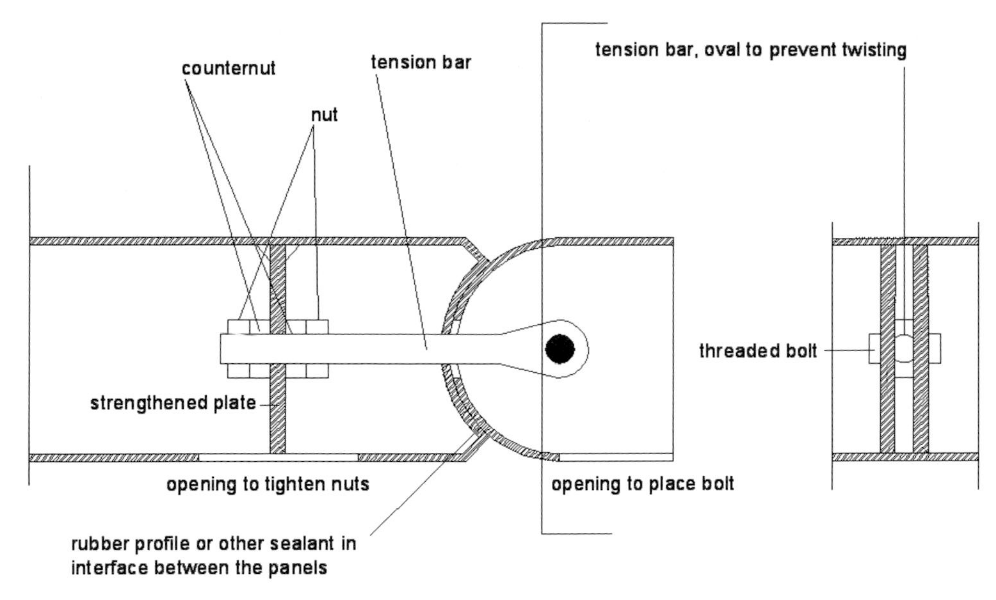

Fig. 6.19 Joint at arch. An innovative concept to mate the two arches and ensure that oxygen leakage is almost nonexistant. Conceived by Joerg Schaenzlin. (Courtesy Florian Ruess and Joerg Schaenzlin)

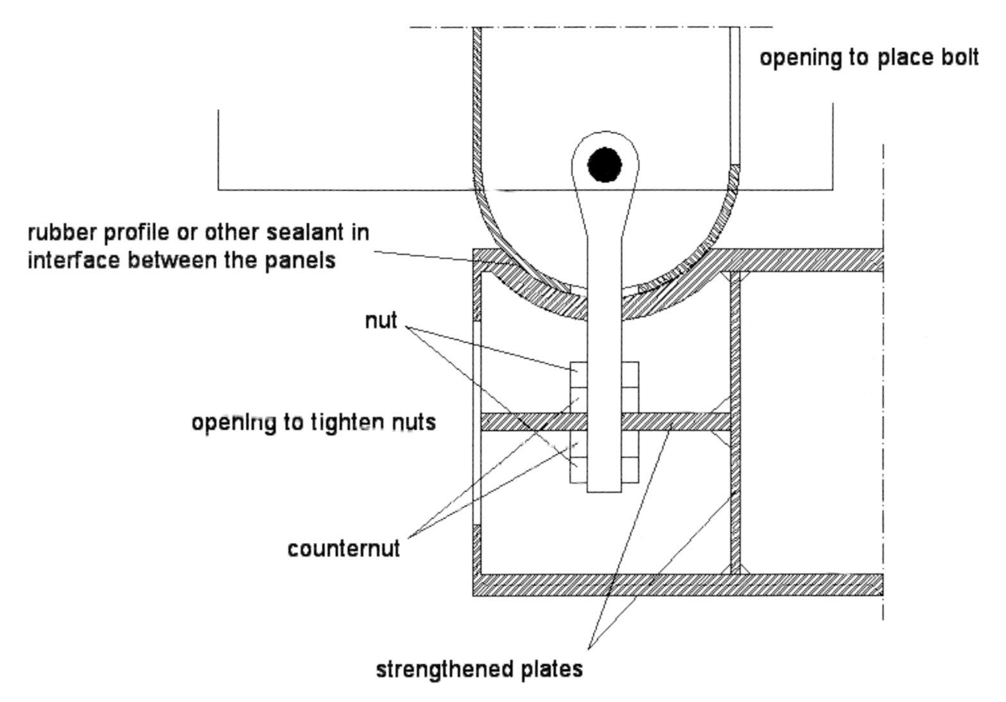

Fig. 6.20 Connection to base tie. A similarly innovative design to connect the arch to the tie, minimizing oxygen leakage. Conceived by Joerg Schaenzlin. (Courtesy Florian Ruess and Joerg Schaenzlin)

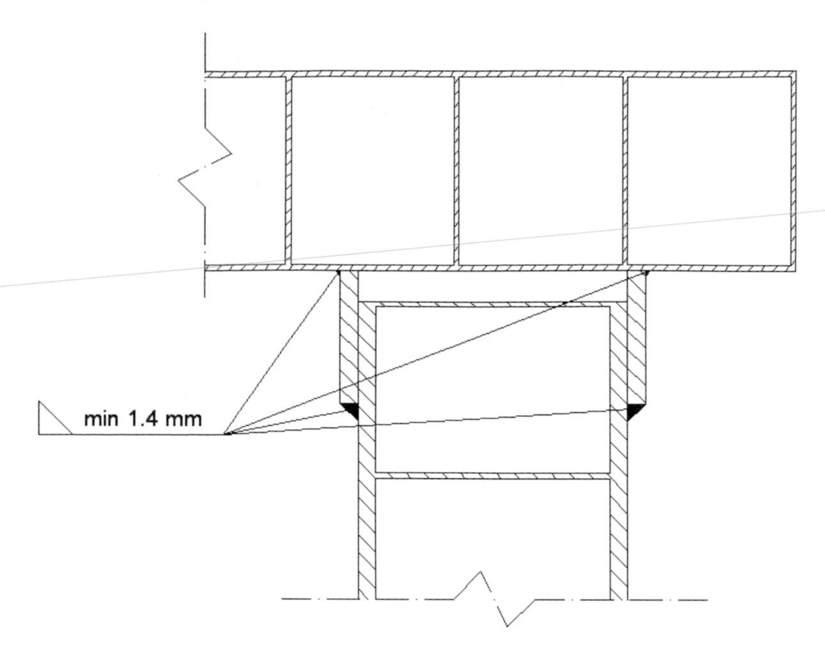

Fig. 6.21 Arch-wall connection. Schematic depicts the connection and weld. (Courtesy Florian Ruess)

Pressure sealing

The nature of the structural system makes it necessary to seal the openings in the structure with an additional element. The area around each of the three hinges has to be sealed airtight with a system that is capable of supporting the pressure loads and flexible enough to allow for the movement of the structure until the regolith cover is finished. The use of membrane fabric strips, glued to the aluminum structure, is a very promising and feasible scenario. Membrane fabrics have been used by the aerospace engineering community.

Openings

Sunlight ingress or access to the support systems will require openings in the structure. They can be at any location on the structure as long as the immediate boundary is strengthened appropriately. Figure 6.22 shows a window opening in the structure.

The vertical boundaries of the window are the web plates of the arch cross-section which can be strengthened if necessary. The horizontal boundaries are additional aluminum plates. It will be a bigger challenge to produce a 3 m deep opening in the regolith cover. Aluminum tubes, for example, could keep the regolith from falling into the opening.

Every manned U.S. space program has included windows in the vehicles. Windows are extremely important to a manned space flight. However, benefits such as

Fig. 6.22 An opening in a panel, shown without glass covering. (Courtesy Florian Ruess)

high crew morale come at a high engineering cost. From Mercury to Apollo to the Space Shuttle and the International Space Station, the use of glass for most of the windows is prominent.

Glass provides many optical benefits for use in space over other transparent material such as plastics. For example, glass is comparatively impervious to attack from atomic oxygen. Most plastics would haze over in a matter of days from atomic oxygen attack if exposed to the space environment. Ultraviolet exposure usually causes embrittlement of plastics. Glass does not experience such degradation.

Glass has many undesirable features when used as a structural material. It experiences static fatigue, has brittle failure modes, and its strength is easily reduced by very small damage. The lunar/habitat environment provides every possible disadvantage to glass. This includes a humid environment, constant loading of the glass in tension, long expected design life, and impact environments on both the internal and external surfaces.

Careful design considers these disadvantageous conditions and ensures the safety and structural integrity of the glass over the desired life span. The International Space Station window design, for example, includes sacrificial panes. On the exterior side of the window a pane is placed for the sole purpose of protecting the pressure panes from impact. It is a completely unloaded pane designed to act as a shield. A similar sacrificial pane is placed on the interior of the International Space Station windows. This pane protects the pressure pane from damage and provides a heater that prevents the window from fogging condensation.[11]

---

[11] L.R. Estes and K.S. Edelstein, "Engineering of windows for the International Space Station," *Space 2004 Engineering, Construction and Operations in Space*, pp. 580–584, 2004.

Problems and outlook

Lunar construction is possible and reasonable. However, the design process for a lunar habitat structure showed several problematic areas:

1. Providing a floor system that is ready to use resulted in high bending moments and therefore high floor mass for the approach taken here. The high interdependency of floor stiffness and bending moments is a result of the soil model that was used. If it is possible to reduce the floor stiffness below a certain level and still make the cross-section capable of resisting the internal forces, the floor could be designed to be significantly lighter.
   Advances in ultra-high strength materials, such as carbon nanotubes, and continued lunar research and exploration that enable the use of lower safety factors will lead to more efficient floor designs in the future. If, for example, a safety factor of only 2 could be used with the same aluminum material, the mass of the floor members would be reduced to about 25% of the floor member mass used for a safety factor of 4.
2. Extreme temperatures may be a problem during the erection of the structure. The chosen sandwich-type cross-section is very efficient for the operational loads but not for the temperature differentials between top and bottom plates while exposed to the Sun. But although the members deform, erection is no problem because of the three-hinge static system. An additional insulation layer is proposed for statically indeterminate structures to protect the structure from the Sun-induced temperature variations. There is much room for further research on the temperature gradient problems on the lunar surface.
3. With every advance in the engineering of lunar construction, different concepts will have to be re-evaluated. Evaluation criteria have to eventually be weighted.

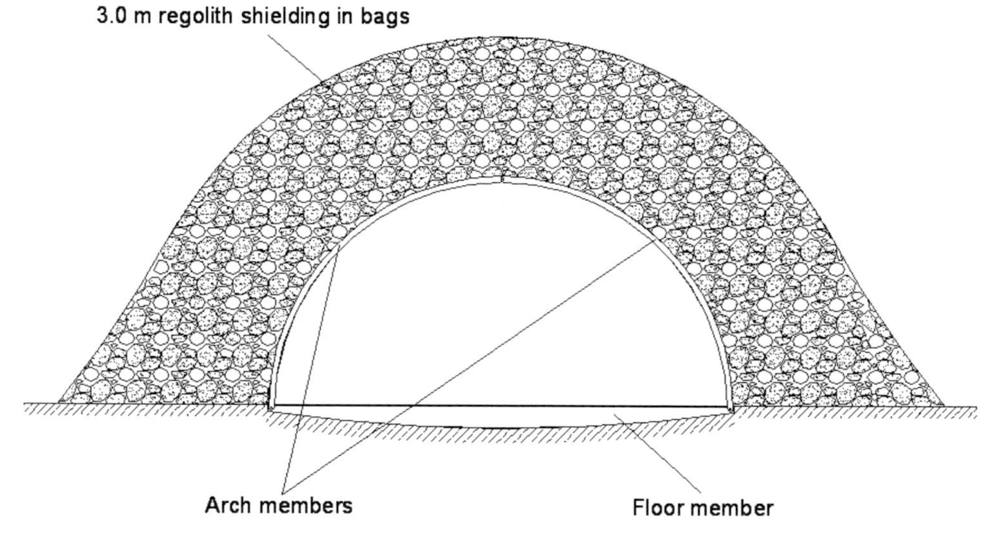

Fig. 6.23 Single habitat with bagged regolith cover. (Courtesy Florian Ruess)

4. A lunar transportation system does not exist. So all dimensions are based on assumptions on what is "transportable." When the transportation vehicle for this purpose is developed and dimension restrictions are available, structural members might be subject to dimension changes.
5. In addition, we have to be aware of the wide range of assumptions and unknowns that cannot be checked now. Many of these assumptions will be verified in the near future during the space agencies' lunar programs. Most important will be data on regolith properties, meteoroid activity, and radiation. The most important point to make is that there are significant but solvable difficulties for our return to the Moon with permanent outposts. This study was primarily focused on a possible structural design, recognizing that the design must be coupled with many other technical and nontechnical aspects of creating a civilization on the Moon.

Figure 6.23 shows the habitat covered by 3 m of regolith. No windows or other details are depicted here. Figure 6.24 depicts an expanded concept. In the background are solar panels.

Fig. 6.24 The look of the igloo base on the Moon. From the scale of the figure, we can see that the shielding is not purely 3 m of regolith. This is a more advanced shielding material that is blended with the regolith. (Illustration by Andre Malok, by permission © 2007 The Newark Star-Ledger)

* * *

Today, in 2169, we see numerous "igloo" structures on the Moon that were built in the first twenty or so years after the permanent settlements began. After that, with the space elevators coming online, first in orbit around Earth and then the Moon, more advanced structures were erected. And then when fully autonomous ISRU robotic construction was feasible, structures became less "tin-can" and more architectural – at least the parts that were above surface.

6.2 Quotes

- "What the nation needs now is not the technology, for we possess that. We have the machines, the people and the know-how to make this journey, to establish a permanent lunar base, and to reach out to Mars." Rod Pyle
- "It's a very sobering feeling to be up in space and realize that one's safety factor was determined by the lowest bidder on a government contract." Alan Shepard
- "What was most significant about the lunar voyage was not that men set foot on the Moon but that they set eye on the Earth." Norman Cousins

7 ISRU for construction

"The freedoms we have here on the Moon are as large as the physical spaces are small."

Yerah Timoshenko

One of the major difficulties in the early settlement of the Moon was the actual process of construction of the structures. Various concepts were explored – from inflatables to self-deploying – in order to minimize the time astronauts would need to spend on the surface doing construction work. In addition to the usual hazards of construction, on the Moon there are the additional worries about radiation exposure.

Automated concepts based on robotics were developed, but in those early years robotics was not yet mature enough to work autonomously. Robots were simple and were not primary actors in space. The use of local resources for the building materials, that is, *in situ* resource utilization (ISRU), was viewed as critical to long-term success on the Moon and Mars.[1] Today we are "living off the land" with an almost fully autonomous fleet of construction robots.

The merging of automated construction and ISRU eventually provided a path to safer and faster lunar and Martian development.

7.0.1 An historical interview with William Siegfried (May 2009)

Can you give us a one or two paragraph bio?

Spent 55 years working in aerospace for the U.S. Navy, Douglas, McDonnell Douglas and Boeing. Worked on programs for the Navy, Air Force, NASA, DARPA[2] and several other agencies. Worked on helicopters (HSS, HR2S, and HSS), aircraft (F2h, F3H), the Thor, Thor Able, Delta and Saturn Launch Vehicles. Participated in Spacelab, Skylab and ISS. Worked missile defense, several classified programs and numerous studies of launch systems, space systems and other types of transportation devices.

[1] S. Matsumoto, T. Yoshida, and H. Kanamori, "Construction of Bases from Lunar Regolith," IAA-99-IAA.13.2.07, 50th International Astronautical Congress, 4–8 October 1999, Amsterdam.

[2] Defense Advanced Research Projects Agency.

Assignments included aerodynamics, flight mechanics, systems engineering, program management, fiscal management and technology leadership reaching as high as Director of Advanced Projects and overall company representation. Scope ranged from individual efforts to leadership of a seven company team successfully competing for a NASA contract.

It looks as though the U.S. will be without heavy launch capabilities beginning in 2010 when the Shuttle is retired. It seems almost unbelievable that we have gotten into this situation. What do you think will happen?

I believe our chosen path will have development and fiscal problems that will force us to rely on other countries for human access to space.

Are you optimistic that President Bush's timeline for the return to the Moon will approximately be kept?

I not only think it will not be kept but fear it may be abandoned!

Is settling the Moon so important for civilization?

Yes. I have written on this. Many believe that extended space travel and colonization will do for the 21st century what aviation did for the 20th. Our current concerns, including terrorism, hunger, disease and problems of air quality, safe abundant water, poverty and weather vagaries tend to overshadow long-term activities such as space colonization in the minds of many. Our leading think tanks do not rate space travel high on lists of future beneficial undertakings even though many of the concerns just mentioned are prominently featured. I contend that space colonization will lead toward solutions to many of the emerging problems of our Earth, both technological and sociological. The breadth of the enterprise far exceeds the scope of our normal single-purpose missions and, therefore, its benefits will be greater.

You have been in this industry for many decades. What are your most vivid recollections of the time of Apollo?

Probably the time I spent at Huntsville as the company rep. during the evolution of the Apollo from EOR to LOR.[3] We were still using slide rules but still managed to construct the correct program.

Do you sense the spirit of Apollo today?

I sense almost NO spirit at all. This includes no one being named to head NASA yet!

What do you see as the major hurdles for our return to the Moon for permanent manned settlements? Are they technical, financial, physiological, psychological?

[3]Earth Orbit Rendezvous and Lunar Orbital Rendezvous

First of all financial. We, in the guise of fiscal restraints have reinvented the wrong version of Apollo and have focused on an approach that falls back to old technology in the interest of savings that will prove to be false. We will be lucky to get back to the Moon, much less populate it. We totally lack the imagination necessary to succeed.

How do we answer critics who say space is too expensive and that there are numerous problems on Earth to take care of first?

Not one dollar spent on space leaves Earth.

One of the reasons given for a robust American manned program is to attract our brightest into engineering and science. Are you concerned about the lack of Americans in these fields? Was this a problem in the aerospace industry?

It was a problem at the start but the space effort energized math and science and engineering education, but many of those people are now close to retirement. [See Figure 3.1.] What is the ratio of foreign born to native U.S. in your school? How many of the foreign born will return home after graduation?

When you envision the future, where do you see us in 50 years? In 100 years? How do you see this evolving from the present?

I really believe it will happen, I just am not sure we will be a part of it!

What do you see as NASA's role?

NASA must get out of program management and become more like they were before they were formed for Apollo. Perhaps they should be a part of an international body like a UN for space that would coordinate a worldwide endeavor.

Can the private sector do this with minimal government participation? You had worked for Boeing – they could probably go to the Moon on their own, couldn't they?

The private sector cannot do it all themselves. They can produce components and systems but management of an overall effort needs a coordinating body to assure all the parts fit together.

Do you think Americans will be the first to send man back the Moon in the 21st century, or will it be the Chinese?

I hope I'm wrong but our approach is not going to beat anyone. The large vehicle (previously known as the Shuttle C) is a faulty design that will not succeed and will drag down the effort. The Chinese effort will be slow but, like the turtle .... .

* * *

7.1 *In situ* resources

The development of techniques for the utilization of lunar resources had a slow start, not because of a lack of interest – interest was keen – but because there were few samples of lunar rocks and regolith upon which to base such procedures.

The majority of lunar rock samples came from the U.S. Apollo missions. The six lunar missions returned 2,415 samples weighing 382 kg (842 lb). Most of these samples came from the last three missions, Apollo 15, 16 and 17. The three unmanned Soviet Union Luna missions returned 326 g (0.66 lb) of samples. This is another example of how human spaceflight and human exploration has an advantage over robotic missions and results in major benefits.

An indirect source of lunar samples were rocks that were ejected naturally from the lunar surface by cratering events and subsequently fell to Earth as lunar meteorites. For example, during approximately 25 years beginning in 1980, over 120 lunar meteorites representing about 60 different meteorite fall events had been collected on Earth, with a total mass of over 48 kg. About one-third of these were discovered by American and Japanese teams searching for Antarctic meteorites, with most of the remainder having been discovered by anonymous collectors in the desert regions of northern Africa and Oman.

Due to the scarcity of such samples, and also since the Apollo samples were stored and secured like the Crown Jewels – which of course they were – simulants were developed. These simulants were designed and fabricated so that they would be representative of the key mechanical and chemical properties of the lunar regolith. Table 7.1 shows a comparison of such properties.[4]

These simulants were used to test concepts based on ISRU. While structures were one potential application for local resources, there were many others – as there had to be. "Silicon and aluminum can be used in the fabrication of lunar based solar cells. In addition, lunar regolith can be melted to form a glassy substrate upon which the silicon solar cells are deposited. The glassy regolith melt is electrically insulating providing good isolation for solar cells. ... In addition, films of the lunar regolith are transparent at thicknesses less than half a micron ... suitable for optical coatings. The regolith film could be used in anti-reflection coatings, transparent protective coatings and possibly even in electronic devices as insulating films."

It was proposed that circuitry and wires be deposited directly onto the lunar surface and then the regolith could be melted thus becoming the substrate. A rapid prototyping process could be used to create such an *in situ* circuit board. Today, in the isolated far-side lunar regions, shielded acre-sized *in situ* circuit boards process astronomical data gathered by telescope arrays.

A 2005 NASA Marshall report summarized these ISRU interests.[5] The program at Marshall also included efforts at *in situ* fabrication and repair. After all, once we learned how to build using local resources, it became possible to also repair utilizing

[4] C. Horton, C. Gramajo, L. Williams, A. Alemu, A. Freundlich, and A. Ignatiev, "Investigations into Uses for Lunar Regolith," CP654, *Space Technology and Applications International Forum* – STAIF 2003, edited by M.S. El-Genk.

[5] M.P. Bodiford, K.H. Burks, M.R. Perry, R.W. Cooper, and M.R. Fiske, "Lunar In-Situ Materials-Based Habitat Technology Development Efforts at NASA/MSFC," 2005.

Table 7.1 Comparison of simulant JSC-1 and lunar regolith.

Compound	JSC-1 (%)	Apollo sample (%)
Silicon oxide	47.71	47.3
Aluminum oxide	15.02	17.8
Calcium oxide	10.42	11.4
Iron oxide	10.79	10.5
Magnesium oxide	9.01	9.6
Titanium oxide	1.59	1.6
Sodium oxide	2.7	0.7
Potassium oxide	0.82	0.6
Chromium oxide	0.04	0.2
Manganese oxide	0.18	0.1
Phosphorus oxide	0.66	0.00

local resources. At Marshall, "these technologies [included but were not limited to the] development of extruded concrete and inflatable concrete dome technologies based on waterless and water-based concretes, development of regolith-based blocks with potential radiation shielding binders including polyurethane and polyethylene, pressure regulation systems for inflatable structures, production of glass fibers and rebar derived from molten lunar regolith simulant, development of regolith-bag structures [as well as] automation design issues."

These technology initiatives formed the backbone of our lunar construction capabilities in the mid-to-late 21st century. As robotic technologies evolved – including the development of self-replicating robotic systems – we were able to construct at a rapid pace, increasing our stock of habitable structures to allow for the addition of hundreds of people per year. Our engineers predict that by the beginning of the 23rd century – only 31 years away – we will have our first completely robotic settlement on the Moon. Its initial purpose will be research, but it is also a test bed as we seek to send robotic teams to some of the less hospitable planetary bodies and moons in the outer and inner Solar System.

7.2 Autonomous construction

7.2.1 Lunar surface habitats

The ideas presented above were later extended to the automated construction of surface structures. This required the development of micro-robotic technologies as well as very high efficiency solar energy converters. The following is an extract from an early report, from about 2006, that proposed these approaches. At that time, only mini-robots were possible. The right ideas were there, the technology had just not caught up to the imagination.

Autonomous construction ideas circa 2006

The next step for manned exploration and settlement is a return to the Moon. Such a return requires the construction of structures for habitation as well as for manufacturing, farming, maintenance and science. The most challenging of these is the construction of structures that can be used for habitation, although the other mentioned applications each offer unique challenges to the design engineer and the astronaut construction team that must erect the structures.

Given the costs associated with bringing material to the Moon (assuming no space elevator exists before man returns to the Moon), how do we design and construct habitable structures on the lunar surface in a feasible way, based on present-day mass and energy constraints? The proposed approach here is to use autonomous solar powered mini-robots – the size of a lawn mower – that can, as a team or a swarm, build a lunar structure for habitation with freeform fabrication technologies using *in situ* resources over a 6–12 month period prior to the arrival of astronauts.

An autonomous system that is capable of constructing such a structure on the surface of the Moon is extremely desirable. This structure will be ready for the first astronaut team that arrives on the Moon that can fit it with the systems necessary for human habitation. In essence, this structure is a "shell" into which pipes, wiring, windows (if any) and equipment can be installed by the astronaut team.

The structure will be human-rated, meaning that it will be shielded and can be pressurized upon arrival of people. The presence of such a structural shell on the Moon, awaiting human arrival, has enormous implications for the logistical planning of Man's return to the Moon. All of the spacecraft volume that would normally be allocated for taking structural materials to the Moon can now be filled with other items. This results in enormous savings in time and money.

Once the mini-robots land, they begin to prepare the regolith, first smoothing out the site where the structure is to be fabricated, perhaps sintering the site. Once the site is prepared, the mini-robots begin to build the lunar structure in layers, a structure that has been designed *a priori* with open volumes for mechanical and electronic equipment that are to be inserted subsequently by astronauts.

The "grand vision" is the design of freeform fabrication machines that operate almost completely under solar power. Of course, this creates limitations due to the reliance on *in situ* resources and the relatively low power availability.

The proposed study to prove feasibility will be comprised of the following aspects:

1. Select a benchmark lunar structure for analysis and construction using the proposed autonomous technologies.
2. Establish the structural strength of "blocks" built from *in situ*/regolith material (what can be expected from such a process ideally and realistically, knowing the kind of efficiencies one can obtain on Earth).
3. Determine energy/power requirements and feasibility (can current and anticipated solar energy conversion technology provide the needed power to drive such a machine, and if not, what are our other options).

4. Rate of construction as a function of the above (how fast can such robots build a facility that can then be prepared for human habitation).
5. Examine limitations on the possible complexity of such a structure (if it is possible to build a habitable surface lunar structure, how complicated can it be – can the fabrication process also include holes and paths for pipes and power lines, for example).
6. Examine the transferability of the technology for the Martian surface.

It is anticipated that the rate at which such an autonomous system can initially operate is low. It may take many months for a team of autonomous systems to erect a simple igloo-like structure.

A benchmark lunar surface structure for autonomous construction circa 2006

Key environmental factors affecting lunar structural design and construction are: 1/6 g, the need for internal air pressurization of habitation-rated structures, the requirement for shielding against radiation and micrometeorites, the hard vacuum and its effects on some exotic materials, a significant dust mitigation problem for machines and airlocks, severe temperatures and temperature gradients, and numerous loading conditions – anticipated and accidental. The structure on the Moon must be maintainable, functional, compatible, easily constructed, and made of as much local materials as possible.

Cast regolith has been suggested as a building material for the Moon. The use of cast regolith (basalt) is very similar to terrestrial cast basalt. The terms have been used interchangeably in the literature to refer to the same material. It has been suggested that cast regolith can be readily manufactured on the Moon by melting regolith and cooling it slowly so that the material crystallizes instead of turning into glass. Virtually no material preparation is needed. The casting operation is simple, requiring only a furnace, ladle and molds. Vacuum melting and casting should enhance the quality of the end product. More importantly, there is terrestrial experience producing the material, but it has not been used for construction purposes yet. Table 7.2 summarizes some key properties.

Table 7.2 Typical properties for cast regolith.

Property	Units	Value
Tensile strength	N/mm^2	34.5
Compressive strength	N/mm^2	538
Young's Modulus	kN/mm^2	100
Density	g/cm^3	3
Temperature coefficient	$10^{-6}/K$	7.5 to 8.5

Cast basalt has an extremely high compressive and a moderate tensile strength. It can easily be cast into structural elements for ready use in prefabricated construction. Feasible shapes include most of the basic structural elements like beams,

Fig. 7.1 Lunar habitat concept based on extruded concrete process. (R.W. Howard, from the National Space and Missile Materials Symposium, Summerline, Nevada, 27 June – 1 July 2005. Courtesy NASA)

columns, slabs, shells, arch segments, blocks and cylinders. Note that the ultimate compressive and tensile strengths are each about ten times greater than those of concrete.[6]

Cast basalt also has the disadvantage that it is a brittle material. Tensile loads that are a significant fraction of the ultimate tensile strength need to be avoided. However, it should be feasible to use cast regolith in many structural applications without any tensile reinforcement because of its moderately high tensile strength. But a minimum amount of tension reinforcement may be required to provide a safe structure. The reinforcement could be made with local lunar materials.

Studies will determine which kind of reinforcing the ISRU-based structure requires – possibilities are a glass fiber-reinforcing or a nano-composite matrix. Some studies have shown that such a layered manufacturing process can be achieved in a number of ways. One promising approach is based on microwave sintering. Much of the reinforcing material can be found in the regolith.

As on Earth, prestressing the concrete by embedding it with high-strength rods broadens its usefulness. On the Moon, tendons for prestressing can also be made from lunar materials.

Since it is extremely hard, cast regolith has high abrasion resistance. This is an advantage for use in the dusty and abrasive lunar environment. It may be the

[6] At the time of this writing (2006) high strength concrete had a compressive strength of about 40 N/mm^2 (about 6000 psi), although some concretes were made with three times this strength. In tension, concrete is much weaker, perhaps 10–15% its compressive strength, or about 4–6 N/mm^2.

ideal material for paving lunar rocket launch sites and constructing debris shields surrounding landing pads.

The hardness of cast basalt combined with its brittle nature makes it a difficult material to cut, drill or machine. Such operations should be avoided on the Moon. Production of cast regolith is energy intensive because of its high melting point. The estimated energy consumption is 360 kWh/MT. It may be possible to slow the construction process such that lower energy levels are required. The fracture and fatigue properties need further research.

A suite of freeform/rapid prototyping manufacturing robots – with some specialization between robots – allows for the construction of more complex structures. Rapid prototyping processes are a relatively recent development. The first such machines were released into the market in late 1987. While rapid prototyping is the term commonly applied to these technologies, the terminology is now a little dated, reflecting the purpose to which the early machines were applied. A more accurate description would be layered manufacturing processes. An alternative term is freeform fabrication processes.

These processes work by building up a component layer by layer, with one thin layer of material bonded to the previous thin layer. There are several different processes. The main ones are: stereolithography, laser or microwave sintering, fused deposition modeling, solid ground curing, and laminated object manufacturing.

In addition there are a number of newer processes that have appeared on the market, such as ballistic particle manufacturing and three-dimensional printing. Some ballistic particle manufacturing technologies use piezoelectric pumps that operate when an electric charge is applied, generating a shock wave that propels particles.

All of these processes essentially start with nothing and end with a completed part. Rapid prototyping processes are driven by instructions that are derived from three-dimensional computer-aided design (CAD) models. CAD technologies are therefore an essential enabling system for rapid prototyping. Other enabling technologies are mini-rovers, energy beams and solar power systems.

The processes use different physical principles, but they essentially work by either using lasers or microwaves to cut, cure or sinter material into a layer, or involve ejecting material from a nozzle to create a layer. Many different materials are used, depending upon the particular process. Materials include thermopolymers, photopolymers, other plastics, paper, wax, or metallic powder, for example. The processes can be used to create models, tooling, prototypes, and even in some cases to directly produce metal components. As such, one may view these capabilities as laying the groundwork for self-repairing systems.

Our vision is that an optimally grouped set of mini-robots will work as a team to autonomously erect a habitable volume that will be finished by the first astronaut team on the Moon. Issues that will be resolved in a preliminary study are the appropriate mix of robots by function, energy/power demands as a function of the rate of construction, and redundancy needs for reliability. Solar and other radiation that hit the lunar surface can be factored into the "curing" process. Also, the hard vacuum on the Moon can be used advantageously.

The appropriate mix of robots refers to the fact that different free-form manufacturing processes will be needed. We know that different parts of the structure require different construction approaches. For example, ballistic particle manufacturing may be appropriate for creating the foundation layer. For such ejection-type systems, a filtering process is needed so that relatively uniform sequences of regolith particles are fed. After a preliminary layer is prepared, some additional sintering may be required, perhaps via microwave.

* * *

Such freeform manufacturing mini-robots work as a team, with the group comprising the skill set of capabilities needed to construct all parts of the structure. In the present, we have been able to miniaturize these robots – some are less than one foot high and we are working on one inch high models – so that it is relatively easy and economical to send a team of these to any body in the Solar System for exploration, sample retrieval and mining.

7.2.2 An historical interview with Kris Zacny (July 2009)

Can you give us a one or two paragraph bio?

I was born in Poland, in an industrial and mining center of the country. As a teenager I moved with my family to Cape Town, South Africa. My Dad was a musician with the Cape Town Philharmonic and my Mother taught statistics at the University of Cape Town. Cape Town is one of the most beautiful cities on Earth with a fantastic climate that makes running, biking, hiking and more possible pretty much all year round. I had a really great time. It's also there where I became interested in astronomy and space exploration. After graduating from the University of Cape Town with a B.S. in mechanical engineering, I moved across the country to a small mining town just south of Johannesburg. For the next two years I worked in a coal mine and also spent some time in the deepest mine on Earth: Tau Tona gold mine, over four miles deep. The rock at this depth is over 50 °C and air has to be cooled to manageable temperatures.

I moved again in 2000, this time to Berkeley, California. As a graduate student at UC Berkeley in the Petroleum Department, I tested new diamond composites used as cutters in drill bits. After the project was completed, I decided to pursue a Ph.D., this time in Mars drilling. This topic allowed me to combine my passion for space exploration with knowledge and experience in terrestrial drilling and mining technologies. Upon graduating, I joined Honeybee Robotics of New York City and have been developing cutting-edge drilling and mining technologies since. Thus far, I have lived in Europe, Africa and in America. Perhaps the Moon will be next?

My work is also my hobby and I have been quite active in publishing on topics related to extreme drilling and excavation. My latest coedited book titled *Drilling in Extreme Environments: Penetration and Sampling on Earth and other Planets* is an extensive summary of all drilling techniques used in oil and gas, underwater, ice, and extraterrestrial drilling.

How did you become interested in robotics and digging technologies? And the connection to space?

Like the majority of us, I was always fascinated by space and at one stage wanted to become an astronaut, or at least an astronomer. My life, though, led me initially in a different direction and instead of watching the stars, I ended up hundreds of feet below the surface working in South African mines. Only now I know that this detour has been very valuable to my career. I eventually ended up at University of California Berkeley, developing new diamond composites used as cutting materials in drill bits. After my M.S., I decided to stay for my Ph.D. and luckily (!) I had no funding. Having no funding, as my advisor Professor George Cooper put it, allowed me to decide on a Ph.D. topic on my own and then try to secure funding for it. I realized then that I had an opportunity to work on my childhood's passion: space exploration. I also realized that no one in the field of space exploration has my type of background: digging and mining and actual experiences from various mines. I jumped on this opportunity and decided that space drilling and mining is what I want to do for my doctorate. My advisor met with Dr. Geoff Briggs of NASA Ames, who has been interested in Mars deep drilling for many years, and we wrote a successful research proposal together. My Ph.D. topic was titled "Mars Drilling," and I believe that this was the very first doctorate given for extraterrestrial drilling work. After graduating, it made perfect sense for me to join Honeybee Robotics, the company that has been developing scoops, drills and grinders for Mars and other planets since the 1990s.

Is settling the Moon so important for civilization?

Although the majority of people are very conservative, in that they dislike any changes in their lives, there is a handful of visionaries or adventurers that are pushing the boundaries of possibilities and dreaming big. This handful of people have been guiding our civilization into unknowns and in turn shaping the world. However, some of these discoveries or accomplishments, such as the Apollo landing on the Moon, died down soon and others like the discovery of America blossomed. The difference was that the former was not sustainable whereas the latter was. It was highly profitable to go to America, not to the Moon (at least at that time). Thus the question should rather be: can settling on the Moon be sustainable? The answer is yes, but not now. The big hurdle is the cost of getting to space and this is what SpaceX is attempting to solve by developing the Falcon family of launch vehicles.

While cheaper launch vehicles are being developed, we must also develop the so-called *in situ* Resource Utilization technologies. These technologies allow future Moon settlers to "live off the land," the way Lewis and Clark did on their maiden voyage through America. ISRU technologies include, for example, producing oxygen from ilmenite, an iron-oxide mineral that makes up a bulk of lunar regolith. These technologies are currently being developed for the Constellation Lunar Surface Systems. Recently launched Lunar Reconnaissance Orbiter and its lunar "impactor" LCROSS will also confirm the presence of water in the lunar poles. If water is indeed present, it can be mined to support future settlers.

Besides the resources that can be mined to sustain human presence on the Moon, there are also some highly valuable elements that can be mined to sustain our presence on Earth. For example, Helium-3 can be used as fuel in fusion reactors on the Earth. A few tons of Helium-3 will be enough to support United States' electricity requirements for a year.

Another way to make lunar settlement sustainable is to develop "lunar tourism." We have seen an advent of space tourism with the launch of Dennis Tito to the International Space Station onboard Soyuz. This trip cost approximately $20 million. Scaled Composites' suborbital flight also showed that one can develop suborbital planes for modest budgets. I'm sure we will see companies like SpaceX or Virgin Galactic launching "tourists" to space hotels and eventually to the surface of the Moon. Thus, tourism and Helium-3 could make lunar settlement sustainable.

Can you describe some of the technologies developed by you and your colleagues at Honeybee that have been used on Mars?

Our flagship Martian technologies included a rock grinder called Rock Abrasion Tool (RAT) and a scoop with a small drill bit called Icy Soil Acquisition Device (ISAD). The two RATs have been operating on the surface of Mars, onboard of Mars Exploration Rovers, since 2003. Their job has been to remove the first few millimeters of dusty, weathered and altered rock and to allow viewing of the virgin rock surface. We have been operating these two RATs from the comfort of our offices in Manhattan since 2003.

The ISAD was also the very first human instrument that touched water-ice on Mars. The purpose of the tool was actually to dig into the Martian soil and also to clear the soil covering the surface of ice. The purpose of the small drill bit mounted to the back of the scoop, called RASP, was to acquire some ice shavings for analysis. This was truly a remarkable device, and we were glad we had it because scoops don't dig frozen grounds very well. Both systems proved invaluable tools enabling further confirmation of ice deposits on Mars and enabling the discovery of salt, called perchlorate. We know from our terrestrial experience that salts not only depress the freezing point of water, hence salty water can remain liquid below 0 °C, but also provide great habitats for bacteria colonies. Consequently, perhaps the next spacecraft should have an ability to look for organics, the signs of life.

What do you see as the major hurdles for our return to the Moon for permanent manned settlements?

As mentioned earlier, the cost of getting to space is prohibitive and this has to be addressed first. Because of the sheer scale of the project (or rather the large cost), no country on its own can afford to develop permanent lunar settlements. This has to be an international project, the way the International Space Station is. Thus the major hurdle for setting up the permanent lunar settlement is its sheer cost. With a healthy budget, all required technologies can be successfully developed.

Robotics is a very exciting area of technology. We have hopes that one day robots will autonomously build structures on the Moon and Mars

and for raw material they will utilize *in situ* resources. How far do you think we are in being able to do that?

Before we can do this on Mars or the Moon, we first need to show we can do it on Earth. Large mining companies, such as Rio Tinto, have been actively developing technologies enabling what they refer to as "the Mine of the Future." This is a vision, slowly becoming reality, of a mine, in the Australian outback, where robots drill and blast, load and haul, crush and pulverize, load onto trains and transport to ports hundreds of kilometers away. This is just one example, but there are many. We need to draw from our terrestrial experience in automating operations that are otherwise hazardous to humans or operations that are in remote locations with scorching heat or freezing cold.

As opposed to the terrestrial robotic systems, planetary robots need to be much more reliable. Planetary robotic platforms, rovers and landers, use double and triple redundancies to increase reliability. This is quite expensive, hence planetary robotic systems cost in excess of $500 million. Redundancy in numbers (sending not one, but two or three robots just in case one breaks down) is also expensive. On Earth, it is more cost effective to spend less money on increasing reliability, and when a robot eventually breaks, send a human to fix it. Of course, on Mars or the Moon, this won't be possible because humans are not there. Thus, a big departure from the terrestrial approach will be in increasing the system's reliability.

Another difference is autonomous operation. Repetitive tasks (welding car frames, for example) can be easily automated. The tasks where the environment changes (e.g., mining), are much more challenging because the robot has to keep adapting to new surroundings. It is easier, however, if a system can be teleoperated. This, for example, has been done by the Soviets on the Moon in the 1970s. Soviet engineers successfully teleoperated two rovers, called Lunakhods, which traversed over 50 km on the lunar surface. The round trip communication delay to the Moon is only 2.5 sec. Thus teleoperation is possible. However, for Mars, where one-way communication delay can reach 20 minutes, teleoperation is not feasible. How soon we can develop technologies for fully autonomous operation will depend mainly on how large the budgets are for these applications.

One of the technologies developed at Honeybee is called percussive drilling. How does that work and why is it so important for digging on the Moon?

On Earth, when we dig in a garden for example, we apply our body weight to counteract reaction forces. But on the Moon, we would weigh six times less, and hence digging would be much more difficult. Apollo astronauts weighing 360 lb on Earth (that includes their spacesuits), could only push with 60 lb of force before they lifted themselves off the surface. Thus to enable digging lunar soil, we needed some innovative technology that reduces the required excavation forces. Lunar excavation is enabled by using a percussive hammer, in the same way we use a jack hammer to break up concrete. A scoop with a percussive hammer can dig deeper and faster than an ordinary scoop, and requires 40 times less push force. Thus, instead of an excavator weighing 4000 lb, now an excavator weighing only 100 lb can do the job. Since placing 1 lb on the surface of the Moon costs around $50,000, reduction of excavator mass translates directly to billions of dollars saved.

Where do you see robotic technologies in, say, 50 years?

With the current progress in technology development, I think we may have an "I Robot" type machine. Some people argue the robots may even have consciousness and in fact demand to have equal rights. Maybe we will be able to even design our "perfect spouses" or companions? One thing is certain: they will be taking over our jobs. Fortunately, I'll be retired by then.

NASA and the manned space program seem to be in limbo right now, especially with a new Administrator, the Augustine Commission still in deliberations, and the Shuttle being phased out. Where do you think all this will go?

It is unfortunate that the NASA vision often changes every time we have a new president. Developing technologies for space exploration takes a long time and there has to be some continuation in achieving a set goal that's longer than four or eight years. Sometimes, cost estimates tend to also be quite inaccurate. This is largely because we are developing brand new technologies and hence there are many unknown problems that often arise and were not accounted for when putting together initial budgets.

I believe that the future rests with international collaboration and the cost of doing it alone will eventually drive us towards this path. The International Space Station is a prime example of international collaboration that worked very well. Another example is the most recent series of talks between ESA and NASA that focused on collaboration on all future Mars missions. The reason? We are after the same science and the cost of doing it alone is prohibitive. There are, however, some obstacles that effectively slow down or even prevent international collaboration. The major ones are International Traffic in Arms Regulations (ITAR) that restrict technology transfer from the U.S. The ITAR was implemented to prevent defense-related technologies from falling into the wrong hands, but the definitions of which technology can be transferred and which cannot are sometimes vague and companies lean towards the safe side and keep information confidential. Thus, to enable international collaboration there has to be an environment that supports doing so. The current environment is very restrictive and I hope this will change.

When you envision human settlement of the solar system, where do you see us in 50 years? In 100 years?

Great question. If I knew the answer I would probably be a very wealthy man. It is very difficult to predict what will happen in even 10 years. New technologies are being developed every year and some of them (e.g., internet) change completely the way we live and function. However, to answer your question, in 50 years, I think we will send the first human to Mars and we will have a hotel in an orbit around the Earth. In 100 years we will have a permanent hotel on the Moon. Unfortunately,

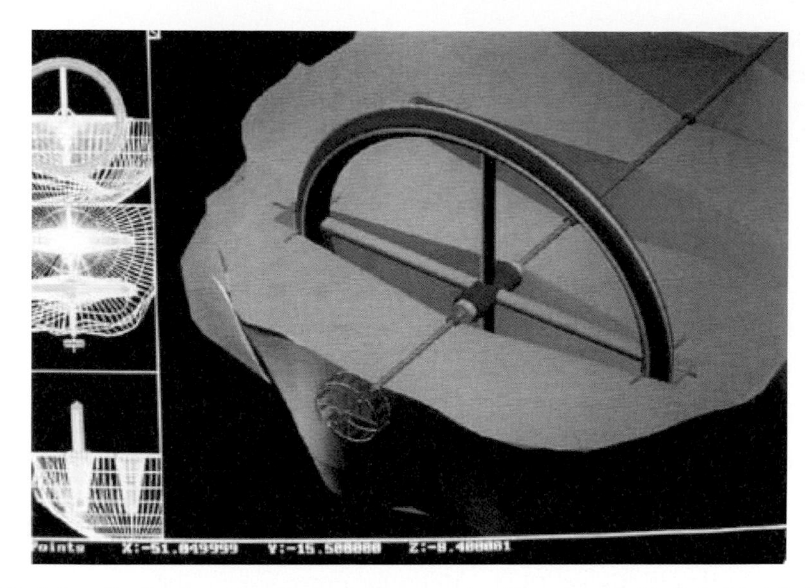

Fig. 7.2 A longitudinal section through the comet, with a toroidal inflated habitat structure inside the ice cavern. (Courtesy Tom Taylor and Werner Grandl)

I won't be around by then, but hope that my kids will take a few days off and go for a quick lunar sightseeing tour!

* * *

7.3 Comet and asteroid mining and living

Ideas on the possibility of utilizing asteroids for mining and habitation were also proposed in the early 21st century.[7] There are two primary reasons for humans to venture to a remote comet or asteroid. The first is a commercial desire to develop a valuable resource, such as water. The potential profits could be significant enough to support the creation of a mining outpost, and perhaps eventually a space colony with as many as 10,000 inhabitants. A second, even more compelling reason is that the knowledge gained through exploration, landing, definition, extraction and ultimate control of these objects might enable us to learn how to control large objects that threaten the destruction of Earth.

Almost 2,300 near-Earth objects are currently known in the 10 m to 30 m size range. It is estimated that there are roughly 25,000 objects larger than 150 m in

[7] T. Taylor, A. Zuppero, A. Germano and W. Grandl, "Commercial Asteroid Resource Development and Utilisation," IAA-95-IAA.1.3.03, 1995.

T. Taylor and H. Benaroya, "Developing a Space Colony from a Commercial Asteroid Mining Company Town," in **Living in Space**, Eds. S. Bell and L. Morris, An Aerospace Technology Working Group Book, 2009.

size. It is estimated that about 250 of these are potentially hazardous to Earth. The number of objects larger than 1 km – the size that is capable of causing a global scale catastrophe – is estimated at between 900 and 1,230. Around 55% of these have been specifically identified. None of these are known to be on Earth-intersecting trajectories. In the event that one of these objects is identified as being a threat to Earth, mastering techniques to relocate these objects is a key side benefit of the mining endeavor.

Figure 7.2 shows a computer-generated cutaway of a concept for digging inside a comet, recovering resources and creating a large volume that can be used for habitation. Figure 7.3 shows a computational model of the facility and Figure 7.4 provides its dimensions. After the main centerline tunnel is complete, one end is designated for discarded materials and the other end is developed for delivery of salable water. A system of machines for ice mining and water recovery will collect and transport materials for use by the colonists and for water ice shipment to customers.

Today, in 2169, we have sent robotic miners very similar to these to a number of large comets and asteroids and have extracted precious metals and water. These resources have been shipped to the Moon for processing and export. The funding for such operations came from Earth- and Moon-based entrepreneurs. Engineers at the lunar mining research facility have developed automated miners that can be sent to potential asteroid sites.

Our prime asteroid farming region is the asteroid belt. The elliptical orbit of the asteroid belt takes three to six years. The asteroid belt has more than 40,000

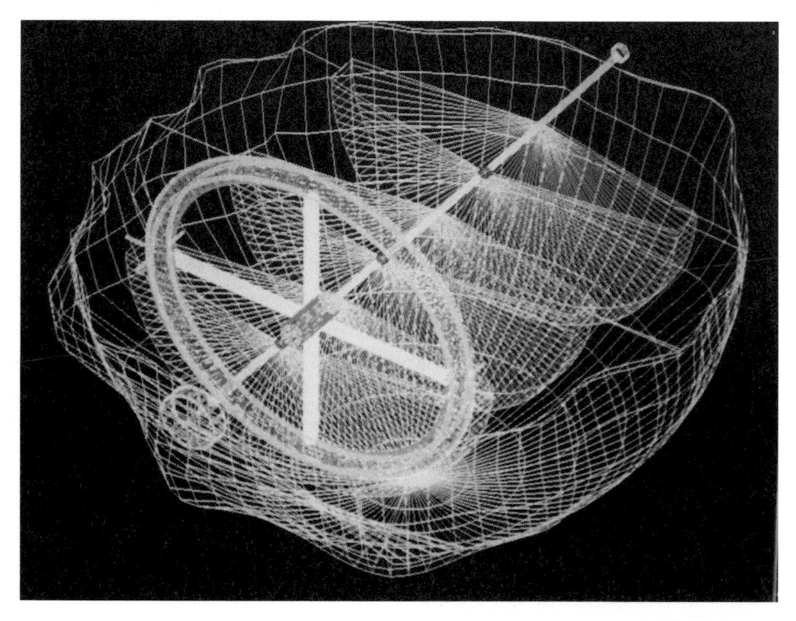

Fig. 7.3 An early digital design (1993) of the comet's interior, including the central tunnel, the caves and the first habitat. (Courtesy Tom Taylor and Werner Grandl) See Plate 7 in color section.

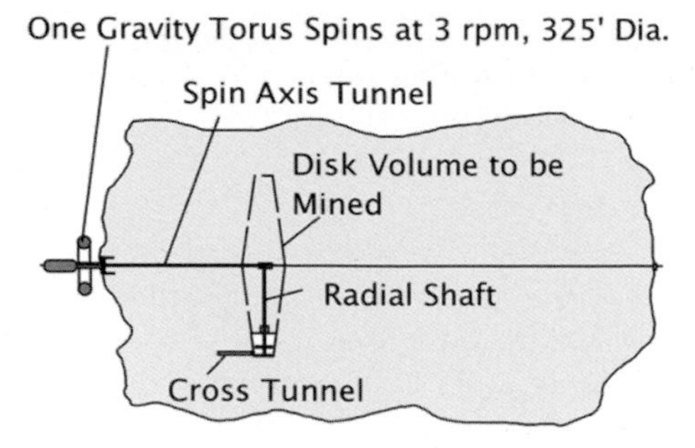

Fig. 7.4 Schematic of an asteroid mining operation. (Courtesy Tom Taylor and Werner Grandl)

asteroids that are half a mile long, as shown in Figure 7.5. The Trojan asteroids lie in Jupiter's orbit, in two distinct regions in front of and behind the planet. If the estimated total mass of all asteroids was gathered into a single object, the object would be less than 1,500 kilometers (932 miles) across, less than half the diameter of the Moon.

We have learned how to shepherd some of the smaller asteroids into lunar orbits for processing. This has many advantages. We don't have to transport mined products large distances and we utilize the lunar space elevators to access the asteroids. As the Martian infrastructure evolves, there is every intent to do the same mining procedures in their orbit. Mars is actually better situated for such mining operations since most of the asteroid belt lies in the region between Mars and Jupiter.

7.4 Lunar concrete

The possibility of creating a waterless lunar concrete had been studied since the late 20th century. Earth concrete is made from a mix of an aggregate of pebbles bound together by water and cement. Maximum strength in this concrete was achieved in one to four weeks depending on the process used. It was proposed[8] that lunar concrete can be made of a mix of lunar dust as the aggregate bound with sulfur, also derived from the regolith. The sulfur would be in a liquid or semi-liquid form so that it can act as a binding agent, requiring a temperature of between 130 to 140 °C. Lunar concrete can attain its ultimate strength of about 5,000 psi in about an hour.

The lunar regolith simulant JSC-1 was used, and in other tests, silica or fiberglass were used to enhance the lunar concrete's strength. The sulfur content was in the range of 12 to 22 percent by mass, with the remaining content being the regolith aggregate. Clearly, the use of lunar concrete was limited to regions where the temperature does not exceed sulfur's melting point. The addition of strength

[8] H.A. Toutanji and R.N. Grugel, "Unconventional Approach," *Civil Engineering*, October 2008.

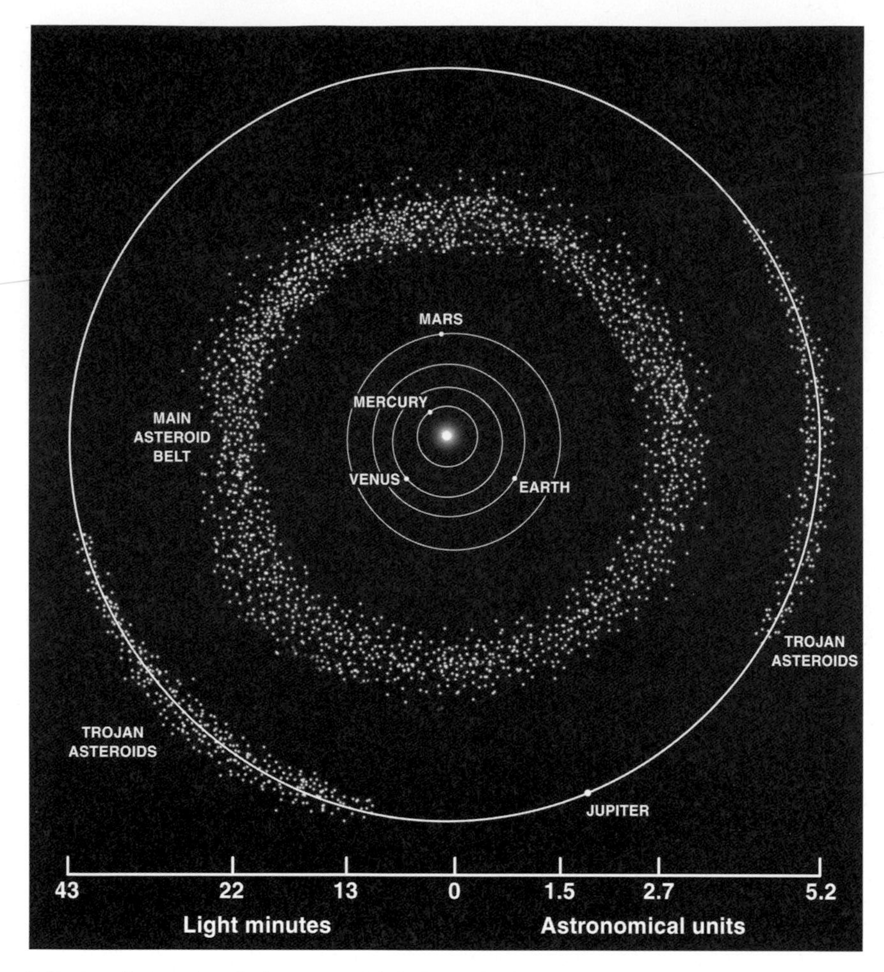

Fig. 7.5 A schematic of the asteroid belt and the Trojan asteroids. An astronomical unit, an A.U., is the distance from the Sun to the Earth, approximately 92,955,807 miles. One light-minute is the distance that light travels in a vacuum, approximately 11,176,943 miles. (Courtesy NASA)

enhancers such as silica or fiberglass can have negative effects. For example, use of silica leads to failures that are more catastrophic – without warning.

Another early effort[9] to create lunar concrete mixed a simulated regolith with aluminum powder at 2,700 °F. The resulting brick had a bearing strength of about 2,500 psi. The strengths attained in these early research efforts varied considerably depending on the process used, the percentages of the components, as well as the other elements introduced into the mix.

Of course, options multiplied suddenly by the discovery in 2009 of water molecules in the polar regions on the Moon by NASA scientists from data gathered

[9]D. Tennant, "Lunar Brick Paves the Way to Moon Colonies," *The Virginia Pilot*, 26 January 2009, online.

Fig. 7.6 This painting shows an asteroid mining mission to an Earth approaching asteroid. Asteroids contain many of the major elements that provide the basis for industry and life on Earth, the Moon and Mars. A NASA-sponsored study on space manufacturing held at Ames Research Center in the summer of 1977 provided much of the technical basis for the painting. "Asteroid-1" is the central long structure and the propulsion unit is the long tubular structure enveloped by stiffening yard arms and guy wires. Solar cells running the length of the propulsion system convert the sunlight into electricity that is used to power the propulsion system. During the mission these solar arrays would be oriented toward the Sun to gather maximum power. In the left foreground is an asteroid mining unit, doing actual mining work. An orbital construction platform in permanent orbit provides power, supplies depot and work volume within which work proceeds. (Artist concept by Denise Watt. S78-27139, June 1977. Courtesy NASA) See Plate 8 in color section.

by the Moon Mineralogy Mapper aboard the Indian Space Research Organization's Chandrayaan-1 spacecraft.

These concrete research efforts – with and without water – were multiplied manyfold in later decades with a resulting significant increase in bearing strengths. Processes were refined and with enhancements in robotic technologies we developed the capabilities to build anything we could imagine by the early 22nd century. This included the construction of very sizable underground facilities. Of particular note is the rapid evolution of our capabilities once self-replicating robotic systems became functional.[10]

7.5 *In situ* lunar telescopes

The idea of placing telescopes on the far side of the lunar surface had has its proponents long before the ability to do this existed. The far side of the Moon is shielded from the "noise" generated by the Earth. The lack of atmosphere greatly improves the resolution of images. The low gravity on the Moon enables us to build much larger telescopes, although the regular strikes by meteorites suggests a trade-off.

A method to make large telescopes for the lunar surface was developed in the early 21st century at NASA Goddard Spaceflight Center in Greenbelt, Maryland. "You can go to the Moon with a few buckets, and build something far larger than anything a rocket can carry," said P.C. Chen, a physicist at the NASA facility.[11] Most of the materials needed are within the lunar regolith. The recipe is a mixture of small amounts of carbon nanotubes and epoxies with lunar dust. A simulated lunar dust called JSC-1A Coarse Lunar Regolith Simulant was used in the mix, since there was not enough regolith available at that time. The result is a strong "smart" material that resembles concrete but can flex or change shape when an electric current passes through it.

Excited, "Chen made a small telescope mirror using a long-known technique called spin-casting. First he formed a 12 in (30 cm) diameter disk of lunar simulant/epoxy composite. Then he poured a thin layer of straight epoxy on top, and spun the mirror at a constant speed while the epoxy hardened. The top surface of the epoxy assumed a parabolic shape [under the centrifugal forces], just the shape needed to focus an image. When the epoxy hardened, Chen inserted it into a vacuum chamber to deposit a thin layer of reflective aluminum onto the parabolic surface to create a 12 in telescope mirror. Figure 7.7 is a photograph of a mirror created in this way.

"Composite materials are synthetic materials made by mixing fibers or granules of various materials into epoxy and letting the mixture harden. Composites combine two valuable properties: ultralight weight and extraordinary strength. The carbon nanotubes make the composite a conductor. Conductivity would allow a large lunar telescope mirror to reach thermal equilibrium quickly with the monthly cycle of lunar night and day. Conductivity would also allow astronomers to apply an electric current as needed through electrodes attached to the back of the mirror, to maintain the mirror's parabolic shape against the pull of lunar gravity as the large telescope was tilted from one part of the sky to another.

[10] J. Suthacorn, A.B. Cushing, and G.S. Chirikjian, "An Autonomous Self-Replicating Robotic System," Proceedings of the 2003 ASME/IEEE International Conference on Advanced Intelligence Mechatronics (AIM 2003), Kobe, Japan, 2003.

[11] J. Hsu, "NASA Envisions Huge Lunar Telescope," *space.com*, posted: 16 July 2008, 06:44 am ET.

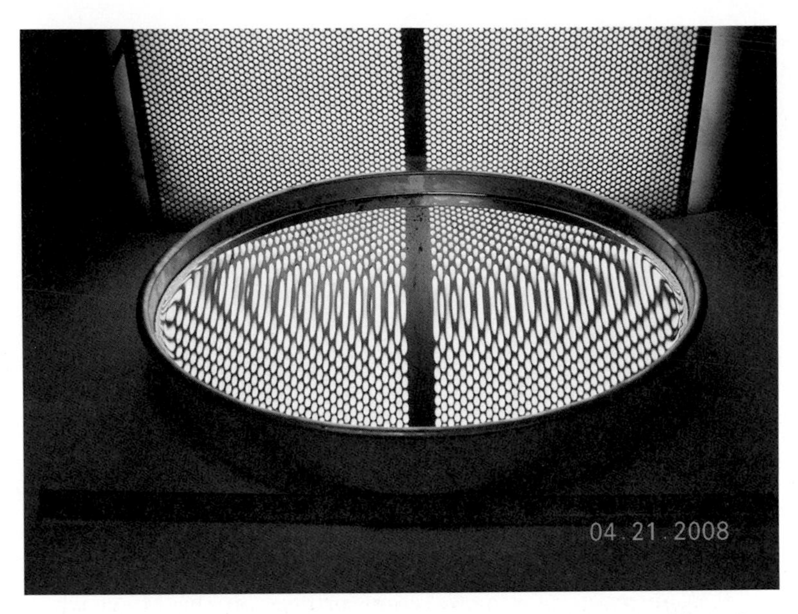

Fig. 7.7 A 12-inch parabolic moondust mirror made by spincasting. The mirror consists of a bottom layer of lunar regolith simulant JSC-1A Coarse mixed with a small quantity of carbon nanotubes and bonded with thinned epoxy. (Courtesy P.C. Chen, NASA/GSFC)

"To make a Hubble-sized moondust mirror, Chen calculated that astronauts would need to transport only 130 pounds (60 kg) of epoxy to the Moon along with 3 pounds (1.3 kg) of carbon nanotubes and less than 1 gram of aluminum. The bulk of the composite material, some 1,300 pounds (600 kilograms) of lunar dust, would be lying around on the Moon for free."[12] The advantage here is that the bulk of the needed materials do not need to be transported from Earth. Figure 7.8 is an artist's rendering of that effort to place telescopes on the lunar farside.

There were many challenges to this innovative process. These included getting the necessary manufacturing equipment to the Moon, such as the spinning table on which the mirror gets created. Astronauts would also have to ensure that none of the free-floating lunar dust contaminates the mirror. Sputtering aluminum vapor onto a large mirror in the presence of ambient dust would be another challenge, because the coating of mirrors on Earth is done in a clean environment. There were practical issues about manufacturability that needed to be resolved.

In the Moon's weak gravity, it was thought to be possible to build a telescope with a mirror as large as 50 m across – half the length of a football field – big enough to analyze the chemistry on planets around other stars for signs of life. These would have resolving power orders of magnitude greater than the largest telescope on Earth or in orbit at that time. There was thought given to constructing them in a crater.

[12] Science@NASA website, http://science.nasa.gov/headlines/y2008/09jul_moonscope.htm, 9 July 2008.

Fig. 7.8 Astronauts constructing a lunar telescope. The bright spot in the figure is a construction light. (Courtesy NASA)

Figure 7.9 is an image of a more recent project on the Moon where a 50-m telescope was built. A closer look reveals a segmented telescope mirror. Since the inception of research in astronomy over 600 years ago, the need to study fainter and fainter objects has naturally led to telescopes of ever larger diameter.[13] But the size of monolithic mirrors for ground telescopes is, in practice, limited to about 8 m. Although monolithic mirrors larger than 8 m are possible, the cost of the required facilities makes this a breakeven point where segmented systems begin to be more economical.

At the beginning of the 21st century, the world's largest telescope was the Keck 10 m telescope, upgraded and still located at Mauna Kea Observatory in Hawaii. Its primary mirror consists of a mosaic of 36 hexagonal segments, and each of them is approximately 2 m across. The 36 segments are adjusted once a minute to compensate for the changing position of the telescope.

The development of structure in the universe – how the first stars and galaxies formed – is a fundamental question that currently engages the attention of many astronomers. Utilizing the telescopes in operation, the first hint of this early structure in the universe is now being discovered. It is clear that the evolution was complex and the objects in question are extremely faint. By pushing the capabilities of all the largest telescopes to their limits, astronomers were just barely able to detect only a tiny portion of astronomical sources in question.

The next generation of larger telescopes was accompanied by advances in the understanding and technology of adaptive optics, optical charge-coupled device arrays, active control of the shapes of telescope mirrors and fast automatic guiding.

[13] S. Jiang, P.G. Voulgaris, L.E. Holloway, and L.A. Thompson, "Distributed Control of Large Segmented Telescopes," Proceedings of the 2006 American Control Conference, Minneapolis, 14–16 June 2006.

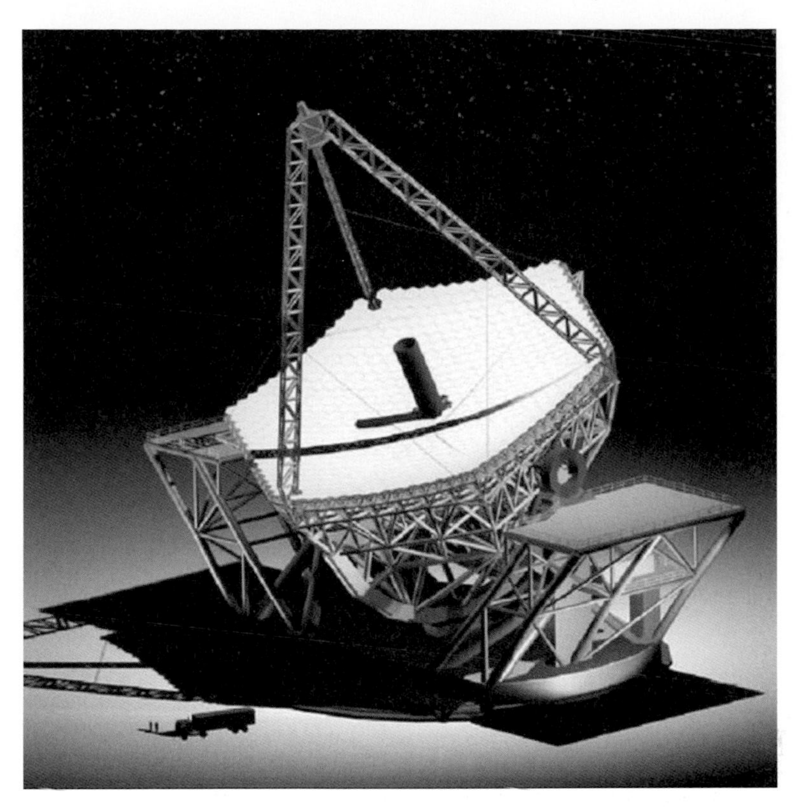

Fig. 7.9 An astronomical telescope. This artist's rendering depicts the recently built 50-m lunar segmented telescope. To show just how large this is, note the 18-wheeler and two people standing in front of the telescope. (Courtesy NASA)

Early examples of such large telescopes on Earth were the Thirty Meter Telescope (TMT)[14] and the European Extremely Large Telescope (EELT)[15]. The EELT was 42 m in diameter. Both of these telescopes came online in mid-decade 2010.

7.6 Lunar ISRU for Martian structures

Even as plans were gearing up for manned missions to the Moon, ideas were being developed on how to get to Mars. Two primary constraints were cost and radiation protection for the crew on the long mission. We summarize an early concept developed in the late 20th century that formed the basis for how we actually went to Mars.[16] In this study, the assumptions were that a manned lunar base was

[14]http://www.tmt.org/

[15]http://www.eso.org/public/astronomy/projects/e-elt.html

[16]B. Tillotson, "Mars Transfer Vehicle Using Regolith as Propellant," 92-3449 AIAA/SAE/ASME/ASEE 28th Joint Propulsion Conference, Nashville, 6–8 July 1992.

operational and could transfer regolith as well as oxygen propellant to low lunar orbit.

The proposed mission profile contained two logistical phases for a space vehicle. "The first involves cislunar operations to load lunar regolith aboard the vehicle. The second includes loading regolith from a Martian moon, perhaps concurrently with Mars surface exploration.

"All elements of the Mars vehicle except the mission payload are assembled in low-Earth orbit (LEO). The vehicle travels to low-lunar orbit (LLO). There it collects lunar regolith from canisters catapulted to LLO by a launcher on the lunar surface. The vehicle then flies to the Earth-Moon Second Lagrange (L2) point, consuming some regolith as propellant during the trip. The transfer habitat, the Mars Excursion Vehicle (MEV) and its payload are assembled in LEO and transported from LEO to L2, where they are attached to the Mars vehicle. The crew joins the vehicle at L2 and the vehicle departs L2 for Mars.

"Arriving at Mars, the vehicle drops the MEV. The MEV aerocaptures into low Mars orbit and lands. Meanwhile the main vehicle spirals in to rendezvous with Deimos, where it acquires a new load of regolith. It then spirals down to low orbit, picks up the MEV ascent stage returning from the surface, and spirals back out to Deimos. The vehicle reloads with regolith and then flies back to the Earth-Moon L2 point.

"At L2, the crew leaves the vehicle and returns to Earth in a cislunar spacecraft. The Mars vehicle proceeds to LLO for reloading. The only equipment to be replaced before the next mission are life support expendables and the MEV."

Materials from the lunar surface are jettisoned into lunar orbit without the use of rockets – the low lunar gravity makes this possible. The transfer habitat stores the regolith propellant on the outside of the structure. The inside of the transfer habitat is in the shape of a racetrack – ellipsoidal – with major dimension 15.2 m and minor dimension 7.6 m. Thrust for the MEV is generated by ejecting regolith at high speed using coil guns. Once the desired speed is attained, the remaining regolith acts as shielding.

* * *

The concept outlined above was developed in advance of the deployment of the space elevators. So while our early trips to Mars were based on this approach, the deployment of the Martian space elevators led to a major increase in traffic between the Moon and Mars without the need for such intricate orbital trajectories.

7.7 Quotes

- "Astronomy compels the soul to look upward, and leads us from this world to another." Plato, 'The Republic,' 342 BCE
- "The contemplation of celestial things will make a man both speak and think more sublimely and magnificently when he descends to human affairs." Cicero
- "SMAISMRMILMEPOETALEUMIBUNENUGTTAUIRAS." Galileo Galilei, an anagram sent to several correspondents. Kepler assumed that Galileo's latest discovery

had to do with Mars, and solved the puzzle as 'Salue umbistineum geminatum Martia proles' (Hail, twin companionship, children of Mars). However the anagram was in regard to Saturn (and what we now see as rings), 'Altissimum planetam tergeminum observavi,' (I have observed the most distant planet to have a triple form), 1610

- "Whether you are an astronomer or a life scientist, geophysicist, or a pilot, you've got to be there because you believe you are good in your field, and you can contribute, not because you are going to get a lot of fame or whatever when you get back." Alan Shepard
- "I have loved the stars too fondly to be fearful of the night." Sarah Williams
- "Those who study the stars have God for a teacher." Tycho Brahe

Fig. 7.10 Full-disk of Neptune, 20 August 1989. This picture of Neptune was produced from the last whole planet images taken through the green and orange filters on the Voyager 2 narrow angle camera. The images were taken at a range of 4.4 million miles from the planet, 4 days and 20 hours before closest approach. The picture shows the Great Dark Spot and its companion bright smudge; on the west limb the fast moving bright feature called Scooter and the little dark spot are visible. These clouds were seen to persist for as long as Voyager's cameras could resolve them. North of these, a bright cloud band similar to the south polar streak can be seen. (Courtesy NASA)

8 Biological issues

"At the core, it is all about people and their emotions."

Yerah Timoshenko

The survival of living things in space and on extraterrestrial bodies was and is the key challenge to a spacefaring humanity. The first priority is living, the next priority is living with sufficient food and comfort, and the next priority is to live and enjoy life.

The Apollo program, and Mercury and Gemini that preceded it, focused on the first priority. When Skylab, Mir and the ISS were built, there was more room. Astronauts could move around – they even had a bit of privacy. Was it comfortable? I would have been claustrophobic.

The design of the chairs and the human/machine interfaces began to look beyond survival and into ease of use. Concepts for space stations and lunar settlements eventually took into account the appearance of the interior – what colors are stress reducers and make people at ease. With each new adventure, and each new generation of that adventure, designs added comfort and took into account the full human being. Layout design facilitated ease of human motility. There were significant efforts given to improving the social and organizational aspects of life in the settlements. Social and psychological issues – effects of stress, recreation and exercise, interpersonal dynamics in space, personal space, privacy, crowding, territoriality – all began to be addressed.

Human physiology took the front row since being alive is a prerequisite for comfort and happiness. But human psychology was close behind. Just being alive was not enough. Once space travelers included people from throughout all of society's strata, not just engineers and scientists with years of training, the whole picture of human survival began to be considered. After all, space tourism had to be an enjoyable adventure, not just an exciting one.

8.1 Human physiology

8.1.1 An historical interview with James Logan (February 2008)

Can you give us a one or two paragraph bio?

I am residency-trained board certified expert in Aerospace Medicine and a 17-year NASA veteran. I have held numerous positions within the agency including Chief

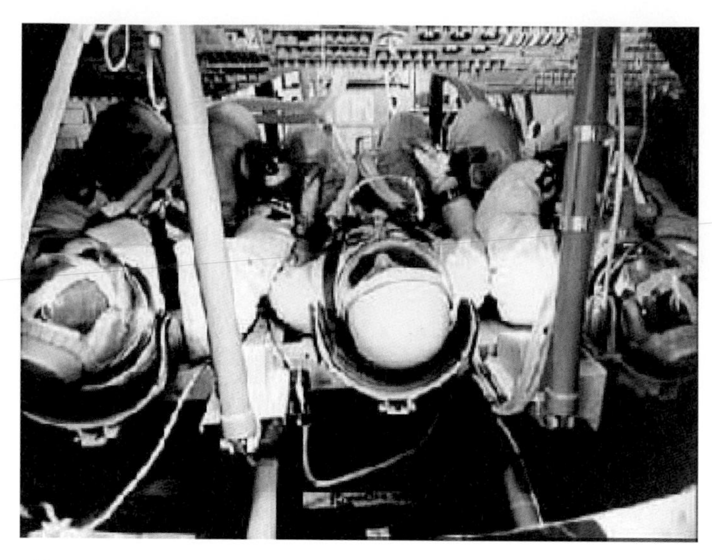

Fig. 8.1 Apollo 8 capsule with astronauts Frank Borman, Jim Lovell, and Bill Anders on the first manned mission to the Moon, 21–27 December 1968. They made a Christmas Eve live broadcast to the people of Earth from lunar orbit. (Courtesy NASA)

of Flight Medicine, Chief of Medical Operations, Chief of Medical Informatics & Health Care Systems and Liaison Officer between Headquarters Life Sciences Division and the Space Station Project Office. In addition I was the only clinical life sciences representative to the NASA Headquarters-chartered Space Station Operations Task Force. I served as crew surgeon, deputy crew surgeon or mission control surgeon for 25 space shuttle missions. I am (and always have been) a maverick.

How did you become interested in the space program?

I was seven years old when the Soviets launched the first Sputnik. I was riveted by the idea a man-made object could circle the Earth every 90 minutes. I knew that wherever machines were sent, man would follow. I never had any desire to be an engineer. I knew my interests would be based in the life sciences because I realized life science barriers would be the most difficult to overcome in the effort to extend human civilization throughout the Solar System.

Why is settling the Moon so important for civilization?

It isn't. Space is essential for ensuring the continuation of civilization but the Moon, *per se*, is an expensive and time-consuming diversion from what should be the prime focus of any rational space policy: Developing the ability to harvest the vast resources of asteroids and comets within our Solar System to create a massive human space-based civilization. A side benefit of that endeavor is that we will learn how to protect the Earth from mass extinctions associated with rogue asteroid and comet impacts. The Moon is a carbon-starved, completely waterless (despite claims to the contrary), utterly hostile, radiation-soaked, gravity-mismatched (too much for rockets, too little for people), dead world. The only things of value on the Moon (other than "science") are massive amounts of oxygen and eventually (once nuclear fusion is perfected) significant reserves of He3. From a resource standpoint, the Moon is entirely uninspiring. If everything of value were extracted from the asteroids and comets, the "slag" that remained would most closely approximate the present composition of our Moon. Planetary scientist Dr. John S. Lewis said it right; the Moon is the "slag heap" of the Solar System.

The Moon was important historically because it was the first step. Once mankind demonstrated the capability to escape the Earth gravitation field, travel to the Moon, land on its surface, take off again and return safely to Earth, it demonstrated that, in theory, it can go anywhere in the Solar System. The Moon is important in the same way the 1903 Wright Flyer was important in proving 'heavier than air human controlled flight' was possible. It served as an excellent "proof of concept." Beyond that, the Wright Flyer had little real effect on the future of aviation. (In fact the method of 'control' the Wright Brothers invented [wing warping] was very soon found to be unreliable and was superseded by Glenn Curtiss' ailerons.)

In my opinion the Moon will never be more than an extended sortie destination (much like Antarctica is today). It will never be a frontier destination. By that I mean it will never be a seat of civilization (men, women and children in multiple generations).

How do we answer critics who say space is too expensive and that there are numerous problems on Earth to take care of first?

First, space is too expensive. NASA was supposed to solve the 'cheap, reliable, robust' access to space problem. It failed miserably and continues to fail. The Shuttle is almost three times as expensive on a cost-per-pound basis to LEO as was the Saturn V ($4166 per pound vs. the Shuttle's $12,500 per pound in real dollars). Everything the Shuttle has ever launched into space is worth more than twice its weight in gold – and that's just the transportation costs!

The fact that we are approaching the 50th anniversary of human space flight without any demonstrable economic benefit of manned space activities whatsoever that can justify (or even partially defray) massive expenditures made in its behalf is an indictment of NASA as well as Congress. Worse, there is precious little attention given to promising technologies that could enable a more self-sustaining human presence in space. I have been a NASA manager for 16 of the last 25 years. Since the end of the Apollo program, NASA's manned program (as opposed to the unmanned program) has been largely underwhelming. Once an ardent supporter, I now find it increasingly difficult to argue that the manned space program has been successful or even worth the investment. We are doing the wrong things in space and that must be changed.

However, the argument that we must solve numerous problems on Earth first before going into space is ludicrous. If we had waited to solve pressing problems before participating in any frontier, mankind would still be living exclusively in Africa.

How is the medical corps at NASA structured? How many doctors support a shuttle launch?

The space medicine corps at NASA is part of the Medical Sciences and Health Care Systems Division at Johnson Space Center. Most of the flight surgeons work in the Medical Operations Branch. The entire medical organization is part of the Space Life Sciences Directorate.

A crew surgeon and deputy crew surgeon are assigned to each shuttle crew and ISS crew. In addition there are launch and landing recovery surgeons that support every launch and landing. The crew surgeon is in the Launch Control Center at KSC of every launch of his/her crew. There are also two or three surgeons assigned to the Mission Control Center at JSC to provide on-orbit support to every crew.

What do you see as the major hurdles for our return to the Moon for permanent manned settlements?

Space radiation, microgravity, toxicology (e.g., lunar dust), isolation and inflight medical care capabilities are the chief life science hurdles. However, the biggest hurdle is logistics. I was a logistics officer in the USAF prior to going to medical school. I am convinced that current technologies (in the absence of a robust and productive ISRU program) are insufficient to support any kind of 'permanent presence' on the lunar surface.

Are human physiological and psychological factors being taken as seriously as the engineering factors in the return to the Moon?

Absolutely not. We are woefully unprepared to return to the Moon for anything other than occasional sortie missions.

As a medical doctor with extensive space experience, are you confident that we know enough about human physiology in low gravity and under space radiation to assure astronaut safety on the Moon for extended periods of time?

No, I am not confident at all. NASA has all but ignored these issues in their haste to build yet another vehicle. The entire NASA program is basically an entitlement program for private sector aerospace engineers living in prominent Congressional districts in Texas, Florida, Ohio, Alabama, Mississippi and California.

We evolved over 3.7 billion years during which the ONLY constant was gravity. It is highly likely that any human civilization (defined as multiple generations of men, women and children) will require Earth normal gravity (or its proxy) for successful gestation, growth and development. Therefore, to thrive in space we will need to take our gravity with us. The only way to create artificial gravity is by rotation. A high rotation rate is destabilizing from a physiological perspective. Minimizing rotation rate will mandate very large structures (made from extraterrestrial materials). The best potential human habitat will be large rotating colonies constructed inside asteroids. The surface of the asteroid will provide required radiation shielding. Hollowing out asteroids will provide resources required for colony construction.

How optimistic are you that President Bush's timeline for the return to the Moon will approximately be kept?

I will put money on the table that the Bush timeline will prove to be utter fantasy.

When you envision human settlement of the Solar System, where do you see us in 50 years? In 100 years?

Without appropriate attention to radiation protection, artificial gravity, extensive *in situ* resource utilization and massive bioengineering, we will still be sending only professional government-supported astronauts into space on nothing more than expensive sortie missions 50 years from now.

As for 100 years from now, who knows? If you had collected the smartest people on Earth in 1908 and asked them for a detailed description of where they thought mankind would be in 2008, I bet they would have missed it by a mile.

* * *

Numerous factors affect living organisms on the Moon. These factors have been known for a considerable period of time.[1] At the lunar surface, the galactic cosmic radiation gives rise to a dose equivalent of about 0.3 Sv/year.[2] On the Earth surface the dosage is 1–2 mSv/year (1/1000–2/1000 Sv/year) due to the protective atmosphere. But at a depth of 1 m of regolith, the annual radiation dose equivalent due to cosmic-ray particles decreases to about 2 mSv/year – comparable to the Earth's surface. Shielding materials can interact with the cosmic-ray particles and release neutrons with an additional dose equivalent of 0.1 Sv/year.

At the Apollo landing site a 282 °C range of temperatures existed, between 111 °C and −171 °C. At a depth of 20 cm, the variation decreases to about 10 °C, and at a depth of 40 cm the variation is 1 °C, where the nominal temperature is approximately −20 °C. The regolith has low thermal conductivity, which is why we see such a drastic reduction in the temperature variations.

"Space flight has been shown to induce varied immune responses, many of them potentially detrimental. Some of these changes occur immediately after arriving in space while others develop throughout the span of the mission. The causal factors include microgravity, the stress due to the high-demand astronaut activities and the social interactions of confinement, diet, lack of load bearing, and radiation."[3]

The criticality of such issues led to the need to be able to monitor astronauts' vital signs so that corrective actions could be regularly implemented. In-flight measuring capabilities were available since the first manned missions, but the level of

[1] G. Horneck, "Life Sciences on the Moon," *Advances in Space Research*, Vol. 18, No. 11, 1996, pp. 95–101.

[2] The sievert (Sv) is the SI-derived unit of dose equivalent. It attempts to reflect the biological effects of radiation as opposed to the physical aspects, which are characterized by the absorbed dose, measured in *gray*. It is named after Rolf Sievert, a Swedish medical physicist famous for work on radiation dosage measurement and research into the biological effects of radiation.

[3] V.M. Aponte, D.S. Finch and D.M. Klaus, "Considerations for Non-invasive In-flight Monitoring of Astronaut Immune Status with Potential Use of MEMS and NEMS Devices," *Life Sciences*, Vol. 79, 2006, pp. 1317–1333.

sophistication needed, beginning with the return to the Moon, led to the development of devices at the micro and nano scale. With time, genetically engineered nano-devices were designed for each astronaut's DNA, providing us with the ability to tailor medicines for them. In the present, in 2169, such technologies are available to all people. The days are past of ingesting mass-market medicines that travel throughout the body in order to cure a problem at one spot. Today, DNA-customized nano-pharmaceuticals deliver their dosage at the molecular level.

Fig. 8.2 During tests conducted for NASA's Desert Research and Technology Studies (RATS) at Black Point Lava Flow in Arizona, engineers, geologists and astronauts gathered to test two configurations of NASA's newest lunar rover prototype. The pressurized version, seen here and called the Lunar Electric Rover, includes a suitport that would allow crew members to climb in and out of spacesuits quickly for moonwalks. Here, Smithsonian Institution geologist Brent Garry dons his suit. (JSC2008-E-139049, 18–31 October 2008. Courtesy NASA)

From the perspective of radiation health, a regolith layer of 2–3 m depth provides radiation shielding of about 40 g/cm^2 whereas the Earth's atmosphere provides a shielding equivalent of 100 g/cm^2. Rare large solar flares require shelters with shielding of at least 700 g/cm^2.

In addition to radiation effects, the lunar regolith dust is abrasive and toxic – bad for machines, deadly for people. Shielding and burial of structures essentially solved the radiation problem early on in our settlement of the Moon, but the dust problem lingered much longer. Sheaths for machines and robots were designed to keep moving parts and joints clear of the dust. Astronaut suits became part of the walls, so we don't have to go through hatches to go to the lunar surface and back.

8.1.2 An historical interview with William Rowe (April 2009)

Can you give us a one or two paragraph bio?

I am a board certified specialist in Internal Medicine. My undergraduate and medical education were obtained at the University of Cincinnati followed by an internship at St. Joseph Hospital, Denver, Colorado. I married toward the end of that year. I then served as a Flight Surgeon for two years which required four hours of flying time per month. This was followed by two years as a resident in Internal Medicine at the University of Iowa, followed by a third year at St. Vincent's Hospital, Toledo, Ohio. I then went into solo practice for 34 years until I "retired" in 1993.

In 1972 I organized the first Treadmill Stress Test Lab of Northwest Ohio at St. Vincent's Hospital, and until retirement monitored over 5,000 symptom-limited maximum stress tests (Bruce protocol) and also served on the hospital electrocardiograph interpretation panel of physicians.

In the 1980s I became involved in the Northwest Ohio Cardiac Conditioning Program at Toledo University and it was there that I began working with Sy Mah, supervising cardiac patients during their exercise conditioning programs. Sy Mah was listed in the *Guinness Book of Records* – having completed 524 marathons by 1988. At that time I was invited to participate in The People to People Fitness Delegation – a group of 75 from North America – and I was selected to present a paper at Sun Yatsen University of Medical Sciences, Guangzhou, China. The circumstances of this participation changed my life since my paper (Fitness Role Models) involved a study of Mah and also a Toledoan and inspired by Mah – U.S.A. 1987 Female Ultramarathoner of the Year – Mary Hanudel, who at the time of my study was training for a 600-mile race in Australia – running 18 miles, seven days a week, and swimming a half hour five days a week.

Both athletes, however, demonstrated abnormal stress tests, in Mah's case by electrocardiograph changes consistent with coronary vasospasm. He died at age 62, 10 months after my study. I sent his heart to Dr. William Roberts, a world-renowned cardiac pathologist at the National Institute of Health. It was Dr. Robert's hypothesis that the scar tissue of the heart muscle was consistent with a "remote ischemic insult" despite perfectly normal coronary vessels. He agreed that this scar tissue could have been triggered by high adrenaline levels over a long period, a discussion about which I subsequently published – emphasizing this hypothesis.

Similarly Hanudel's exercise performance was sub-standard and she was found to have only half her blood volume with subsequent studies consistent with chronic gastrointestinal bleeding. In both cases I postulated that there must have been chronic severe elevations of adrenaline compromising the circulation of the heart and in Hanudel's case shunting of blood away from the GI tract. My studies of Mah were subsequently published in the journal *Chest* and later in *The Lancet* followed by a guest editorial in *Sports Medicine.*

I became interested in stress-related complications involving chronic elevations of adrenaline not only related to the two athletes whom I studied but stemming from my own stress-related problems which I postulated were related at least partially to my running. In my case I had frequent facial skin cancers, some quite severe, requiring extensive plastic surgery and finally in 1995 when I developed cancer of the vocal cord I gave up my running program entirely.

My third role model was Lorna Michael, also influenced by Mah. When I visited Michael in 1993 she had moved to Wisconsin and was training for a 2900-mile, 64-day staged race across the U.S. from Huntington Beach California to Central Park, New York, including the crossing of the Mojave desert (28 miles/day) and the Continental divide. I instructed Lorna to begin taking magnesium supplements after finding in her a significant reduction, along with a high magnesium diet, and requested that her physician inject her with intramuscular magnesium twice a week beginning three weeks before the race. By that time she was up to 225 miles of running/week. Michael came in third among 13 finishers; the others were all males; there were no complications of any kind.

How did you become interested in the application of medicine to the space program?

After publishing the *Lancet* paper in 1992 I was struck with the concept that there were ongoing vicious cycles, triggered by persistent elevations of adrenaline and reductions of magnesium ions and this in turn could damage the lining of the blood vessels (endothelium). With space flight, vicious cycles can occur between magnesium ion deficits and adrenaline elevations and between the latter and ischemia, i.e., an imbalance between oxygen supply and demand. Furthermore, vicious cycles can also develop between reductions of vessel dilators, spasm of the coronary blood vessels (for example) with turbulence, and injuries to the lining of the blood vessels.

In the average man, the endothelium has the surface area of seven tennis courts and weighs about 2 kg. When it is damaged there is the potential for injuries to any of the organs of the body. I soon discovered that the world's leading authority on magnesium was Mildred Seelig, M.D. and I went down to the University of North Carolina where she was a professor to make sure my thinking regarding magnesium ions was sound. With little experience in publishing, I needed reassurance; Seelig provided it and became my mentor, subsequently inviting me to present papers at Magnesium Conferences in Austria, Australia and Japan.

Then my life took another path when I came across a book review of *Space: the Next 100 Years* by Nicholas Booth, an Englishman. I was intrigued by his statement that in order to reduce the likelihood of bone and muscle loss with space flight astronauts would have to exercise as much as four hours a day. Immediately

I knew I was on to something potentially very important regarding my interest in the vicious cycles that I've described above involving adrenaline and magnesium, and wondered whether astronauts would end up with cardiac and gastrointestinal complications similar to those of my two extraordinary "role model" subjects. I arranged to pay a visit to Booth, who lived in London.

At our first meeting, Booth had arranged for me to meet with Anders Hansson, Ph.D., a brilliant biophysicist and one of the world's leading experts regarding the risk to the body due to space flight. Hansson, the author of *Mars and the Development of Life*, was also a Senior Science Consultant at the International Nanobiological Testbed. After a very long discussion during our first London meeting, Hansson informed me that, in his opinion, I was the first to show how the endothelium could be injured – complicating the vicious cycles I described pertaining to space flight – and that the stress encountered by extraordinary runners would also apply to astronauts engaged in long space missions. He suggested that upon my return to the States I should immediately meet a deadline in three weeks for submitting an abstract to present the following Spring at the 1995, 11th IAA Man In Space Symposium in Toulouse, France. Subsequently, Hansson became my space mentor, and we visited each other across the Atlantic several times over the next 10 years.

When I presented my paper in Toulouse, the moderator interrupted me after my having spoken for less than two minutes and stated something to the effect that I had a very provocative paper and to be sure to leave plenty of time for questions. It was my very first presentation of a paper and I felt fortunate that I had a voice – not because of anxiety so much but because I had just completed a course of radiation therapy for cancer of the vocal cords 10 days before my departure for France and had very little voice until miraculously my voice returned the day prior to my presentation. I'm convinced that the stress-related adrenaline elevations from running played a role in my cancer along with sleep deprivation; the average duration of sleep in space is six hours – conducive to oxidative stress.

Is settling the Moon important for civilization? How do we answer critics who say space is too expensive and that there are numerous problems on Earth to take care of first?

I believe that settling anywhere away from Earth is important for civilization because of the risk eventually of a large meteorite destroying our civilization, but not for the purpose of avoiding destruction from a nuclear attack. I believe we need perhaps at least another 200,000 years of evolution before we are rid of our savage instincts, making us safe from our own annihilation; this encroaches somewhat on Fermi's question as to why we haven't been paid a visit from elsewhere. On the other hand, it may be because interstellar travel is impossible.

Certainly the Moon should be the place to settle rather than Mars, for example, because of its proximity. If some day we do settle on the Moon wouldn't it be great if we could eventually raise cattle? I believe that early man survived by ingesting blood providing both oxygen delivery and at the same time taking advantage of red blood cells' capability as a vitally important antioxidant. Blood ingestion – rather than iron ingestion in the form of tablets or iron injections – may provide exactly the right amount of iron absorption.

What do you see as the major hurdles for our return to the Moon for permanent manned settlements?

I have published numerous papers addressing the medical problems, particularly the potential hazards to the endothelium, stemming from microgravity and hypokinesia. I have stressed the fact that pharmaceuticals cannot be used in space because of invariable malabsorption, deterioration of some to the point that they no longer meet U.S. Pharmacopeial standards thought to stem possibly from radiation, and potential impairment in metabolism and excretion from ultimately progressive loss of hepatic and renal tissue respectively. In order to colonize I believe, for example, gene therapy will be required to provide a vitally important peptide (atrial natriuretic peptide) which is a vessel dilator and clot buster. I believe that with research over the years, and depending upon how much money is available, and the integrity of NASA medical investigators, we can develop gene therapy to replace a very large number of vital enzymes and peptides, for example. This will require many decades of research. I believe very strongly that those who argue that we can get by with less than 1 g are wrong. Why? Because we have 1 g genes.

Are human physiological and psychological factors being taken as seriously as the engineering factors in the return to the Moon?

I am certain that human physiological factors are not taken as seriously as the engineering factors regarding return to the Moon. As to the psychological factors, it is much easier for all of us to understand and with which to identify.

As a medical doctor with space experience, are you confident that we know enough about human physiology in low gravity and under space radiation to assure astronaut safety on the Moon for extended periods of time?

No.

You have studied and written about the health of the Apollo astronauts after their return to Earth. Can you summarize your conclusions and their implications?

My main concern has been the risk to the endothelium, a delicate structure, and with limitations as to the repair process potential in space because of deficits in magnesium, for example, and development of collateral circulation. It has been shown that after age 30 the endothelium is not adequately repaired. This has been emphasized by Vanhoutte *et al.*[4] Therefore we should consider selection of

[4] P.M. Vanhoutte, L.P. Perrault, J.P. Vilaine, "Endothelial Dysfunction and Vascular Disease" in *The Endothelium in Clinical Practice*, eds. G.M. Rubanyi and V.J. Dzau, p. 273, Marcel Dekker, New York, 1997.

astronauts at a young age to begin training – as was the case of Olympic athletes by the East Germans.

In addition, since there is invariable malabsorption with all medications, and magnesium will have to be given by means of a subcutaneous delivery device, dependable in the presence of vigorous movements during a space walk and one which can be replenished once it leaves the manufacturer; such a device does NOT exist. Furthermore, there is at this time no significant experience with subcutaneous magnesium except in experimental animals.

A problem which I believe has not been addressed pertains to the use of 100% oxygen for up to two hours prior to a space walk to prevent decompression sickness whereas during the early Apollo missions this was used for three hours. During my first meeting with Hansson, he expressed his concern regarding the fact that this is conducive to oxidative stress and has the equivalent effect of reperfusion.

I have stressed the fact that since our genes haven't changed much in the past 50,000 years, we can't expect to survive for very prolonged periods in space without paying a price; we must try to duplicate the lifestyle which our genes demand. I believe that without gene therapy – decades from now – it is conceivable that all of those existing for very prolonged periods in space will ultimately develop renal vascular hypertension with progressive renal failure. The fact that, in space, astronauts are "non-dippers" – with adrenaline remaining elevated at night, i.e., not dipping to lower levels with sleep as on Earth, along with invariable space-related anemia of about 10% – is conducive eventually to renal vascular hypertension, complicating oxidative stress. In addition there are at least two other major space flight-related antioxidant deficiencies: water (hydrogen) and magnesium ions.

When you envision human settlement of the Solar System, where do you see us in 50 years? In 100 years?

I have stressed our genetic limitations regarding specifically your question. I can't answer that other than to quote a highly respected anthropologist, Richard Lee, who states: "The upshot of all this is to inject a somber note in an otherwise optimistic proceeding. We need to learn a little humility and at least to become aware of our own unexamined assumptions. These scientists aspire to the stars and yet their vision is profoundly limited by the blinds of one culture at one point in history. History is messy, and the human material we are working with is messy. Let us at least try to be aware of the triumphalism that I hear again and again: 'We're going to space, its our destiny.' " Such sloganeering strikes a hollow note for those of us who are far from sure that technology will solve all of our problems.

* * *

8.1.3 Lunar dust

We have discussed lunar dust throughout this volume. It is ubiquitous and dangerous to human and machine. Lunar dust began as a problem when the Apollo astronauts found the grey powder clinging to everything during their sojourns on the Moon. Even the vacuum cleaners designed to clean the spacesuits and craft of the dust choked.

From a physiological point of view, the concern about the dust was that it would enter the lungs and interact with the blood and be carried throughout the body. The reduced gravity would keep the fine particles suspended in the airways longer and allow them to get deeper into the lungs and reach the bloodstream. Particles that are smaller than 2.5 microns (0.0025 mm) can cause the most damage and are also most affected by the reduced gravity on the lunar surface. Lunar dust is the portion of the regolith that is less than 20 μm in size and it makes up about 20% of the volume by weight. A special simulant of the lunar dust – known as JSC-1Avf – was created to study its potential effects.[5]

NASA had formed the Lunar Airborne Dust Toxicity Advisory Group (LADTAG) of experts in September 2005 to address the problem of setting health standards for astronaut exposure to lunar dust. Expertise came from astronauts, flight surgeons, inhalation toxicologists, particle physicists, lunar geologists, and pathologists. The group was evenly balanced between NASA experts and others from academia or industry. The goal of LADTAG was to identify research questions, suggest a means of answering those questions, guide research to answer the questions, and then set short-term and long-term human exposure levels that are safe. These environmental standards would guide vehicle engineering decisions for the Crew Exploration Vehicle and the Lunar Surface Access Module, form the basis of flight rules, and dictate the need for environmental control and monitoring.

The group had "concerns about the toxicity of the chemically reactive lunar dust grains, which also contain nanoparticles of natural metals and glass shards formed from a combination of chemical reactions, meteorite impacts and solar wind bombardment."[6] The goals, quoted below, were to assess human exposure and potential health effects:

- Human exposure
  - create activated simulant and/or lunar dust and characterize their passivation in a life support habitat
  - review data on lunar dust with size < 10 μm for surface area, mineralogy, size distribution, surface morphology, chemistry, and electrostatic properties, with special attention to differences from Earth analogs
  - *in situ* assessment of dust at proposed landing site(s): size distribution, chemical composition, chemical reactivity, and passivation in a habitat atmosphere
  - how different is the dust at the south pole compared to lunar dust and simulant samples that we have?

---

[5] J. Park, Y. Liu, K.D. Kihm, and L.A. Taylor, "Characterization of Lunar Dust for Toxicological Studies. Part I: Particle Size Distribution," *Journal of Aerospace Engineering*, October 2008.

Y. Liu, J. Park, D. Schnare, E. Hill, and L.A. Taylor, "Characterization of Lunar Dust for Toxicological Studies. Part II: Texture and Shape Characteristics," *Journal of Aerospace Engineering*, October 2008.

[6] J. Hsu, "Scientists to Set Lunar Health Standards," space.com, 10 June 2008.

- Potential health effects
  - review database on human and animal exposures to materials similar to lunar dust: volcanic ash, mineral dusts, and occupational dust exposures
  - review consequences of Apollo astronauts' exposure to lunar dust. Can we look at filters from Apollo capsules and learn anything?
  - critical review of existing lunar dust studies to deduce what we can learn
  - conduct in-vitro studies of cellular response to simulants and lunar dusts
  - conduct intratracheal instillation studies of simulants and lunar dusts in one rodent species
  - conduct 6-hr inhalation study of simulant and one lunar dust in one rodent species
  - conduct subchronic (28-day) inhalation study using simulant or lunar dust in one rodent species
  - conduct brief human exposure to activated simulant or lunar dust to assess the acute response of man, contingent on the above results.

* * *

The Moon's hard vacuum leaves the dust grains covered with chemically reactive bonds that would be neutralized on Earth by interaction with atmospheric oxygen. On the Moon, the dust can react in toxic ways when breathed into lungs.

The dust is electrostatically active and sticks to everything. Any piece of equipment will be covered in a matter of days. Due to the charge, the dust migrates and crawls up surfaces. Today, we place our structures and delicate equipment at least a meter above the surface and we microwave the surface to create a melt. Our habitats are overpressured around the openings to prevent ingress of dust.

8.2 Human Psychology

In addition to physiological challenges, the psychological pressures of lunar habitation can be severe. The early settlers were truly isolated in close quarters. Over the decades we built larger facilities and residences. Currently, we have as much interior space, both private and public, as the average Earth dweller. Of course, we have no outside space. Many of the public spaces we are currently building – 140 years after the founding of the first permanent settlement – have large atriums, few hallways except as necessary, and very high ceilings. Our public space, while interior, is our outside space.

We had a lot to learn about the psychological aspects of long-duration space flights, such as the almost year-long flight to Mars in 2033–34. And the isolation felt by the first people back on the Moon beginning in 2024, but especially those who settled permanently on the Moon beginning in 2029. These were true pioneers in the full sense of the word – isolation, danger, close quarters – what a challenge to the human body and mind!

Mixed crews were used for such long-duration missions. Couples were sent, along with emotionally mature individuals with training in conflict resolution and an ability to work through issues such as jealousy. Many of these early problems disappeared as the settlements grew larger and the populations went into the hundreds. Our greenhouses helped many of us during our depressions. Not only did the plants provide us with all forms of sustenance, they helped us keep our emotional balance and put our feelings into perspective, especially during the times that we were living one day at a time, when our main goal was to make it to the next day.

This is not to say that we do not have our daily challenges in our lunar cities. We do. They are just much more manageable than they were a hundred years ago.

8.2.1 An historical interview with Sheryl Bishop (January 2009)

Can you give us a one or two paragraph bio?

I am a Social Psychologist and Associate Professor at the University of Texas Medical Branch, Galveston, Texas and currently hold the post of Senior Biostatistician for the School of Nursing. I was the founding curriculum director for the new Space Life Sciences Ph.D. curriculum in UTMB's Department of Preventive Medicine and Community Health from 2004–2008. I have served as lecturer, faculty and SSP co-chair for the Space Life Science and Space and Society Departments at the International Space University, Strasbourg, France, since 1996. For the last 20 years, I have investigated human performance and group dynamics in teams in extreme environments, including deep cavers, mountain climbers, desert survival groups, polar expeditioners, Antarctic winter-over groups and various simulations of isolated, confined environments for space. I routinely present my research at numerous scientific conferences, and have over 60 publications and 50 scholarly presentations in both the medical and psychological fields on topics as diverse as psychometric assessment, research methodology, outcomes research, stress and coping in extreme environments, psychosocial group dynamics and human performance in extreme environments. I am often requested to be a content expert by various media and have participated in several television documentaries on space and extreme environments by the Discovery Channel, the BBC and 60 Minutes.

How did you become interested in the adaptation of humans to extreme environments?

In the mid-1970s I was trying to settle on a career in order to finish my bachelor's degree. As an avid science fiction reader, I decided to visit Johnson Space Center in Houston, Texas, to see what career opportunities were available. They advised me to look into closed loop environments as a engineer. As engineers were plentiful at that time, I decided to project a bit further into the future and it occurred to me that there would come a time when someone would need to address the myriad issues surrounding human adaptation to the space environment. So I took up social psychology with a specialty focus on space. To my chagrin, I found access to actual space crews nonexistent and, once again, realizing that there was work that needed to be done before we would ever have large numbers of actual crewmembers in space, I sought the next best thing ... analog environments. Since any analog to space needs to mirror some aspects of confinement, isolation, and, possibly, threat, this led me to the various extreme environments I've studied over the last 20 years.

Do you think that settling the Moon is important for civilization beyond economic benefits? Is there an epic vision for human exploration and growth that postulates the need for such endeavors?

I do think the Moon is important if for no other reason than the fact that setting up large-scale habitats that could grow into real communities is logistically more easily achieved with the Moon's proximity. I want to go to Mars as badly as anyone but recovery from the inevitable failures and breakdowns inherent in any exploration process is a magnitude more difficult to accomplish. As an 'Apollo Orphan' I'm more vested in any effort that moves us beyond our near Earth orbit accomplishments and breaks the ennui that strangles mankind's journey outward since Apollo.

Is there an epic vision for human exploration and growth that requires such endeavors? I believe that there is a positive cycle that is generated from great exploration challenges. It is the nature of nations to be concerned with known geopolitical conditions and maintenance of each nation-state's self-interests. But any system that is limited in its ability to expand will soon generate conflict. Movement outward has always been the evolutionary and social release for humankind. There is no reason to believe that movement off-planet will not provide the same stimulus for growth and development that any other exploration cycle in human history has provided.

How do we answer critics who say space is too expensive and that there are numerous problems on Earth to take care of first?

First, you must distinguish those naysayers who will not be convinced by any amount of evidence from those who are open to possibilities. The former are a waste of time to confront and simply must be out-lived and out-endured. If the historical evidence regarding the failure of all previous societies to stagnate and collapse once growth and expansion is halted is not enough, then the latter should be directed to the overwhelming evidence of technological innovation and economic stimulus that our space program has provided over the years. As a scientist, I always believe that data takes precedence.

What are the psychological and physiological traits of a person who you think would be suited to make a new life on the Moon or Mars?

In truth, there doesn't appear to be a single "right stuff" profile. Rather, there are characteristics of the environment that would be a better fit for individuals that fall within a certain range of characteristics. For instance, all space environments will entail a great degree of enclosure and habitat confinement since we must create a sustainable environment. In such closed loop environments, the ability to tolerate confinement is absolutely essential. However, a person that pilots a shuttle from Earth to Moon is faced with short duration confinement compared to a person that is actually living and working on the Moon. For Mars, additional factors involve the magnitude increase in isolation and distance from other humans. In general, individuals that tend to do well in isolated, confined extreme environments tend to be self-sufficient, mature, flexible, comfortable in working with others, self-motivated, conscientious, agreeable, resilient, achievement oriented (but not competitive) and both moderately task and interpersonally oriented.

Are human physiological and psychological factors being taken as seriously as the engineering factors in the return to the Moon?

"Taken seriously" is a relative term. There is certainly an acknowledgement by all concerned about the importance of attending to physiological factors and to a lesser extent, psychological factors. However, when it comes to committing dollars to programs, engineering activities take precedence. The space arena is still very much dominated by engineers who only understand an engineering perspective. So the pragmatic answer is "no," it's not given the same weight or importance as "bending metal."

How can engineers become more aware of the importance of the non-technical issues that are really as critical to our success in this journey?

I've had success in heightening awareness in the engineering community by making the linkage between group and individual functioning and how all those engineered systems are used explicitly. For instance, when I demonstrate that understanding how spatial configuration of workspace affects mood, concentration and group functioning, and that subtle changes to accommodate how humans use space instead of insisting that humans conform to how machines "use" space improves overall performance, engineers typically "get it." In fact, they see the resolution to a lot of downstream user-interface "problems" by fitting systems to how humans function at the beginning. But it's taken a while to convince the community that people can't simply be "forced" to fit the system. The best and most effective opportunity for making change is for psychologists and social scientists to be at the table in the design phases ... not be brought in only when the problems start piling up to "fix" things.

Are you confident that we know enough about human physiology in low gravity and under space radiation to assure astronaut safety on the Moon for extended periods of time?

Wow ... we can't "assure" the safety of the average citizen in one of our metropolitan areas with its attendant risks from environmental pollution, overpopulation, etc. Why should there be any expectation that we "KNOW" all of the myriad unforeseen risks from space radiation to assure astronaut safety? At best, all risk evaluation is probabilistic. We "know" some of the characteristics of the radiation profile on the Moon but there are so many unknowns that are possible, the best we can say or do is to provide what can be done given what is known.

How does our experience in Antarctica prepare us for long duration space travel and settlement on other planetary bodies?

Of all the "analog" environments here on Earth, Antarctica exemplifies many more of the elements that characterize space environments, most specifically the risks of an extraordinary environment with extreme temperatures, structures, isolation, requirements for life support and habitat sustainability, confinement, circadian dysrhythmia, difficulty of rescue, etc. As a laboratory for distilling the critical fusion and fission factors that promote or undermine effective individual and group functioning, the presence of an environment that poses high threat and a substantial degree of uncertainty is of essence.

From a psychosocial perspective, do you think Americans will be the first to settle the Moon in the 21st century, or will it be the Chinese?

It's not a psychosocial issue but, rather, a political issue. The nation with the political will to pursue a sustained effort to colonize the Moon will be the first to achieve that milestone. Given the differences in our political systems, it is difficult to see America sustaining such a program over the length of time it will take to succeed. However, I think you've overlooked a "dark horse" in this race and that of commercial space interest may beat them all.

When you envision human settlement of the Solar System, where do you see us in 50 years? In 100 years?

50 years – I would hope that we would at least have one manned permanent outpost on the lunar surface. We might have the equivalent of a lunar surface ISS. There may be some initial commercial small-scale manufacturing facilities operating but they will most likely still be proof of concept operations rather than viable commercial facilities. We hopefully will have at least made some initial exploratory class missions to Mars (flags and footsteps) with, at best, some proof of concept power generation, water production, etc., small scale facilities.

In 100 years, the lunar surface should be reminiscent of early Antarctica while Mars will still be characterized by few established exploration-class bases. The wild card will be commercial space organizations. If they can find a viable combination of space tourism and commercially profitable manufacturing interests, things may accelerate to a degree.

* * *

Evolutionary psychology helped us understand what to expect as we went into space to colonize the Moon. Our environment causes our body and our mind to change. This change can be rapid or take years and generations. We are always interested in the role that the environment plays – especially an extreme environment – in our adaptation and evolution. Gravity, or its lack thereof, shapes our lives. Earth gravity is directly related to what we have become physically and psychologically. Therefore, as we began to move into space in the early 21st century we were concerned about how microgravity in space and low gravity on the Moon and Mars would impact our daily and long-term lives. In those early years, Valeri Polyakov held the distinction of spending the most time in space, 438 consecutive days on the Russian space station Mir.

Some of the effects of changes in gravity are overcome by the body after a period of time. Some are temporary and others may become permanent. On shorter missions astronauts reported severe motion sickness. This is because "the vestibular system relies on gravity, so as the head tilts, hair cells in the inner ear are displaced, creating a signal to the brain regarding the head position and balance. When these signals from the vestibular system and the brain are incongruent, an individual feels nausea."[7] Astronauts on longer missions can adapt to this incongruence.

On longer missions, more serious effects are noticed. Due to the lower gravity, blood and body fluid shift, there is muscle atrophy and bone density loss. In microgravity, fluids shift upward in the body – in Earth gravity we walk and the mechanical loads result in leg muscle contraction, arterial and venus constriction, and a resulting blood flow through one-way valves. "However, without gravitational opposition, blood pressure decreases causing the potential of cerebral ischemia (blood loss in the brain) and neuronal (brain cell) death."[7]

Additionally, the lack of gravitationally-induced signals leads to bone loss and slows biological growth. Microgravity resulted in a loss of 1–2% bone loss in astronauts in weight-bearing areas such as the pelvic bones, lumbar vertebrae and femoral neck. As bones in the lower body atrophy due to lack of use, upper body skeletal regions grow in density.

Research was performed on the utility of artificial hypergravity in countering these effects. In the present, we have not totally resolved these issues. Exercise is a significant fraction of our days. Pregnant women spend more than half their days in rotating habitats to help reduce much of the low gravity effects on them and their fetuses. But we envision that as we become "permanent" residents of the Moon (beyond the hundred years-plus that we have spent here) we may decide that we will let evolution takes its course. Very few of us visit Earth. Our lives are here and outward. We are becoming prepared to see subsequent generations with less development of their lower bodies.

However, some of the effects can be temporarily pleasing, as this early quote shows: "Former Astronaut Susan Helms liked how her body changed while she was on the International Space Station for six months. 'I got taller. I shed about 20 pounds. Your legs get very skinny. If you have varicose veins, they'll go away. Wrinkles go away because the fluid shifts. You're getting the idea that they could build a spa in space and there would be a lot of people paying money to go,' she jokes. 'It's almost like a fountain of youth in a sense. Your body goes back 20 years and sheds the effects of gravity. It's pretty amazing.' she says. Unfortunately, it all reverses in a short time once you return to Earth."[8]

Some of us have plans to move to other planetary bodies with larger-than-Moon gravity fields. Those that do are undergoing hypergravity training to build up their ability to survive in gravity fields several times that of Earth.

Another effect of microgravity and low gravity is on our ability to judge size and distance.[9] Apollo astronauts reported difficulties judging distances and features. On

[7] S.E. Bell and D.L. Strongin, "Evolutionary Psychology and its Implications for Humans in Space," in *Beyond Earth: The Future of Humans in Space*, ed. B. Krone, pp. 78-83, CG Publishing, Apogee Space Press, 2006.

[8] L.S. Woodmansee, *Sex in Space*, p. 48, CG Publishing, 2006.

[9] R. Courtland, "Zero-gravity may make Astronauts Dangerous Drivers," *NewScientistSpace.com*, 22 September 2008.

the Moon, vision can be distorted making it difficult to judge the speeds of objects, which we estimate by observing how an object changes in size.

We learned much about the psychological effects even during the short spaceflights during the Apollo and Shuttle eras. An early study explored how the value hierarchies of people in the space program changed after a trip to space.[10] "Transcendence[11] leapt from last place to second place among the men and from second to unrivalled first place among the women." Such changes did have a significant impact on the lives of astronauts and their close associates. We now know that among the populations on the Moon and Mars, a higher purpose and spirituality are very common. Of course, these personality traits could be predicted as being dominant for those who leave the comforts and beauty of Earth, as well as family and friends, for a life on planetary bodies that can never feel like home in the case of the Moon, and will take centuries to have some semblance of Earth in the case of Mars.

8.3 The role of plants

Green has always been a part of the vision for space – not money, that is, but plants. The role of plants in a manned facility on the Moon and on Mars is multi-faceted. Some of us have plants in our homes and offices because their presence makes us feel better. They make the rooms feel more in tuned with our psychological needs. We like the colors of plants. They become part of the atmosphere of the room in numerous ways.

Plants were as crucial to human survival in space as oxygen. They are "complex eukaryotic[12] organisms that share fundamental metabolic and genetic processes with humans and all higher organisms, yet their sessile nature requires that plants deal with their environment by adaptation *in situ*."[13] This is very different than how humans deal with their environments, which is by creating a protective structure, by changing the environment or by leaving.

Plants address their environment by adjusting their metabolism. In this way, plants can be viewed as biological sensors that can be monitored to report on their environments. They are sensitive to fluctuations in gravity, radiation, temperature and pressure – as are humans. "Truly insightful experiments that address fundamental questions about biological adaptation and responses to extreme extrater-

[10]P. Suedfeld, "Space Memoirs: Valuable Hierarchies Before and After Missions – A Pilot Study." *Acta Astronautica*, Vol. 58, 2006, pp. 583–586.

[11]Transcendence – a combination of spirituality (unity with nature, inner harmony, detachment from material desires) and universalism (protecting the environment, wisdom, peace).

[12]Eukaryote – any organism having as its fundamental structural unit a cell type that contains specialized organelles in the cytoplasm, a membrane-bound nucleus enclosing genetic material organized into chromosomes, and an elaborate system of division by mitosis or meiosis, a characteristic of all life forms except bacteria, blue-green algae, and other primitive microorganisms.

[13]R.J. Ferl and A.-L. Paul, "Plants in Long Term Lunar Exploration," 2159.pdf, NLSI Lunar Science Conference (2008).

restrial environments can be answered using plants. They are easy to transport in spaceflight in their seed, under complete vacuum and in extremely low temperatures and are fundamental components of any long-term bio-regenerative life support system."[13]

Plants and animals have complementary existences as "plants recycle human wastes and provide human nutrition, while humans recycle plant wastes and provide plant nutrients. As biosensors and due to their adaptation to the space environment via their ability to alter their gene expression, plants are viewed as an extremely useful indicator of how humans would evolve over the generations in space and on the planetary bodies." Currently, in 2169, our knowledge of plants has evolved to the point where there is an almost symbiotic relationship between plants and humans.

We would not have survived on the Moon without our tremendous greenhouses. Also, the interiors of our habitats have plants distributed throughout. While we do not speak to our plants – that is, most of us don't – we communicate with them psychologically, and through sounds and music. Some of the flowering plants are provided appropriate light so that they can bloom. In the early years on the Moon, those blooms gave more joy to our pioneering settlers than anything else.

We have been able to genetically engineer certain plants so that they can fully survive in the lunar environment via *lunaponics*. Recall hydroponics from Earth, where the cultivation of plants was accomplished by placing their roots in liquid nutrient solutions rather than in soil, leading to the soilless growth of plants. Lunaponics is similar – plants are coupled to bioengineered vines that provide the plant access to cyanobacteria that can break down lunar rocks into constituents that the plants can use for food. These evolved plants require minimal shielding on the surface.

In the early 21st century there were significant initial discoveries about plants. It was found that marigolds could grow in crushed rock very like that found on the lunar surface, with no need for plant food.[14] The marigolds were planted in crushed anorthosite – a rock very similar to much of the lunar surface – and with added bacteria the plants thrived.

"The medicinal significance of marigolds is high, and a marigold may be recommended for antiulcerogenic action, for prevention of liver and kidney inflammation, and for oncoprotection."[15] The first plants on the Moon were utilized to create a fertile soil that could grow second generation plants for human consumption. The pioneer plants survived on lunar regolith, were resistant to diseases, little light, low gravity and all of the harsh conditions found on the Moon. The microbial community infused into the regolith was critical in allowing the plants to survive.

Experiments also showed that cyanobacteria can grow in lunar soil, if provided with water, air and light. When cyanobacteria were placed in a container with water and simulated lunar soil, they produced acids that broke down tough minerals

[14] R. Black, "Plants 'Thrive' on Moon Rock Diet," BBC News website, 17 April 2008.

[15] N.O. Kozyrovska, T.L. Lutvynenko, O.S. Korniichuk, M.V. Kovalchuk, T.M. Voznyuk, O. Kononunchenko, I. Zaetz, I.S. Rogutskyy, O.V. Mytrokhyn, S.P. Mashkovska, B.H. Foing, V.A. Kordyum, "Growing Pioneer Plants for a Lunar Base," *Advances in Space Research*, Vol. 37, 2006, pp. 93–99.

including ilmenite, which is relatively abundant on the Moon. The cyanobacteria use only sunlight for energy. "Cyanobacteria in growth chambers, where water, sunlight and lunar soil are provided, ... can be harvested and further processed to make use of the elements they extract from the lunar soil ... that can be used as fertilizer for food plants grown in hydroponic greenhouses ... [and] ... methane given off by the breakdown of the cyanobacteria could be used as rocket fuel."[16]

"The most challenging technologies for future lunar settlements are the extraction of elements (for example, iron, oxygen, and silicon) from local rocks for life support, industrial feedstock and the production of propellants. While such extraction can be accomplished by purely inorganic processes, the high energy requirements of such processes drive the search for alternative technologies with lower energy requirements and sustainable efficiency. Well-developed terrestrial industrial biotechnologies for metals extraction and conversion could therefore be the prototypes for extraterrestrial biometallurgy. Despite the hostility of the lunar environment to unprotected life, it seems possible to cultivate photosynthetic bacteria using closed bioreactors illuminated and heated by solar energy. Such reactors might be employed in critical processes such as air revitalization, element extraction, propellant (oxygen and methane) and food production."[17]

These pioneering studies have led to our ability today to utilize bacteria in all of the industrial and life support processes on the Moon and Mars. As already mentioned, plants and humans form a partnership in space. We are really parts of a whole biosystem that is beginning to flourish.

8.3.1 An historical interview with Robert Ferl (July 2009)

Can you provide us with a brief bio?

I am the Director of the Interdisciplinary Center for Biotechnology Research at the University of Florida and a Professor of Plant Molecular Biology. I have a B.A. in Biology from Hiram College and a Ph.D. in Genetics from Indiana University in 1980. Directly after my Ph.D., I spent time in Australia at the CSIRO[18] in Canberra to work on some of the earliest plant gene sequencing. I have been a P.I. for two NASA Flight payloads, one that launched in 1999 (STS 93) and one that is scheduled to launch as an ISS payload in November 2009. With my colleague, Anna-Lisa Paul, I have conducted a variety of experiments that explore the molecular responses of plants in other extreme and novel environments to understand the potential responses to extraterrestrial environments. In addition to spaceflight, we have examined changes in gene expression patterns in response to parabolic flight environments, extreme low atmospheric pressures and to mineral features of the Haughton Impact Crater on Devon Island.

[16] D. Shiga, "Hardy Earth Bacteria can Grow in Lunar Soil," *NewScientistSpace.com*, 14 March 2008.

[17] I.I. Brown, J.A. Jones, D. Garrison, D. Bayless, S. Sarkisova, C.C. Allen, and D.S. McKay, "Possible Applications of Photoautotrophic Biotechnologies at Lunar Settlements," Rutgers Symposium on Lunar Settlements, 3–8 June 2007.

[18] Commonwealth Scientific and Industrial Research Organisation.

Your research within molecular biology specializes in the molecular mechanisms involved in gene expression and the genetic responses of plants to harmful environments. You also study genetically engineered plants for their responses to microgravity and space. Much has been written about man's risks in space. Can you summarize the risks to plants? Are they better suited for space, radiation and microgravity than is man?

In the main, the risks to plants would be very similar to the risks to man. In fact, we are among those folks who look more at the parallels among plants and humans because both are complex eukaryotic organisms that make complex developmental and metabolic choices depending on their environment. Spaceflight offers several environmental complications to those organisms that evolved on Earth. Radiation is a big risk shared by all, if the flight gets anywhere near high Earth orbit or beyond. Lack of gravity is another impact that is shared, with clear biological effects on both plants and humans. Plants also have incredibly sensitive reactions to the environment and those reactions influence growth, flowering, and developmental choices.

Now, are plants better suited? Very interesting question. In some ways no – mostly because plants cannot take medicine, put on clothes, or breathe through a tube in an emergency. Therefore plants have a more difficult challenge in most spaceflight scenarios. On the other hand, plants are incredibly capable of modifying their growth to meet environmental circumstances and scientists and engineers have learned over time how to improve plant growth hardware on orbit.

Can you tell us a bit about the Arthur Clarke Mars Greenhouse and the NASA Haughton Mars Project Base Camp on Devon Island?

The ACMG greenhouse was established in 2002. In general the ACMG is not a full-featured, high fidelity simulation of a greenhouse to be established on Mars. It does, though, support scientific and operations research in an environment which is operationally relevant to Mars or the Moon in unique ways – each at a specific level of fidelity and complexity. The main greenhouse systems include the plant growth system, environmental control system, power system, communication system and local network and the data acquisition and control system.

The Haughton-Mars Project (HMP) is an interdisciplinary and international field research project making use of the Haughton crater impact structure and surrounding terrain as a terrestrial environmental analog for Mars and the Moon. The Haughton Mars Project Research Station (HMP RS) is located in a remote polar desert on Devon Island, an uninhabited island in the territory of Nunavut in the Canadian High Arctic. The geologic features and biological attributes of Devon Island in general, and the HMP RS site in particular, offer a unique research and operations environment appropriate to the development of approaches, technologies, and field-based operational methodologies that might be critical to successful extraterrestrial human missions. The HMP RS is managed and operated jointly by the Mars Institute and the SETI Institute with support from the Canadian Space Agency and NASA.

How does it feel to live on Devon Island? Does it help you imagine how life on the Moon or Mars might feel?

As part of the whole project experience on Devon Island, the answer is clearly YES. First off, the project is on the edge of a large impact crater, one that presents a large amount of impact breccias and in that regard it is very similar to what one might experience landing near a crater on the Moon or Mars. The biggest life analogs are operational – the logistics, communications and the simple fact of living in a place with no water in the soil, rocky and a difficult terrain to move about and manipulate. As a laboratory scientist who is used to a clean and easy operations environment, dealing with the operational realities of living in the field and dealing with the limited logistics of a field deployment are big eye-openers. So, again as scientists, what we propose to do and accomplish on the Moon or Mars can be tremendously informed by first-hand experience in these difficult operational deployments.

Is there a significant difference for plants on the Moon versus Mars? Can we learn from plant survival and growth on the Moon and apply that knowledge to Mars?

There are certainly differences. The rocks are different, the gravity is different, and the vehicles will be similar but different. But having said that, it is also clear that learning how to deal with impact breccias and craters, how to deal with dust and non-terrestrial soils, how to encourage plants to be productive and informative in an extraterrestrial environment – all of these things learned on the Moon would be huge in buying down the risks on a Mars mission.

In what ways do you see plants supporting humanity's expansion into space?

Ahhh, much has been written on this subject but for me all we have to do is look at history. Human movement around the Earth occurred first through small explorations but civilizations only expand where their agriculture can take them. It is very clear to me that true expansion into space (as opposed to short-term exploration) will absolutely require plants and agriculture to provide life support.

What are some remarkable facts about plants that most people do not know?

Foremost – plants are more like humans than you might ever imagine. All of your central metabolism is shared with plants. All of the key metabolic pathways are shared. Plants have DNA and complex developments and behaviors that make them pretty close relatives in the grand scheme of things.

Is settling the Moon important for civilization?

In my opinion, yes. Again all we need to do is consult history, and also the writings of the past few generations. It is clear that our imaginations take us there, it is clear that we can envision ourselves living there. To me, this suggests strongly that we want to settle there, and if we want to then we essentially need to.

How do we answer critics who say space is too expensive and that there are numerous problems on Earth to take care of first?

Show them the reaction of the Earth to the first Moon landing. Show them the reaction of kids to Hubble images. Show them the excitement that is created in humans that get any experience at all with space. To me, the answer is not in economics or spin-off benefits to problematic things on Earth. To me the answer is pure excitement, pure exploration and, most importantly, pure knowledge of where we as humans fit into the grand scheme of the universe. It is too expensive? It surely is expensive, and yes there must be some balances between space exploration and local terrestrial needs. But it is not an either/or situation. Never has been. Rather, I see how much we spend on space (money, lives or simple emotional energy) is a reflection of how broadly we are thinking and how deeply we are considering our role in the universe.

What do you see as the major hurdles for our return to the Moon for permanent manned settlements?

From an engineering perspective, radiation safety. Providing a safe haven for periods of high solar activity is a big, big construction issue. The issues of habitat design, growing plants to support life, conquering the problems associated with using resources on the Moon, these are all a matter of cost/benefit. But I am not at all confident that we know how to protect terrestrial biology when it is outside of the van Allen belts – on the Moon, Mars or in transit.

How optimistic are you that President Bush's timeline for the return to the Moon will approximately be kept?

Not very. Sorry, but we have proven to be not very good at aggressive timelines. Witness the ISS.

* * *

8.4 Quotes

- "As soon as somebody demonstrates the art of flying, settlers from our species of man will not be lacking [on the Moon and Jupiter] ... Given ships or sails adapted to the breezes of heaven, there will be those who will not shrink from even that vast expanse." Johannes Kepler, letter to Galileo, 1610
- "I think there's a supreme power behind the whole thing, an intelligence. Look at all of the instincts of nature, both animals and plants, the very ingenious ways they survive. If you cut yourself, you don't have to think about it." Clyde Tombaugh
- "Space flights are merely an escape, a fleeing away from oneself, because it is easier to go to Mars or to the Moon than it is to penetrate one's own being." Carl Gustav Jung

– "There is a theory which states that if ever for any reason anyone discovers what exactly the Universe is for, and why it is here, it will instantly disappear and be replaced by something even more bizarre and inexplicable. There is another that states that this has already happened." Douglas Adams

9 Getting there

"It IS rocket science."

Yerah Timoshenko

Getting there has always been viewed as the largest obstacle to a significant presence in space. The cost of launching payloads into Earth orbit had always been prohibitively expensive, in the order of $10,000 per pound. The cost to launch an early 21st-century Space Shuttle was $450,000,000. Research efforts on rockets were significant, but the science of escape velocities and orbital mechanics put up walls that seemed insurmountable.

A number of different methods for propelling spacecraft beyond conventional rockets had their hopeful proponents, for example, ion propulsion and anti-matter/matter collisions. One idea that was often discussed was solar sailing. Light has momentum, so a giant mirror can be used to reflect sunlight, thus imparting a small thrust to the spaceship. Over a long period of time, the thrust accumulates and results in large velocities. However, the amount of sunlight drops off inversely proportional to the square of the distance from the Sun. The same sail would receive only 4% as much sunlight at Jupiter as when near the Earth, so solar sailing would only be useful for exploration within the inner Solar System.

Another idea that drew much attention was the mass driver. Just as a rocket ejects oxidized fuel to provide thrust in the opposite direction, a mass driver ejects mass. On an asteroid, if we set up a launch rail to throw pieces of the asteroid off in one direction, then we will move it in the other direction at a speed that is in proportion to the relative masses. Today, we do this quite often to bring asteroids into lunar orbit so that we can mine them within the reach of our elevators.

Once in space, aerobraking and gravitational assists are used to alter spacecraft orbits. Aerobraking consists of using a planet's atmosphere to slow down a spacecraft when it arrives at its destination. Since we still send rockets directly to planets that have atmospheres but no elevators, the use of aerobraking instead of standard retrorocket systems saves a lot of fuel and reduces launch costs.

Gravitational assists have been used by a great number of different missions. In this scenario, a space probe approaches a planet from a carefully-planned orbit such that the planet's gravity transfers some of the kinetic energy of its orbit to the probe. Some examples: Voyager 1 (1977) came in behind Jupiter and got an extra kick of energy while slowing Jupiter's orbit around the Sun. Since Voyager was miniscule compared to Jupiter, the difference in Jupiter's orbit afterwards was

Fig. 9.1 Astronaut at far end of mass driver. Deflection plates near the end of the mass driver make minute adjustments to the trajectory of the launched ore to ensure that it reaches its target: a mass catcher at the L-2 point. (Courtesy Space Studies Institute)

extremely small – not measurable. But the extra velocity the probe gets can be very significant, trimming many years off the voyage time to more distant places. Both the Galileo (1989) and Cassini (1997) missions were planned so as to swing those probes around planets in the inner Solar System (Earth and Venus in both of these missions) to work up enough velocity to orbit out to the much more distant planets Jupiter and Saturn.

9.0.1 An historical interview with Lee Morin (May 2009)

Can you give us a one or two paragraph bio?

My parents were in the Foreign Service, and I grew up in Washington D.C. area and abroad in Kobe, Japan, Baghdad, Iraq, and Algiers, Algeria. Attended Western Reserve Academy in Hudson, Ohio, and then the University of New Hampshire (Math/Electrical Science). Spent some time at MIT with the group that became the Media Lab. Attended the M.D.-Ph.D. Medical Scientist Training Program at NYU and was trained as a DNA chemist and molecular biologist. My dissertation was "Synthetic Deoxyoligonucleotides and Prophage Phi-80 Induction." This work identified the molecular trigger for error-prone "SOS" DNA repair. (It is the fragment of DNA that has the sequence GG.) Dr. Michio Oishi was my advisor.

Medical school was completed at Bellevue Hospital, and then completed two years of General Surgery residency at Montefiore and the Bronx Municipal Hospitals. Decided I didn't want to spend my life in rooms without windows, joined the Navy, and spent the next two years in the Submarine Service, aboard the USS Henry M. Jackson (SSBN-730). Then went to Pensacola and became a Naval Flight Surgeon. Left the Navy after several years as a Flight Surgeon in Pensacola, became Medical Director of the largest Occupational Medical practice in Florida, in Jacksonville. Was a reservist with the U.S. Marine Corps, got mobilized and recalled to active duty for Desert Storm, served in Bahrain for 15 months. Decided to stay in the Navy and returned to Pensacola as a Flight Surgeon and Diving Medical Officer. Completed residency in Aerospace Medicine. Applied to NASA and was accepted in 1996. Flew aboard Atlantis in 2002 on mission STS-110, performed two EVAs to attach the first piece (called S-Zero) of the truss that holds the solar arrays onto the space station. Am currently working on development of the Orion Spacecraft as Deputy of the Cockpit Working Group and lead of the displays Rapid Prototyping Team.

How did you become interested in the space program?

First became seriously interested when I first joined the U.S. Navy and saw a personnel announcement on applying to become a naval astronaut, but did not have the five years' minimum naval service at that time. Did not seriously (i.e., actually) apply until a decade later, when the stars aligned. I was interviewed on my first application and was accepted. I figured I would always regret not having applied at least once, and it would be interesting to get the interview process if I got that far, especially the flight physical. (I was an aerospace medicine specialist and had performed many specialized physicals for the U.S. Navy.)

Is settling the Moon so important for civilization?

If we are to become a spacefaring civilization we must have sustained presence off the Earth. If we can't get that done on the Moon, we certainly aren't going to get it done anywhere else.

You are both an astronaut and a medical doctor (and Ph.D.). That must be very rare in NASA. It must give you a unique perspective?

About 10% or so of the astronaut corps are physicians or life scientists (biochemists, vets, etc.). Life Science is a selection category; applicants are screened within their categories for initial selection. Since higher education is considered strongly during selection, it is not surprising that M.D.s, Ph.D.s, and M.D.–Ph.D.s are not uncommon in the astronaut corps. Right now there are two M.D.–Ph.D.s in the corps, five other M.D.s, and two biomedically-related Ph.D.s. (nine of about 90 active astronauts)

Do you interact with the medical corps as a colleague? Are you involved with some of the medical deliberations and research?

We do work with the Life Sciences Directorate on medically-related issues, often in a liaison role, as representatives of the crew office to various panels, boards, or development teams. Some examples are the radiation protective measures for Orion, and crash impact attenuation when the capsule returns to Earth under parachutes.

What do you see as the major hurdles for our return to the Moon for permanent manned settlements?

Heavy lift to get stuff there. Political will to actually do it. Radiation when we get there. The real key is identifying an economic driver that works to pay the way. This will turn out to be something unexpected.

Are human physiological and psychological factors being taken as seriously as the engineering factors in the return to the Moon?

Yes, for the sortie missions. We have a strategy that works for older adults for six months, i.e., ISS exercise program and post flight rehab. But long-term radiation exposure to a fetus or a child or a young woman is a much harder issue.

As a medical doctor with extensive space experience, are you confident that we know enough about human physiology in low gravity and under space radiation to assure astronaut safety on the Moon for extended periods of time?

We know what to do, basically monitor and shelter using the mass of lunar material as a radiation shield. Being in deep space for a long time is the problem – you may need strong magnetic fields to stop the ions.

Can you put into words the feelings that go through you when you blast off into space? And then view the Earth from several hundred miles up?

On my flight, there was a computer glitch on some computer I never heard of before or since in the LCC[1] just before liftoff. We heard that they were rebooting it. After what seemed like a long time, my Commander asked how long was rebooting going to take? He was told a couple of minutes. He said, well, you only have 50 seconds left in the launch window. I was ready to release my seatbelt and get ready to climb out. (However it takes hours to get out after a launch abort – once you are on the pad, the quickest way to get to the bathroom is to get into space and to set up the WCS[2] (shuttle bathroom). All of a sudden, the call came that the computer was rebooted and was go. The Flight Director then polled the room (FIDO[3] is GO, Booster is GO, Surgeon is GO, ...) and Flight Director said GO! And the main engines light and the shuttle "twangs" – rocking back and forth about $1\frac{1}{2}$ feet – then you get rear-ended by the semi and a deep rumble and you are pressed into your seat and you know you are going somewhere and you are glad you did not release your seat belt and you rumble and get heavier and heavier and then there is a loud bang and the pressure gets less for a little while and the ride is smoother and the pressure builds back up and it starts to get hard to breathe and it reminds you of when you were a kid and the bully on the playground was beating you up and it gets really hard to breathe and then it gets real still and your pencil is floating up on its string and your experienced crewmate is out of his seat and says "Welcome to space, rookie!" as he pats you on the back while he floats by and you open your seatbelt and the straps float up and you feel like the bottom is coming out of the elevator at the Empire State Building and you start to float forward out of your

[1]Launch Control Center.
[2]Waste Collection System.
[3]Flight Dynamics Officer.

seat and you can't seem to stop drifting forward and you find you are tensing your abdomen to catch yourself and you remember that they told you that people get backaches in space because they tense their abdomen too much and you float up against the wall because anything you touch you move away from and you feel like a cat with her claws out, slipping on a glass table only in 3-D . . . .

NASA and the manned space program seem to be in limbo right now, especially with the lack of an Administrator and the Shuttle being phased out. Where do you think all this will go?

The new Administrator will get confirmed and the program will move forward and the Shuttle will stop flying in 2011 and Orion will get built and will fly probably in 2015.

When you envision human settlement of the Solar System, where do you see us in 50 years? In 100 years?

I am a proponent of telepresence robotic industrial development and material processing of lunar material as the only practical path to having enough stuff in space to stay there. The telepresence has to turn lunar material into the material needed to sustain the presence off Earth – everything has to be built in space from stuff that is already there. When we can do that, we will be a spacefaring civilization. The key is to have economic products that pay the way. The only thing that you can afford to return to Earth from the Moon is information products, such as intellectual property, images, telepresence experiences, monuments – those we can move between Earth and Moon at the speed of light without heavy lift or any lift once you are established. The key is to construct an economy around these products to achieve sustainability.

This approach provides a path to an exponential growth model. If such a path is realized, the space business can look very different in 50 or 100 years, with as much change as other aspects of our lives have changed in the last 50 years. If it doesn't, it will probably still look like it does now, only more expensive because energy will be more expensive.

* * *

Addendum: A New NASA Administrator

At the time of the above interview, President Obama had nominated on 23 May 2009 General Charles Bolden for Administrator of NASA. Charles Bolden retired from the United States Marine Corps in 2003 as the Commanding General of the Third Marine Aircraft Wing after serving more than 34 years, and was at the time of his nomination the CEO of JackandPanther LLC, a privately-held military and aerospace consulting firm. General Bolden began his service in the U.S. Marine Corps in 1968. He flew more than 100 sorties in Vietnam from 1972–1973.

In 1980, he was selected as an astronaut by NASA, flying two space shuttle missions as pilot and two missions as commander. Following the *Challenger* accident in 1986, Gen. Bolden was named the Chief of the Safety Division at the Johnson Space

Fig. 9.2 Astronaut Charles Bolden, confirmed as NASA Administrator on 16 July 2009. (Courtesy NASA)

Center with responsibilities for overseeing the safety efforts in the return-to-flight efforts. He was appointed Assistant Deputy Administrator of NASA headquarters in 1992. He was Senior Vice President at TechTrans International, Inc. from 2003 until 2005. Gen. Bolden holds a B.S. in Electrical Engineering from the U.S. Naval Academy, Annapolis and an M.S. in Systems Management from the University of Southern California.

The U.S. Senate unanimously confirmed General Bolden on 16 July 2009.

* * *

9.1 Rockets

The history of rockets is almost as old as mankind. Rockets are remarkable collections of human ingenuity that have their roots in the science and technology of the past. They are natural outgrowths of literally thousands of years of experimentation and research on rockets and rocket propulsion.[4] Robert Goddard designed, built and flew the first liquid-fueled rocket in 1926.

The Saturn V

The Saturn V was a multistage liquid-fuel expendable rocket used by NASA's Apollo and Skylab programs from 1967 until 1973. In total, NASA launched 13 Saturn V rockets with no loss of payload. It remains the largest and most powerful launch vehicle ever brought to operational status from a height, weight and payload

[4] http://www.grc.nasa.gov/WWW/K-12/TRC/Rockets/history_of_rockets.html

Fig. 9.3 Robert H. Goddard, one of the founding fathers of modern rocketry, was born in Worcester, Massachusetts in 1882. As a 16-year-old, Goddard read H.G. Wells' science fiction classic the *War of the Worlds* and dreamed of space flight. By 1926 he had designed, built, and flown the world's first liquid fuel rocket. Launched 16 March 1926 from his aunt Effie's farm in Auburn, Massachusetts, the rocket, dubbed "Nell," rose to an altitude of 41 feet in a flight that lasted about 2 1/2 seconds. Pictured here, Goddard stands next to the 10-foot-tall rocket, holding the launch stand. To achieve a stable flight without the need for fins, the rocket's heavy motor is located at the top, fed by lines from liquid oxygen and gasoline fuel tanks at the bottom. During his career, Goddard was ridiculed by the press for suggesting that rockets could be flown to the Moon, but he kept up his experiments, supported in part by the Smithsonian Institution and championed by Charles Lindbergh. Widely recognized as a gifted experimenter and engineering genius, his rockets were many years ahead of their time. Goddard was awarded over 200 patents in rocket technology, most of them after his death in 1945. A liquid fuel rocket constructed on principles developed by Goddard landed humans on the Moon in 1969. (Courtesy NASA Marshall Space Flight Center)

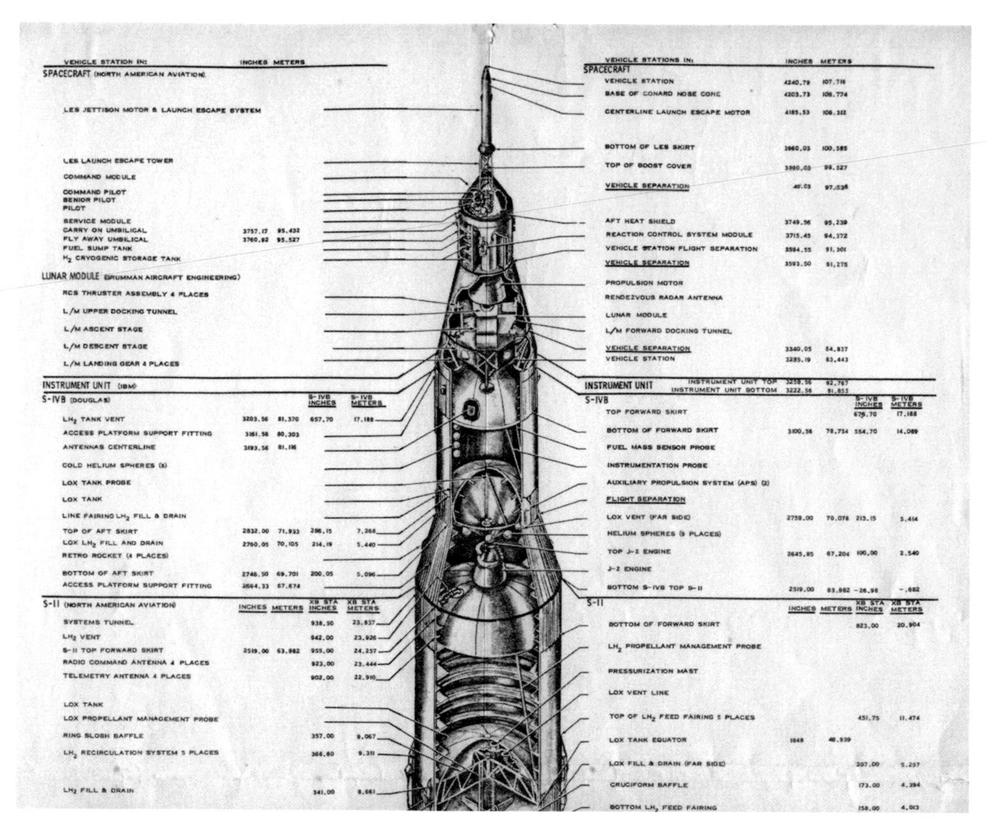

Fig. 9.4 The Saturn V – top half. The whole rocket was 363 ft tall. (Courtesy NASA)

standpoint. The Soviet Energia, which flew two test missions in the late 1980s before being canceled, had slightly more takeoff thrust.

The largest production model of the Saturn family of rockets, the Saturn V, was designed under the direction of Wernher von Braun at the Marshall Space Flight Center in Huntsville, Alabama, with Boeing, North American Aviation, Douglas Aircraft Company, and IBM as the lead contractors. The three stages of the Saturn V were developed by various NASA contractors, but following a sequence of mergers and takeovers all of them are now owned by Boeing.

9.2 Space nuclear power

While solar power is an important part of our energy mix on the Moon, we could not survive without nuclear power. With our built-up infrastructure, energy-demanding industries, and sizable population base, solar power cannot do the job alone. Nuclear fission reactors were brought to the Moon very soon after the first permanent settlement was founded in 2029. These reactors were physically small units and, even though none never failed, they were placed quite far from the settlements

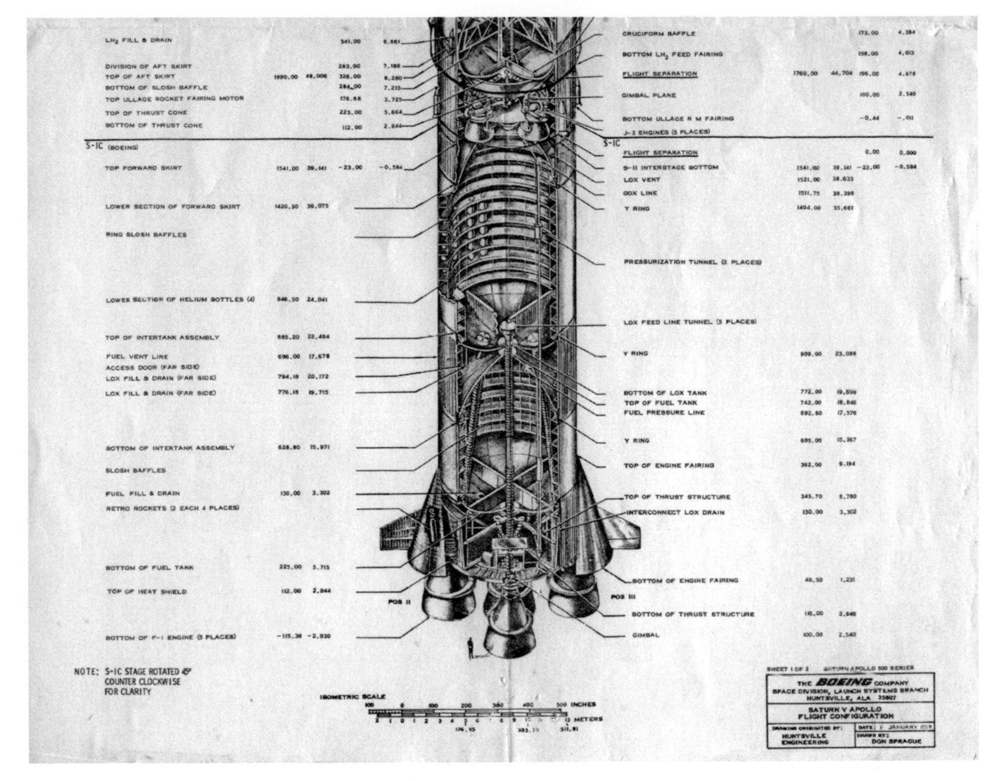

Fig. 9.5 The Saturn V – bottom half. (Courtesy NASA)

Research on fusion reactors took place on the Moon. We always believed that the locally abundant Helium-3 would eventually provide all the energy we could ever want. And that is what happened. The first fusion reactor came online in 2070.

It took a few more years before such reactors were used on Earth. People there wanted to make sure that all the engineering kinks had been removed before the fusion reactors were hooked up to the power grid on Earth. There were no problems of significance. Mars adopted reactors, fission and fusion, as soon as possible. Solar power is not as potent there as on the Moon. Within a year after the first permanent Martian colony in 2041, nuclear fission power was pumping power into the evolving infrastructure. And as soon as the first lunar fusion reactor came online, one was sent to Mars along with the Helium-3 needed to run it for a decade. That was when Helium-3 became our most profitable export.

Nuclear power in space began in 1965 with the launch of SNAP-10A,[5] shown in Figure 9.6. NASA's Project Prometheus, announced in 2003, renewed compre-

[5]The Systems Nuclear Auxiliary Power Program (SNAP) was a program of experimental radioisotope thermoelectric generators (RTGs) and space nuclear reactors flown during the 1960s by NASA. Odd-numbered SNAPs were RTG tests and even-numbered SNAPs were compact reactor system tests. One even-numbered unit, the SNAP-10A, had the distinction of being the only nuclear reactor launched into space by the United States.

Fig. 9.6 The world's first nuclear reactor power plant to operate in space, SNAP-10A, was launched into Earth orbit on 3 April 1965. (Courtesy U.S. Department of Energy)

hensive efforts to develop advanced technologies for space use, in particular, the development of electric propulsion technologies that would be combined with power generated by a space nuclear reactor. Ion thrusters would be used.

These technologies would allow missions with more sophisticated active/passive remote sensing, greater launch window flexibility, spacecraft maneuverability at the destination planet or moon, and greatly increased science data rates. The first proposed mission application, the Jupiter Icy Moons Orbiter mission, focused on the 200-kW Prometheus 1 spaceship, shown in Figure 9.8, utilizing electric propulsion.

Fig. 9.7 Several dozen nuclear powered spacecraft had been launched by the early 21st century. The multi-billion dollar Cassini mission to Saturn, launched on 15 October 1997, carried the largest amount of Plutonium ever sent into space (72 lbs) for this nuclear power plant. (Courtesy NASA)

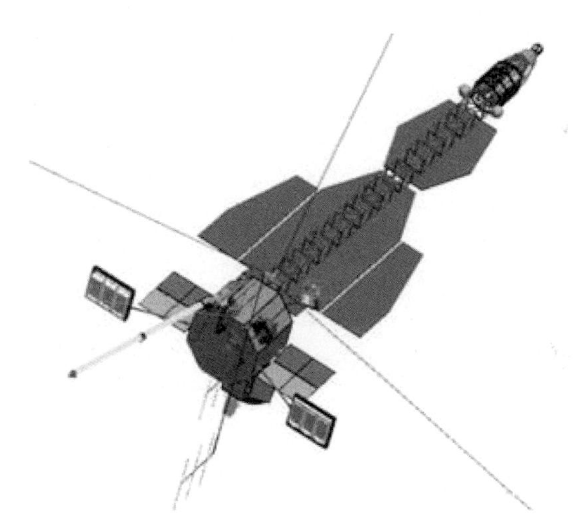

Fig. 9.8 Spacecraft concept proposed for the Jupiter Icy Moons Orbiter. (Courtesy NASA Glenn)

The proposed mission had two principle objectives:

1. Tour and characterize three icy moons of Jupiter: Callisto, Ganymede, and Europa.
2. Demonstrate nuclear electric propulsion flight system technologies that would enable a range of revolutionary planetary and Solar System missions.

NASA's primary research program for developing these technologies was Prometheus Nuclear Systems and Technology, named for the ancient Greek god of fire and craft. Prometheus 1 would launch using conventional chemical rockets. But once in Earth's orbit, a nuclear electric propulsion system would propel it through space. Nuclear electric propulsion was over 10 times more efficient than chemical rockets and produced 20 times more power than the generators used on space probes such as Voyager and Cassini or solar-powered systems like Deep Space 1.

Coupled with traditional rocket launchers, nuclear electric propulsion would allow spacecraft to travel farther and faster and to perform in-flight course changes and precise maneuvers. It would also carry heavier, power-hungry equipment, including precise cameras, sophisticated scientific instruments, high-speed computers and advanced communication systems – at least ten times as much payload science as other systems. As a result, scientists could closely study the surfaces of the outer planets and their moons as well as the surfaces and interiors of comets.

In a nuclear electric propulsion system, a nuclear reactor produces heat, a power-conversion system converts the heat to electricity, and an ion thruster uses the electricity to propel the spacecraft.

Fig. 9.9 Ion engines emit only a faint blue glow of electrically-charged atoms of xenon, the same gas found in photo flash tubes. (Courtesy NASA)

Deep Space 1, which launched in 1998, was the first spacecraft to use an ion thruster as its primary propulsion system. But that thruster, developed at the NASA Glenn Research Center, drew its electricity from solar panels. Solar power is efficient and lasting, but it cannot maneuver spacecraft at locations in the outer Solar System which are in orbits beyond that of Mars. The only way to study a planet is to land on it or travel in its orbit, and the only way we can do that in the outer Solar System is with nuclear propulsion.

Scientists and engineers from the NASA Glenn Research Center and the Jet Propulsion Laboratory had worked for more than two years to build a high-power ion thruster that would have helped NASA realize the Vision for Space Exploration.

Named *Herakles*, this ion thruster descended from the Deep Space 1 system, but it could have used electricity from a reactor to produce plasma from xenon gas. Ions from the plasma were to be extracted and then accelerated at extremely high speeds. In space, as the xenon ion beam rushes from the thruster, it would propel the spacecraft in the opposite direction.

Although Herakles' thrust was designed to be as gentle as the force of 10 U.S. quarters in your hand, the speed of the spacecraft would constantly increase as it travels. That means Prometheus 1 could eventually have reached speeds greater than 200,000 mph – 10 times faster than the speed of the Space Shuttles.

Herakles and other components of Prometheus 1 were originally scheduled for advanced flight development in 2006, but NASA concurred with a General Accounting Office (GAO) recommendation and deferred the Jupiter Icy Moons Orbiter mission based on concerns over cost and technical complexity. NASA shifted the focus of Project Prometheus from developing nuclear power and propulsion systems needed for deep space probes like the Jupiter Icy Moons Orbiter to the development of space-qualified nuclear systems, such as surface nuclear power, nuclear thermal, and nuclear electric propulsion systems needed to support the near term goals of NASA's Vision for Space Exploration.

In the present day, Prometheus-like Nuclear Systems and Technology help us explore the outer planets to search for signs of life and clues to the origin of the Solar System.

Also very exciting to us are the robotic nano-craft that we have designed for extra-Solar System travel at near-light speeds. With these spacecraft, within a matter of years we will begin to get glimpses of other planetary systems around stars that are relatively few light years away.

9.3 The space elevator

The space elevator, like many engineering marvels, was born in science fiction and grew up in science and engineering fact. The idea of the space elevator is attributed to Konstantin Tsiolkovsky (1857–1935). While visiting Paris in 1895, the remarkable Eiffel Tower made him think of a similar structure that could reach all the way into space. In Tsiolkovsky's vision, a "celestial castle" would be built at the end of a cable 35,790 kilometers long. This put the terminus of the structure in geostationary orbit.

He was a visionary and a pioneer of astronautics. He theorized many aspects of human space travel and rocket propulsion decades before others, and he played an important role in the development of the Soviet and Russian space programs. Tsiolkovsky was certain that the future of human life would be in outer space, so he decided that we must study the cosmos to pave the way for future generations.

Later, he proved mathematically the possibility of space flight, and wrote and published over 500 works about space travel and related subjects. These included the design and construction of space rockets, steerable rocket engines, multi-stage boosters, space stations, life in space, and more. His notebooks are filled with sketches of liquid-fueled rockets, detailed combustion chamber designs with steering vanes in the exhaust plume for directional control, double walled pressurized cabins to protect from meteorites, gyroscopes for attitude control, reclining seats to protect from high g loads at launch, air locks for exiting the spaceship into the vacuum of space, and other accurate predictions of space travel. Many of these were done before the first airplane flight.

He determined correctly that the escape velocity from the Earth into orbit is 8 km/s, and that this could be achieved by using a multi-stage rocket fueled by liquid oxygen and liquid hydrogen. He predicted the use of liquid oxygen and liquid hydrogen or liquid oxygen and kerosene for propulsion, spinning space stations for artificial gravity, mining asteroids for materials, space suits, the problems of eating, drinking, and sleeping in weightlessness, and even closed cycle biological systems to provide food and oxygen for space colonies.

Fig. 9.10 Konstantin Tsiolkovsky (1857–1935). (Courtesy New Mexico Museum of Space History)

In 1926 Tsiolkovsky defined his "Plan of Space Exploration" consisting of 16 steps for human expansion into space:

1. Creation of rocket airplanes with wings
2. Progressively increasing the speed and altitude of these airplanes
3. Production of rockets without wings
4. Ability to land on the surface of the sea
5. Reaching escape velocity and the first flight into Earth orbit
6. Lengthening rocket flight times in space
7. Experimental use of plants to make an artificial atmosphere in spaceships
8. Using pressurized space suits for activity outside of spaceships
9. Making orbiting greenhouses for plants
10. Constructing large orbital habitats around the Earth

11. Using solar radiation to grow food, to heat space quarters, and for transport throughout the Solar System
12. Colonization of the asteroid belt
13. Colonization of the entire Solar System and beyond
14. Achievement of individual and social perfection
15. Overcrowding of the Solar System and the colonization of the Milky Way Galaxy
16. The Sun begins to die and the people remaining in the Solar System's population go to other suns.

We have yet to make our way through three quarters of this list, even 250 years after it was created.

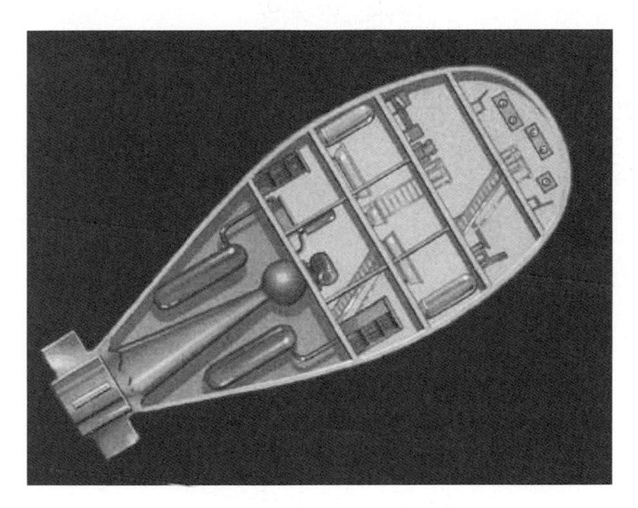

Fig. 9.11 A spaceship concept by Tsiolkovsky. (Courtesy David Darling)

We can summarize for the reader a few of the elements of space elevator design concepts and some of the difficult issues that had to be considered as part of a complete design.

There are a number of key elements in a space elevator. The most important is the ribbon, upon which the climber makes its way into space. The ribbon was the fundamental stumbling block to a reliable and practical elevator design. Until carbon nanotubes (CNT) were discovered in the very early 1990s, there were no materials that could be utilized to make a ribbon. Imagine this ribbon of fantastic length, stretching from the surface of the Earth to a point more than a third of the way to the Moon, a length of 144,000 km (89,477 miles). The average distance from the center of the Earth to the center of the Moon is 384,403 km (238,857 miles).

Carbon nanotubes are a type of material that are very much stronger than the strongest material known at that time. CNTs are less dense than other very strong materials such as steel and Kevlar while having a tensile strength of over 25 times stronger.

The essential idea was to deploy a cable/ribbon from geosynchronous orbit in two directions, "up" and "down." It would be deployed down to the Earth's surface

Fig. 9.12 A space elevator being propelled up the ribbon by an ocean platform-based laser. Present-day climbers are being propelled by very small fusion reactors. These also power the shielding devices on board each elevator. (Courtesy Space Elevator Visualization Group and Alan Chan) See Plate 9 in color section.

and up well beyond geosynchronous orbit (GEO). Both parts would be in tension, with the equilibrium position at GEO. If the upper part needed to be shorter, then a counterweight had to be placed at the end, keeping the structure in mass balance. The highest tension would be in the middle and the least tension at both ends. Due to this, the ribbon design needs to be tapered from the middle to the two ends as the stress in the cable decreases.

A climber moving to the outer end can be given enough energy to escape from Earth's gravity well and proceed to the Moon or another destination.

In addition to the ribbon, there are the climbers, and an anchoring structure on the Earth – although anchor is not the proper term since the space elevator is an orbital structure, orbiting over a fixed spot on the surface of the Earth. Key to the success of the space elevator is a powerful and reliable energy source to drive it along the ribbon.

There were many engineering aspects to the design of a space elevator. Two early discussions of the design of a space elevator were given by Jerome Pearson[6] (1975: pre-CNT) and by Bradley Edwards[7] (2000: post-CNT).

There was a significant list of design considerations that needed to be addressed before the early space elevator pioneers built their first prototype.[8] For a structure of this length, there are numerous environmental conditions that act upon it. Some of the most serious concerns exist within the Earth's atmosphere and low Earth orbit. In that region there are weather considerations, and at the orbital altitudes the possibility of debris. Lightning strikes on the tether could potentially destroy it due to the extreme heating. Atmospheric considerations placed limitations on where over the Earth the tether could be orbited.

Additionally, there was the possibility of micrometeorite impacts. The cable was susceptible to radiation damage in the Earth's radiation belts as well as atomic oxygen erosion. Due to the electromagnetic properties of the CNT ribbon, it is heated by magnetic field induced electrical currents.

Structural oscillations of the tether were a design concern. The tether oscillates along its axis as well as in a direction perpendicular to the axis. The frequencies and periods of such oscillations depended on the tether's geometric and material properties. These values directly affected the allowable speeds of the climber because of resonance concerns. Resonance in this instance is an undesirable vibration characteristic where large amounts of energy can be transmitted to the ribbon from the climber inducing very large and dangerous oscillations.

As stated by Pearson, "the 'orbital tower' could be built only by overcoming the three problems of buckling, strength and dynamic stability. The buckling problem could be solved by building the tower outward from the geostationary point so that it remains balanced in tension and stabilized by the gravity gradient until the lower end touches the Earth and the upper end reaches 144,000 km altitude. The

[6] J. Pearson, "The Orbital Tower: A Spacecraft Launcher Using the Earth's Rotational Energy," *Acta Astronautica*, Vol. 2, 1975, pp. 785–799.

[7] B.C. Edwards, "Design and Deployment of a Space Elevator," *Acta Astronautica*, Vol. 47, No. 10, 2000, pp. 735–744.

[8] *The Space Elevator: A Revolutionary Earth-to-Space Transportation System*, B.C. Edwards and E.A. Westling, B.C. Edwards Publishers, 2002.

Fig. 9.13 Space elevator being propelled further up the ribbon by an ocean platform-based laser. Here the elevator has passed through an orbital station. (Courtesy Space Elevator Visualization Group and Alan Chan) See Plate 10 in color section.

strength problem could be solved by tapering the cross-sectional area of the tower as an exponential function of the gravitational and inertial forces, from a maximum at the geostationary point to a minimum at the ends. The strength requirements are extremely demanding, but the required strength-to-weight ratio is theoretically available in perfect-crystal whiskers of graphite. The dynamic stability is investigated and the tower is found to be stable under the vertical forces of lunar tidal excitations and under the lateral forces due to payloads moving along the tower. By recovering the excess energy of returning spacecraft, the tower would be able to launch other spacecraft into geostationary orbit with no power required other than frictional and conversion losses. By extracting energy from the Earth's rotation, the orbital tower would be able to launch spacecraft without rockets from the geostationary orbit to reach all the planets or to escape the Solar System." Pearson called his structure an orbital tower, and referred to perfect-crystal whiskers of graphite because CNTs had not yet been discovered.

As an economic driver, the space elevator design teams projected a drop in cost to low Earth orbit of three orders of magnitude, that is, from about $100,000 down to $100 per pound when compared to the cost of rocket launches. Edwards[9] predicted the following possible markets would be developed and created:

- solar energy satellites (clean, limitless power from space)
- space-system test-bed (universities, aerospace)
- environmental assessment (pollution, global change)
- agricultural assessment (crop analysis, forestry)
- private communications systems (corporate)
- national systems (developing countries)
- medical therapy (aging, physical handicaps, chronic pain)
- entertainment/advertising (sponsorships, remote video adventures)
- space manufacturing (biomedical, crystal, electronics)
- asteroid detection (global security)
- basic research (biomedical, commercial, university programs)
- private tracking systems (Earth transportation inventory, surveillance)
- space debris removal (international environmental)
- exploratory mining claims (robotic extraction)
- tourism/communities (hotels, vacations, medical convalescence).

"We expect solar power satellites to be one of the major markets to develop when we become operational and have begun dialogues with British Petroleum Solar about launch requirements and interest. Solar power satellites consist of square miles of solar arrays that collect solar power and then beam the power back to Earth for terrestrial consumption. Megawatt systems will have masses of several thousand tons and will provide power at competitive rates to fossil fuels, without pollution, if launch costs get below $500/lb to GEO.

"Another market we expect to emerge is Solar System exploration and development. Initially this would be unmanned but a manned segment, based on the Mars Direct scenario, could emerge early after elevator operations begin. The exploration market would include:

[9] B. Edwards, *The Space Elevator NIAC Phase II Final Report,* 1 March 2003.

- exploratory and mining claims missions to asteroids, Mars, Moon and Venus
- science-based, university and private sponsored missions
- *in situ* resource production on Mars and Moon
- large mapping probes for Mars and the asteroids
- near-Earth object catastrophic impact studies from space."

These predictions were accurate. We have implemented all these systems and programs. And as mentioned earlier, we have expanded the scope of the elevators so that they can haul very large objects into space.

9.3.1 Possible problems

The space elevator, while a beautiful concept – one that we witness at many locations around Earth, the Moon and Mars – was a challenging one to the engineers of the 21st century. It was a long ribbon extending from the surface of the Earth to one-third of the way to the Moon! It made its way through Earth's atmosphere of wind, rain, hurricanes, lightning, wayward aircraft and terrorists. As it extended beyond the atmosphere it became vulnerable to Earth's van Allen radiation belts, recurring orbital debris and micrometeorites, and subjected to solar radiation pressure. The solar radiation pressure resulted in slow oscillations of very large amplitudes – on the order of 200 km – that could hamper ribbon control. Climbers rode up and down the ribbon causing some damage with each ride.

The Earth's natural van Allen radiation belts offered an early and serious hazard to space elevator designers. Since the elevator traveled slowly up the ribbon as compared to rocket speeds, people inside would be exposed to potentially fatal doses of radiation.[10] Additionally, the radiation effects on the equipment in the elevator was problematical without some shielding. Early possible solutions included additional passive shielding such as aluminum – with final weight becoming a burden on the ribbon – and active magnetic fields to negate the radiation belts – where providing the necessary power became the issue. These were much like the concepts proposed for the active shielding of lunar surface bases.

Ribbon damage due to debris and micrometeorite impact also challenged the engineers. In an early study,[11] a group of engineering students was interested in the reliability of a prototype ribbon. In particular, they were trying to understand how topology, or the configuration of the ribbon, can affect the survivability of the ribbon to multiple hits by micrometeorites. Figures 9.14–9.16 show the chosen ribbon configuration in three states: new condition, slight damage, and failure-in-progress. The prototype ribbon was placed in an Instron machine to test the yielding characteristics of the specimen.

In this experiment, the premise was that micrometeorites would sever single strands of the ribbon randomly. The ribbon is under tension, as the real elevator

[10] A.M. Jorgensen, S.E. Patamia, and B. Gassend, "Passive Radiation Shielding Considerations for the Proposed Space Elevator," *Acta Astronautica*, Vol. 60, 2007, pp. 198–209.

[11] M. Antal, A. Little, E. McIntyre, L. Schaper, B. Vogeding, and H. Benaroya, "The Space Elevator: A Case Study," Department of Mechanical and Aerospace Engineering, Rutgers University New Jersey, 2003.

ribbon is when deployed. As the strands are severed, the tension force is redistributed among the remaining strands, eventually leading to a failure when the remaining strands cannot resist the force. One of the conclusions from this study was that the ribbon is remarkably robust and can survive with many cut strands. This is a good thing because there would then be enough time to make repairs after the first one or two micrometeorite hits.

Fig. 9.14 Scale model of intact ribbon under tension.

Fig. 9.15 Ribbon with a few strands punctured and severed by simulated meteorite impacts.

Fig. 9.16 Severe ribbon damage with failure imminent.

What happens if a space elevator ribbon breaks? There was great concern about the damage the lower part of the ribbon could cause as it wraps itself around the Earth.[12] No matter where the elevator breaks, the top segment is permanently ejected from Earth orbit. The lower segment will fall back to Earth and as the Earth rotates the ribbon will wrap as far around the Earth as ribbon length allows – most of the ribbon survives reentry. "Once decelerated, the ribbon falls slowly to the Earth and imposes a significant but probably not destructive force to objects on the ground. ... [It is] an insignificant threat to satellites, but is a large risk to other space elevators."

There were also concerns that when a ribbon fiber snaps a large amount of energy is converted from the internal strain of the carbon nanotube into significant amounts of thermal and kinetic energy. Both of these energy forms can damage the surrounding ribbon fibers creating an obstacle to the elevator.

Fortunately, we have not experienced a catastrophic failure of an elevator – they are properly spaced. We have emergency rockets placed along our ribbons in case of a failure and are ready to shoot any wayward ribbon into space. Our elevators are also able to propel themselves to nearby orbital platforms under their own emergency power.

9.3.2 An historical interview with Jerome Pearson (May 2009)

Can you give us a one or two paragraph bio?

I was born in the Southern U.S. in 1938. I grew up in Kansas City, not far from Robert Heinlein when he was writing his early science fiction, like "The Man Who Sold the Moon." In the early 1950s, I got stars in my eyes about astronomy and space travel, inspired by Edgar Rice Burroughs and the 1952 Collier's Magazine

[12]B. Gassend, "Fate of a Broken Space Elevator," *Space Exploration 2005*, SESI Conference Series, Vol. 1, 2005, Eds. J.A. Smith and B.E. Laubscher.

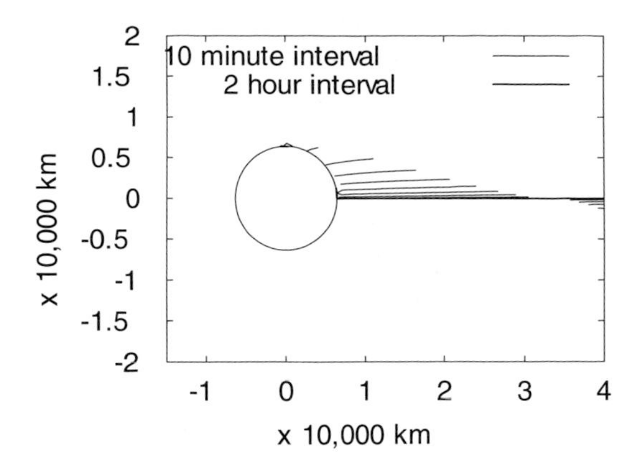

Fig. 9.17 Early stages of space elevator break. (Courtesy Blaise Gassend)

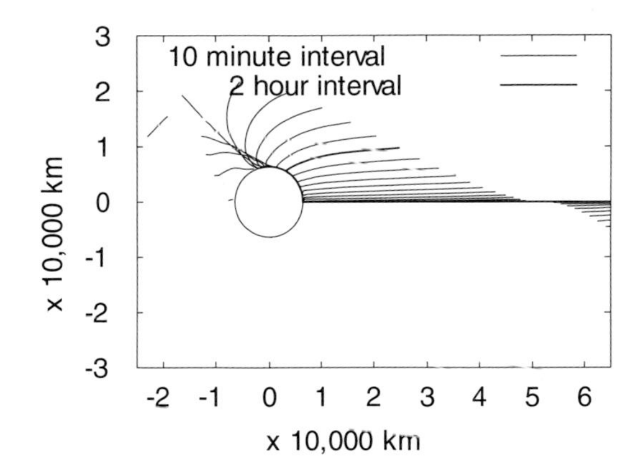

Fig. 9.18 Intermediate stages of space elevator break. (Courtesy Blaise Gassend)

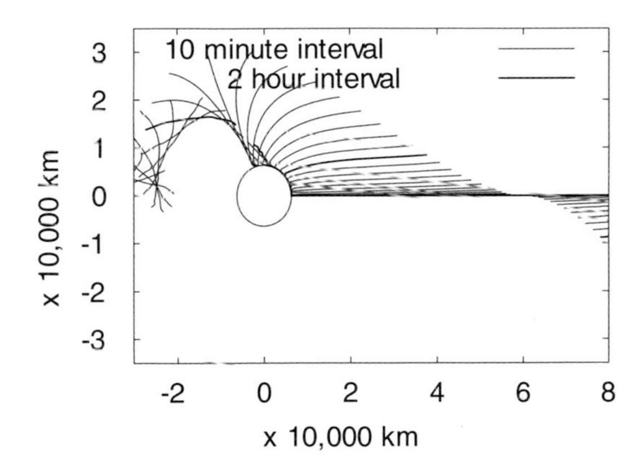

Fig. 9.19 Later stages of space elevator break. (Courtesy Blaise Gassend)

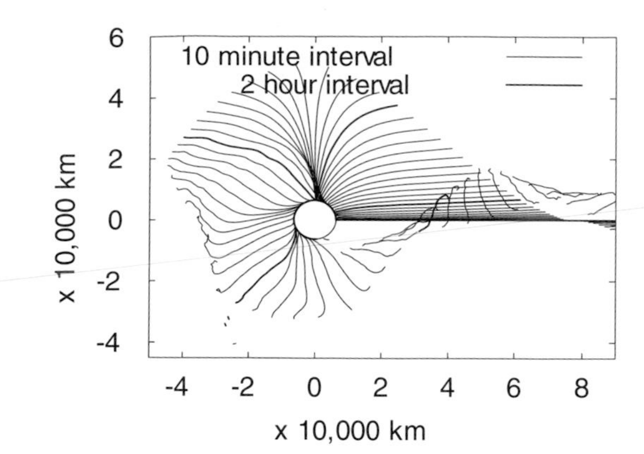

Fig. 9.20 Space elevator break and wrapping around the Earth. (Courtesy Blaise Gassend)

articles about the Mars Project by von Braun. When I read Isaac Asimov's Foundation series, I was hooked; I studied astronomy, read science fiction, and my first job out of engineering school was with NASA, working on the Apollo program. I even applied to be an astronaut, but they turned me down because of my glasses. But I have been very fortunate in my career, working for NASA during the glory days of Apollo, working for the Air Force during the heady days of Strategic Defense Initiative (SDI), and now working for my small business developing advanced aircraft and spacecraft. And I got to meet my boyhood science fiction heroes Bob Heinlein, Isaac Asimov, and especially Sir Arthur Clarke, whom I visited in Sri Lanka.

You were the first person to do a dynamic feasibility study of the space elevator cable. This culminated in your paper of 1975, which was long before the discovery of carbon nanotubes. What drew you to this engineering problem?

I invented the space elevator, and was the first to publish a technical article on a feasible space elevator system. My 1975 paper in *Acta Astronautica* gave the idea to the world spaceflight community, and led Arthur Clarke to contact me for information on his space elevator novel, *The Fountains of Paradise*, published in 1979; he gave me a nice credit in the Afterword. Actually, it was Arthur who led me to create the idea of the space elevator in the first place. In 1969 I saw his testimony to Congress about geosynchronous communication satellites, and he described them as "perched on imaginary towers 35,000 km above the equator." I thought: why not drop a cable down and make a real tower? That's why I called my invention the "Orbital Tower." Since I was an aerospace engineer specializing in dynamics, the vibration modes and such were the aspects that concerned me most.

Did you have to contend with skepticism?

Boy, did I ever! I finished the calculations and wrote my paper by 1970, but it took me five long years to get it published. I was turned down by the AIAA, and even by the British Interplanetary Society. Finally, I found the courageous editor Antoni Oppenheim at *Acta Astronautica*, the *Journal of the International Academy of Astronautics*. He is still a professor at UC Berkeley, and back then he had the self-confidence to publish the article. An editor faced with an outrageous idea can safely turn it down, and no one will ever know; but if he takes a chance and publishes something unsound, he ends up with egg on his face – it took real courage by Antoni to publish my paper. (At a 1978 conference where I presented my lunar space elevator for the first time, the AIAA editor who had turned down my original space elevator paper came up to me and apologized for not accepting it!)

Were there others who also developed some of the concepts for the space elevator?

Yes, the concept of the space elevator has a long history. Konstantin Tsiolkovsky did a famous thought experiment in his 1895 book *Dreams between Earth and Sky*, in which he imagined a tall tower on the Earth and looked at the apparent gravity on the tower. He found the point where gravity disappeared (which is the stationary altitude), and calculated this altitude for all known bodies of the Solar System (including the Sun; after all, it was just a thought experiment!), but concluded that a real tower could not be built on the Earth. The early Russian researcher Tsander conceived a lunar space elevator of iron in about 1930, but didn't publish it. Two groups published papers in the 1960s about elongating GEO satellites to support a payload at a lower altitude, but didn't extend the concept to the space elevator. John Isaacs published a space elevator concept in *Science* magazine in 1965 (with a caveat from a cautious editor!), but he lacked the nerve to do a massive space elevator, and proposed such a microscopically thin diamond wire that it would be severed by meteoroids in an instant, and was completely impractical. The only person other than me who invented a functional space elevator was Yuri Artsutanov, who did not publish, but wrote a popular article in *Pravda* in 1960. It was unknown outside the Soviet Union, and was not available to me when I invented the space elevator. Yuri and I met in St. Petersburg in 2006 (see the STAR, Inc. space elevator page at http://www.star-tech-inc.com/id4.html). Bradley Edwards took the Pearson/Artsutanov space elevator concept and developed practical means to build and operate it. There are also dynamic space elevators that depend on mass in motion to keep them aloft. These include the orbital rings by Paul Birch, the rotating tethers of Yuri Artsutanov and Hans Moravec, the launch loop of Keith Lofstrom, and the space fountain by Rod Hyde.

Where are we today (2009) with the space elevator? Is it an area of study that still draws you?

Yes, I still think about space elevators and how to make them. In my original paper, I surmised that the building material would need the strength-to-weight of what

I called "perfect-crystal whiskers of graphite," which is a good description of the carbon nanotubes discovered two decades later. (The enormous material strength required for the Earth space elevator led me to invent the lunar space elevator, because the Moon's mass is so much smaller.) But even assuming the availability of carbon nanotubes as the building material, there is another huge problem facing the space elevator – we will have to clean up space of all debris, and even all non-geostationary satellites, to keep it safe from collisions. Arthur Clarke described such a clean-up in his *2061: Odyssey Three*, and *3001: The Final Odyssey*. Actually, as a result of the recent loss of the Iridium satellite in a collision with a defunct Cosmos satellite, I expect to see a government program to remove space debris, and my company actually has a propellantless space vehicle that could do the job. There are annual space elevator conferences, and lots of work being done in Japan, and a logical application of their high-speed trains would be to short space elevators attached to Birch's orbital rings in LEO. I'd like to do a paper on that, if I had the time!

Based on what we know today, is there one key hurdle to building the space elevator?

Actually, there are still four: the material, which is still not available; the radiation effects of having the terminal in GEO, above the protection of the van Allen belts; the collision risk from meteoroids and space debris; and interference with and from other satellites. Since every Earth satellite other than the geostationary satellites eventually intersect the space elevator, the elevator cable must be wiggled to avoid them, or they must be controlled to miss it, or they must be removed from orbit. Also, after 15 years of research into carbon nanotubes, we still do not have a usable nanotube composite, and it's not clear when that might happen. In addition, passengers going through the radiation belts on the elevator would receive high radiation doses. One way to solve all these problems is to build the space elevator in LEO, based on the Paul Birch orbital ring concept. The orbital ring is a hollow ring around the Earth, at about the altitude of the Space Station. It is gravitationally neutrally stable, so a wire conductor is placed inside it and accelerated to move faster than orbital velocity; this produces an upward force that is balanced by the weight of short space elevators from the ground to the ring. Electrifying the ring for accelerating the conductor would also enable electrifying the short vertical space elevators, and allow the use of vertically oriented shinkansen-type high-speed trains to move up the cables, as envisioned by Shuichi Ono, chairman of the Japan Space Elevator Association. This concept would not be bothered by orbital debris or other satellites, it would have lower radiation levels, and the short space elevators could be built of existing composites, just like the lunar space elevator. The orbital ring could support several short space elevators, so there would be many spots on the Earth from which to climb an elevator into space.

What would it cost to put up a prototype of such a system? Can we do it today?

At current launch costs of $2,000 per kg, the cost would be enormous; so we have to find a way to reduce launch costs. Actually, the carbon nanotube composites could be used to build reusable single-stage-to-orbit vehicles that could take off and land at conventional airports, greatly reducing the cost. We could also use electromagnetic guns or gas guns to launch the materials into orbit. Once an initial strand is in place, we could use the space elevator cables to carry up the additional material more cheaply. We can build this space elevator today, by putting a minimum strand in place and bootstrapping the rest, so it could actually lower the launch costs and reduce its own construction costs. We might be able to do it for some tens of billions of dollars, less than the International Space Station, at $50 B. It needs to be done now, and we could actually do it now using the orbital ring approach.

How would the space elevator affect humanity becoming a spacefaring civilization? What role do you see for the space elevator?

I believe the one key thing preventing us from becoming a spacefaring civilization is the high cost of getting from the ground into orbit. The sooner we overcome this limitation, the sooner we can move into space, with private companies, explorers, adventurers, and settlers moving into space and on to the Moon and Mars. The space elevator could be the catalyst that lowers launch costs and frees us from the bonds of gravity. Even if we go with the low-altitude space elevator, the carbon nanotube research being done could lead to high-strength composites that would help in making spaceflight routine for many private citizens, who could visit orbiting hotels riding in composite nanotube SSTOs (single-stage to orbit rockets).

Can you provide us with a cost comparison between using rockets to go back to the Moon and using space elevators?

Rockets can launch payloads to the Moon at a cost of about $5,000 per kilogram at current prices. The return trip, to bring resources from the Moon to the Earth, costs even more, because we would have to either develop lunar propellants or carry all the return-trip propellants from the Earth's surface. Space elevators could reduce costs in both ways. The Earth space elevator could launch people, equipment, and supplies to the Moon at perhaps $100 per kilogram, and the lunar space elevator could launch lunar materials to Earth space elevator rendezvous at perhaps twice that price. So space elevators might be just 6% of current costs.

What do you see as the major hurdles for our return to the Moon for permanent manned settlements? Are they technical, financial, physiological, psychological?

I'm a technological optimist, so I think the technology problems can be solved; however, they may be solved in ways that are unexpected.

There are physiological problems involved in lunar habitation, and those are dust, gravity, and radiation. The lunar dust is quite unlike Earth dust, being very jagged, small, charged, and pernicious. Inhaling it can be as deadly as inhaling asbestos particles, and it sticks to your space suit every time you go out on the lunar surface. NASA is now thinking about having space suits permanently mounted on the outside of airlocks so that you can get into them without bringing lunar dust into your habitat. We may have to do other things like paving the entire lunar surface around all habitats and rocket landing pads, probably by just microwave sintering of the top surface, as proposed by Larry Taylor of the University of Tennessee at Knoxville. This would really cut down the prevalence of dust near habitats.

Long-term exposure to lunar gravity is still a big unknown. Since NASA has not done studies of lunar or Martian gravity on animals, we don't know the long-term effects, and we certainly don't know about the effects on animals or humans gestating and being born on the Moon under $1/6\ g$. This is something that needs to be studied and understood before we build habitats on the Moon. I have proposed to NASA that we build a rotating gravity laboratory in Earth orbit, with white rat experimental subjects in capsules at the ends experiencing long-term lunar and Martian gravity. This critical experiment is decades late.

Radiation is also a big problem. The exposure on humans on the surface is really scary, without any protection from an atmosphere or radiation belt. All of the lunar habitats will probably have to be built underground, with at least a meter of regolith above them.

This may result in psychological problems like claustrophobia. Perhaps outside views can be provided electronically to help alleviate this. Good communications with other habitats on the Moon and with people on Earth should also help; easy access to the Lunar Internet would be necessary also.

But I think the major hurdles now are cultural. NASA has become very safety-oriented, quite unlike the way it was during Apollo. This emphasis on safety has not yet improved safety, but it has slowed progress and raised costs. And NASA is not emphasizing cost reduction of getting into space. The current Ares-Orion system is based on many elements of the Space Shuttle, and will unfortunately not produce great reductions in launch costs. Even under Ares-Orion, most of the NASA budget will go into transportation to the ISS, as it does now for the Space Shuttle.

In the larger picture, the United States is no longer as excited about space as it was during Apollo; in response to this, the government may turn its attention to more immediate social needs because it's overwhelmed with welfare costs. These two aspects were summed up succinctly by the NASA Administrator James Fletcher at a conference in the late 1980s. He described the then-current NASA plan to return to the Moon, taking 12 years, and I pointed out that the original Apollo program took less than that, and build three generations of spacecraft in doing so. His response was, "Yes, but NASA was a different agency in the 1960s, and the United States was a different country." How true!

How do we answer critics who say space is too expensive and that there are numerous problems on Earth to take care of first?

Actually, progress has been the greatest when a civilization addresses the tough challenges. Building great cathedrals in the 12th and 13th centuries, sailing over the oceans in the 15th to the 17th centuries, and overcoming nuclear, air, and space challenges in the 20th century were accompanied by great advances in the general welfare. The U.S. had great poverty in the 1930s when we were spending very little on defense and training our soldiers with wooden rifles; but poverty greatly decreased in the 1980s despite the costly Reagan defense buildup.

When you envision the future, where do you see us in 50 years? In 100 years? How do you see this evolving from the present?

Technologically, we could be controlling the Earth's climate and developing Mars in 50 years, and could be a multi-planet species in 100 years. But culturally, I'm less confident. In the 1960s when I worked on Apollo, I fully expected us to be on Mars by the 1980s. But it has been 40 years since the first lunar landing, and now we cannot even get people beyond low Earth orbit. I still see possibilities of space development and expansion, even in spite of weakened governments, or governments so occupied with social services that they turn inward and ignore space development. But we may have to depend on the vitality of lunar colonists pioneering John Kennedy's "new frontier" to get us through this phase.

Can the private sector do this with minimal government participation?

Yes, but the one requirement we need is low-cost access to space. With low-cost launchers, such as SSTOs made of nanotube fibers that could take off and land at conventional airports, we might have a new era of space "barnstormers" like the aviation barnstormers of the early 20th century. Small companies and space entrepreneurs could even build their own spaceliners and space hotels, and provide transportation and services without government.

Do you think Americans will be the first to send Man back the Moon in the 21st century, or will it be the Chinese?

I certainly hope the U.S. is the first, but our hopes are sinking fast. After next year, when the Space Shuttle fleet is retired, the United States will not even have the capability to launch humans into orbit, and that will last until a new launch system comes online. The Ares/Orion system won't be ready until 2015 or 2016, so we will be dependent on Vladimir Putin and the Russian Soyuz vehicles to launch astronauts to the space station from 2010 to 2016, even if the new administration reaffirms the program. Since the President has not appointed a new NASA Administrator, and has asked for a summer-long study before he does, the outlook is not good.

In the longer view, Western civilization seems to be losing its vitality. Most of Europe has less than replacement birth rates, and all of the population growth is coming from non-assimilating Muslims who do not share our Western views and ideals. The U.S. seems destined to be overwhelmed by the costs of our retirement and medical care systems, and may also be overwhelmed by so many immigrants seeking work, but coming with low skills and high social needs. It is not clear to me that the fragile Western liberal culture will survive another century in Europe. The Chinese also have a long way to go to get humans to the Moon, and they still lack a rocket big enough to do it, like the Apollo's Saturn 5. But they are determined, and they may be able to do it by, say, 2020. If we lag, they could beat us. However, there are cracks in the Chinese Communist model as well, and they may undergo a collapse like the Soviet Union's in the next two decades. The current worldwide recession might be the catalyst for that collapse if it continues for a few years, as some are predicting.

But I'm still an optimist, and surely mankind will find a way through our current rough times to see a bright future as we move from Earth to Moon to Mars, and then onward to the stars.

* * *

9.3.3 Other space elevators

While the "traditional" concept involved using rocket propulsion or laser light pressure to propel loads up a cable anchored to Earth, an early study showed that a rotating space elevator could do away with engines or laser light pressure application completely. Instead, the unique double rotating motion of looped strings could provide a mechanism for objects to slide up the elevator cable into outer space.[13]

Theoretical physicist Leonardo Golubović and his graduate student Steven Knudsen at West Virginia University explained this possible concept. Golubović and Knudsen introduced the Rotating Space Elevator (RSE), a rotating system of a floppy string that forms an ellipse-like shape. Unlike the traditional Linear Space Elevator (LSE) made of a single straight cable at rest, the RSE rotates in a quasi-periodic state. As the scientists explained, RSE motion is nearly a geometrical superposition of two components: its geosynchronous rotation around Earth (which has a one-day period), and the internal rotation of the string system that goes on around the axis perpendicular to the Earth (about a 10-minute period). This internal rotation of the string is the fundamental mechanism allowing objects to freely slide along the string, and also providing the dynamical stability needed to maintain the elevator's shape.

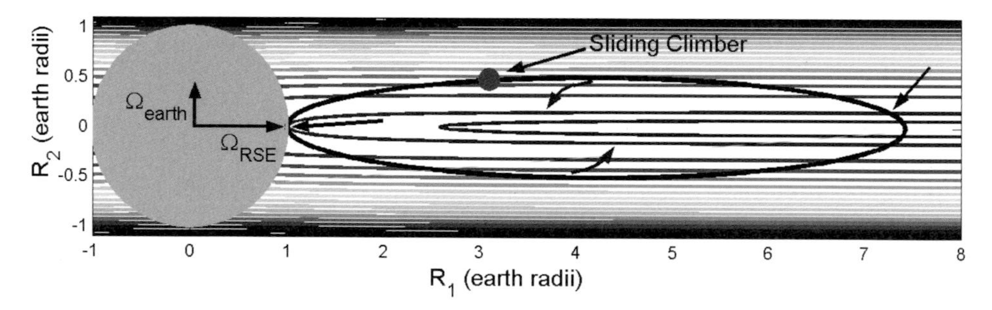

Fig. 9.21 Rotating space elevator shown as a gray circle on the rotating elliptical ribbon. (Courtesy Leonardo Golubović and Steven Knudsen)

[13] L. Golubović and S. Knudsen, "Classical and Statistical Mechanics of Celestial-scale Spinning Strings: Rotating Space Elevators," *European Physics Letters*, Vol. 86, 2009, pp. 34001.

Another elevator concept was proposed where a nonstationary space elevator (also known as a skyhook) is timed to rotate synchronously with a maglev-type rail system on the Moon. By matching velocities at the lower terminus of the elevator, a maglev train riding on the surface of the Moon on the same path as the elevator terminus can vertically launch a payload to be grappled by the elevator.[14] This was viewed as a more effective option than the stationary elevator, primarily due to the variations in the gravity field around the Moon.

A maglev is a magnetically-levitated vehicle. A maglev track is one where the train in motion is suspended above the track by a magnetic field and propelled along the track by the moving field. Some have suggested the idea of a maglifter. In this case, a spacecraft is magnetically levitated and accelerated along a maglev track that is sufficiently long so that the speed needed to release the craft into orbit is achieved. At that speed, the spacecraft is released and onboard rockets take the craft into orbit. On the Moon, this was viewed as a less expensive way to carry payloads into orbit. We found uses for both elevator and maglev on the Moon.

9.4 Quotes

- "The space elevator will be built about 50 years after everyone stops laughing." Arthur C. Clarke
- "This is no longer science fiction. We came out of the workshop saying, 'We may very well be able to do this.' " David Smitherman (2000) NASA/Marshall Advanced Projects Office
- "We will never be an advanced civilization as long as rain showers can delay the launching of a space rocket." George Carlin
- "What is it that makes a man willing to sit up on top of an enormous Roman candle, such as a Redstone, Atlas, Titan or Saturn rocket, and wait for someone to light the fuse?" Thomas Wolfe

[14]G.L. Matloff and P. Roseman, "Lunar Partial, Nonstationary Space Elevators and Maglevs: A New Lunar Launch Option," *Acta Astronautica*, Vol. 65, 2009, pp. 599–601.

10 Science and commerce

"Commerce is life. Without commerce, human civilization would still consist of isolated small groups of cave dwellers."

Yerah Timoshenko

Science and commerce go hand-in-hand. While space has been somewhat of an exception given the unmanned explorations of the Solar System going forward without obvious commercial benefits, there really were such benefits. Dual-use technologies have worked their magic in both directions. The engineering of the space probes, their cameras and instrumentation, the mini-chemistry labs that landed on numerous extraterrestrial bodies – all of these filtered back to the Earth economy. But in the same vein, developments in materials, microminiaturization of electronics, nanomechanisms, computational power – these technologies allowed for the development of the space probes, the airbags for landing, and the capabilities to explore the Solar System. Science and commerce are symbiotic – it is not science vs. commerce.

10.0.1 An historical interview with Alan Hale (Jan 2009)

Can you give us a one or two paragraph bio?

In a nutshell: raised in southern New Mexico; attended the U.S. Naval Academy and spent some time thereafter in the Navy; worked for $2\frac{1}{2}$ years at the Jet Propulsion Laboratory for the Deep Space Network; attended graduate school at New Mexico State University, earning a Ph.D. with a thesis on extrasolar planet detection; briefly worked at (what is now) the New Mexico Museum of Space History; formed what is now the Earthrise Institute; discovered a comet somewhere along the way; presently involved in scientific research/education and what I call "science diplomacy" (using science as a tool for building bridges between nations and cultures).

You are a discoverer of the Comet Hale-Bopp, discovered on 23 July 1995. It must be an amazing feeling to know that a comet bears your name. What other feelings did you experience as you realized that you had actually discovered a new astronomical body?

Fig. 10.1 Airbags inflate to allow for a soft landing over the hard rocks on Mars. The airbags used in the Mars Exploration Rover mission were the same type that Mars Pathfinder used in 1997. Airbags must be strong enough to cushion the spacecraft if it lands on rocks or rough terrain and allow it to bounce across Mars' surface at freeway speeds after landing. (Courtesy NASA)

It was an exciting feeling, certainly. I suspected something "interesting" almost immediately, and within an hour had verified that it was a comet and that there were no known comets in that location, so in pretty short order I knew I had discovered one.

In addition to excitement, I felt a pretty strong sense of irony; I had spent several hundred hours over many years trying to discover a comet and had been unsuccessful, and now I had one drop into my lap without even looking for one.

It should be remembered that Hale-Bopp was a rather dim object at the time of its discovery, and there was no way to know that it would later become the bright and dramatic object that it subsequently became.

How do you view the major efforts by many nations to reach for the stars?

Ultimately, I believe it is important that, as a species, we expand into space; if we don't, we eventually either stagnate or go extinct (either by asteroid impact, or self-destruction, or something). It will take us a long time to accomplish and I'm sure there will be quite a few detours along the way – but I think the overall goal must stand.

With President-Elect Obama taking office in less than two weeks, how do you see President Bush's vision for the Moon evolving?

I'm not sure I can give a reasonably intelligent answer to this; this will probably have to be a "wait and see" endeavor. I'm aware that the transition team has

Fig. 10.2 "I'm sending along the 'standard' H-B photo that I normally use; I took this from the deck of my house on the day of the comet's perihelion (1 April 1997)." Alan Hale in response to a request for a photo of Comet Hale-Bopp. (Courtesy Alan Hale)

been looking at the program – a process I support – and may recommend some changes in focus or approach; I do hope that the overall objectives of the program are retained.

Is settling the Moon so important for civilization? How do we answer critics who say space is too expensive and that there are numerous problems on Earth to take care of first?

I will concede that those who make these types of arguments have a point, and it would be wise for us space enthusiasts to realize that. For a starving family in some third-world nation it would probably be pretty difficult to get excited about settling the Moon or exploring space in general.

Perhaps if all the money and resources that are spent on developing new and creative ways of blowing each other up were instead spent on eradicating hunger and disease, and on education and raising the worldwide standard of living, it would be easier to make a convincing argument that settling the Moon and exploring space are vital to our future.

While arguments about "the need to explore," etc., are relatively convincing to space enthusiasts, I believe that ultimately we're going to have to make economic arguments as a primary rationale for going into space. There are plenty of natural resources on the Moon and in near-Earth asteroids, for example, that can be utilized for developing a vibrant global civilization here on Earth (and also aid in preserving Earth's natural environment) – and this can ultimately produce a positive feedback loop and stimulate additional expansion into space.

Fig. 10.3 Image of comet C/1995 O1 (Hale-Bopp), taken on 4 April 1997 with a 225 mm f/2.0 Schmidt Camera (focal length 450 mm) on Kodak Panther 400 color slide film with an exposure time of 10 minutes. At full resolution, the stars in the image appear slightly elongated, as the camera tracked the comet during the exposure. E. Kolmhofer, H. Raab; Johannes-Kepler-Observatory, Linz, Austria (http://www.sternwarte.at), Creative Commons Attribution, ShareAlike 3.0.

As a scientist how do you view the debate of manned versus robotic exploration of the Solar System?

There is room for both. We are a very long way from being able to send humans to outer Solar System destinations like Jupiter, Saturn, Uranus, Neptune, Pluto – all of which either have been visited, or will be visited, by robotic missions. Our knowledge of these places has been and is being enormously increased as a result of such probes.

On the other hand, there are numerous investigations that only analysis by trained human beings can accomplish; for example, it will probably not be until humans have visited Mars that we'll be able to make a definitive statement as to whether or not life has ever existed there.

It is, of course, much more expensive and logistically complex to send human beings to these places than it is to send robotic probes; but if we are ever to expand into space, we have to go there.

You have founded the Earthrise Institute. From the Institute's website there is the following mission statement: "Simply stated, the mission of the Earthrise Institute is to use astronomy, space, and other related endeavors as a tool for breaking down international and intercultural barriers and for bringing humanity together." Can you tell us a bit more about your hopes and dreams for the Institute, and how it is being received?

I should probably state that Earthrise is presently in a state of flux, as we are presently reevaluating some of our activities and goals, and there will probably be some change in focus over the coming months. That being said, the overall mission should probably remain the same.

That mission was inspired in part by two delegations of Americans that I led to Iran (for astronomy-related purposes) in 1999 and 2000, and the fact that we were able to utilize astronomy for building bridges between the American and Iranian peoples. Changes in the political leadership of both countries have stalled these efforts during recent years, but I'm cautiously optimistic that upcoming changes in political leadership will help in reviving these efforts.

Fig. 10.4 The explorer in the foreground, wearing a constant-volume, hard space suit with rotating joints, is a representative of a commercial enterprise that intends to develop and exploit extraterrestrial resources. A lunar oxygen production plant, set between the two large solar panels, is generating a supply of rocket fuel that will be used for journeys to Mars. The lunar base can be seen in the distance. (Artwork by Mark Dowman and Mike Stovall of Eagle Engineering, Incorporated. (S89-25055) Courtesy NASA)

As an astronomer, what do you see as a benefit of having a presence on the Moon?

The Moon has no atmosphere or clouds to get in the way, and offers a stable platform for astronomical instruments; plus, objects can be observed for two weeks at a time. There are some drawbacks, including the "clingy" nature of Moon dust, the fact that we lose objects for two weeks at a time, and that there are significant temperature extremes. But by being creative we should be able to get around these drawbacks.

I'm not a radio astronomer, but a radio telescope located on the Moon's far side would be shielded from all the radio noise emanating from Earth and would accordingly be much more sensitive than anything we could build here.

When you envision the future, where do you see us in 50 years? In 100 years? How do you see this evolving from the present?

A lot depends upon how successful we are at eliminating poverty and disease, and in whether or not we're able to find a way to avoid blowing each other up and instead resolve disputes peacefully. If we can accomplish these things, then I'm reasonably optimistic that perhaps we can achieve great things over the coming decades and centuries, including expanding out into space. If we can't accomplish these things, then it all might become pretty moot.

If we do survive these challenges, then I see us becoming more and more globalized, and political boundaries becoming more and more a thing of the past in any practical sense. We are already communicating globally via the Internet and other recent technological advances, and our children are growing up in a world where this is all normal. As they reach adulthood and have children I see this process only accelerating, and perhaps within a couple of generations people will be communicating with their counterparts all over the world via implanted devices – almost a form of telepathy.

I'd also like to think we'll develop more robust ways of global transportation, and also make major steps towards eliminating the many diseases that afflict us. And, certainly, I'd like to see us expanding into space – in part to obtain the resources that will drive all this activity.

There are, again, many challenges that we'll have to overcome, in addition to those I mentioned earlier. These include controlling our population growth (especially as disease is eliminated and people live longer) and arresting and reversing environmental degradation.

Given the recent financial meltdown worldwide, does the return to the Moon become more tenuous?

In the short term, very possibly. If the world economy can recover at some point in the not-too-distant future, then perhaps any current slowdown in space efforts may not be much more than a temporary glitch. We'll have to see, though.

Do you think Americans will be the first to send Man back the Moon in the 21st century, or will it be the Chinese?

No way to know at this point. It depends upon who has the most will, and upon who has the resources to make it happen.

* * *

Fig. 10.5 An artist's concept of an Independent Lunar Surface Sortie (ILSS), one of the potential future options for space activity studied at the time this image was created. The picture appeared in a September 1977 publication from the NASA-JSC Program Planning Office entitled "A Compendium of Future Space Activities." The ILSS configuration consisted of a crew and equipment module; a lunar transport vehicle (LTV) for landing and takeoff, and two experiment/exploration payloads. The LTV would be used for descent from lunar orbit, landing and ascent back to lunar orbit. Its single-stage booster uses LOX and LH2 propellants. During an ILSS the LTV would be abandoned after the crew and equipment module is joined with the orbit transfer vehicle for return to Earth. The term "independent" signifies that each mission is self-supporting, as were the Apollo lunar missions. (S76-24320, September 1976. Courtesy NASA)

The debate of robotic vs. manned exploration was hot throughout the pioneering days of space. Apollo of course had no such debate because the Soviets had moved into space with men and the United States could do no less. Probes, however, were part of our space program from the beginning. They were born with our rocket programs.

There were the solar probes. Pioneer 5 measured magnetic field phenomena, solar flare particles, and ionization in the interplanetary region between March and April 1960. The Advanced Composition Explorer (ACE) satellite began making solar wind observations in August 1997.

There were the Mercury probes Mariner 10 and Messenger. The Venus probes Mariners 2 and 5 and the Soviet Venera series, of which Venera 7 landed. And

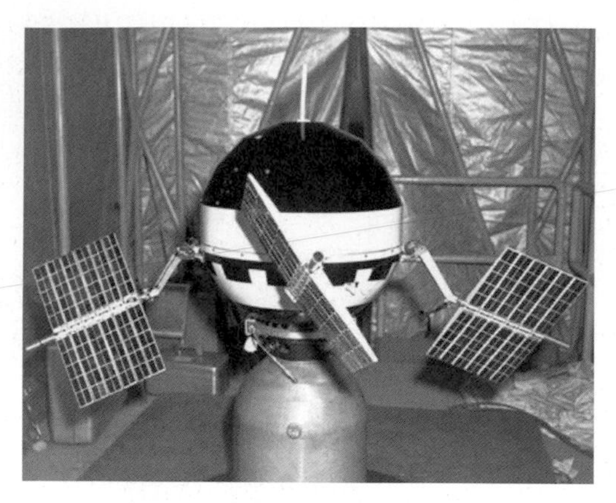

Fig. 10.6 Pioneer 5, launched 11 March 1960. This space probe was used to investigate interplanetary space between the orbits of Earth and Venus. The probe was 0.66 m in diameter and weighed 43 kg. (Courtesy NASA)

Fig. 10.7 Advanced Composition Explorer. This probe was launched on 25 August 1997 with enough propellant to operate through 2024. Its purpose was to study low-energy particles of solar origin and high-energy galactic particles. It was 1.6 m across not including the solar arrays and 1 m high. It weighed 785 kg. (Courtesy NASA)

numerous lunar probes: Luna, Pioneer, Surveyor, Explorer and Zond. Some flew by the Moon, others orbited, and others landed. They were in preparation for the manned missions.

There were Mars probes, those that went to Phobos, the asteroids, Jupiter, Saturn, Titan, Uranus, Neptune, Pluto, the comets, and some that left the Solar System. The Dawn probe, launched in September 2007, visited the asteroid Vesta and the dwarf planet Ceres in the asteroid belt.

Fig. 10.8 Dawn Probe. It was launched on 27 September 2007 with the purpose of investigating protoplanets Ceres and Vesta between Mars and Jupiter. It arrived at Vesta during August 2011, and left for Ceres during May 2012 and arrived at Ceres in February 2015, orbiting it until July 2015. (Courtesy NASA)

In the later part of the 21st and early 22nd centuries, probes of many nations embarked on trips throughout the Solar System. Most landed on their targets and began to beam scientific data back to Earth. Today we have a full spectrum of data on each significant body in the Solar System. Environmentally, we know what to expect on all of the planets, their moons, the dwarf planets, and the major and minor asteroids. We have enough information to send robotic missions to any of these and set up autonomous scientific and observation stations with very high reliability.

We have recently sent nanoprobes[1] – traveling at almost light speed – to HR 8799, shown in Figure 10.9. It was the first extra-Solar System with planets that we discovered in the early 21st century. HR 8799 is a young (approximately 60 million year old) main sequence star located 129 light years (39 parsecs) away from Earth in the constellation of Pegasus, with roughly 1.5 times the Sun's mass and 4.9 times its luminosity. It is part of a system that also contains a debris disk and at least three massive planets which, along with *Fomalhaut b*, were the first

[1]D.H. Wilson, "Near-lightspeed Nano Spacecraft Might be Close," *msnbc.com*, 8 July 2009.

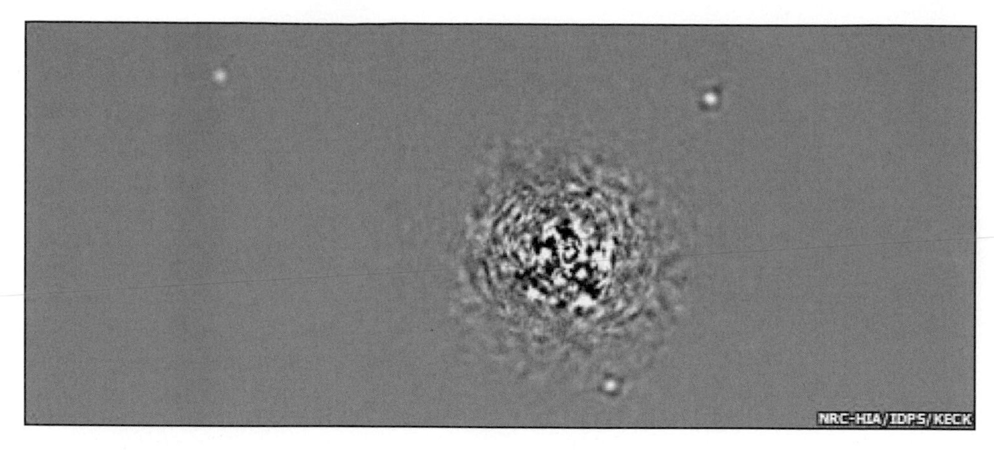

Fig. 10.9 HR 8799 (center blob) with infrared images of planets HR 8799d (bottom), HR 8799c (upper right), and HR 8799b (upper left). (Courtesy C. Marois/National Research Council of Canada)

extra-Solar planets with orbital motions that were confirmed via direct imaging. The designation HR 8799 is the star's identifier in the Bright Star Catalogue.

There is no doubt that the probes have been extremely successful on all counts. We continue – in 2169 – to rely on them. Today they are not passive – they all have self-replicating capabilities and have been planted on some of the more important bodies in the Solar System. The most recent of these probes have undergone 10 generations' self-replication in about four years' time.

But as good as the unmanned missions are, those with people are a thousand times more scientifically (and economically) profitable. This has always been true, not because we cannot develop massive scientific missions, but rather that we cannot garner the political support for massive scientific missions.

The most massive such missions of the past have been the NASA Great Observatories: the Hubble Space Telescope, the Compton Gamma Ray Observatory, the Chandra X-ray Observatory, and the Spitzer Space Telescope. These unmanned observatories brought us to a new level of understanding of the universe around us.

As miniaturized electronic and mechanical systems evolved, especially with nanotechnology continually in revolution, it made less sense to launch massive systems into orbit when a much smaller system would be just as effective at a thousandth the weight and ten-thousandth the cost. Micro- and nano-satellite technology evolved rapidly in the 21st century and the most pressing issue, after studying the extraordinarily large datasets that they sent back to their masters, became tracking them and making sure that they did not interfere with each other.

For historic interest, we have started sending robotic recovery missions to bring back the very old non-operational probes. We can learn from their state of repair and perhaps some never-transmitted data can be accessed. In some instances we have sent robotic repair missions to probes that are too far away to bring back – these are fixed up and programmed for new missions. These robotic repair missions are essentially perpetual missions since they can refuel and take on stock using any body in space.

The debate between manned and unmanned/robotic missions faded away naturally because of these technological innovations. The impetus then focused on manned missions. There was a clear case for human space exploration, and the political will materialized in the early to mid-21st century in response to serious efforts by a number of nations for manned returns to the Moon: by the U.S. with Canadian participation, the Chinese, the Europeans, an Indian–Israeli collaboration, the Russians, and a Japanese–Korean effort. Other nations had initiatives as well that fed into these major programs. Competition has always been and continues to be the source for invention.

Science has been and is a major beneficiary of having people in space. Thousands of references to the primary research literature verifies the scale of scientific bounty that resulted from Apollo, the scientific legacy of which can be summarized as follows:

1. 382 kg of lunar rock and soil samples including 24 drill cores, some to a depth of 3 m. "The analysis, and especially the dating, of this material has had a major impact on our understanding, not only of lunar history, but of the origin and evolution of the Solar System as a whole.
2. "Information on the lunar interior obtained by the Apollo seismology experiments. It is noteworthy that the Moon is still the only planetary body other than the Earth whose interior has been probed in this way.
3. "Information gleaned from *in situ* measurements of the gravity, heat flow, charged particle environment, and 'atmospheric' composition at some or all of the six landing sites."[2]

While robots could gather specimens, as did the Soviet Luna probes, the quantity thus collected comprised less than 0.1% of the quantity returned by Apollo, which were "intelligently collected." The political will and economic costs of travel in space are of such magnitude that science has not and will not be the justification. It was not for Apollo. It has not been for our developing cities on the Moon or Mars or our manned missions to the outer planets and moons.

Similarly, during the construction of the International Space Station in the period between the last decade of the 20th century and the first decade of the 21st century, many lamented at the costs of construction and the lack of use of the final product. Only a few at the time realized that the process was more important than the product – even though the product was worth its weight in gold.

Of greatest value to us today as we evolve our space-based human civilization is the ISS experience of space construction. "The ISS [was] by far the largest and most complex structure yet assembled in space. Once developed for the ISS, this experience [has been available] for the construction of ... large space facilities, some of which (e.g., astronomical instruments and lunar or planetary outposts)

[2]I.A. Crawford, "The Scientific Case for Human Space Exploration," *Space Policy*, Vol. 17, 2001, pp. 155–159.

are themselves [of] large scientific impact."[2] In fact many of our interstellar probes were constructed in space, in space docks orbiting the Earth and the Moon. These probes could be optimized in a more effective way than were the early Earth probes that had to be launched into space through Earth's gravitational field.

The European Space Agency's X-ray telescope (XEUS) was regularly refurbished on the ISS, as were many scientific instruments. The construction of the ISS necessitated the development of new concepts in the management of complex projects – as did the NASA Apollo program – that have benefited us to this day, in particular the international agreements. A quote from that era is quite succinct: "If experience building and operating the ISS helps lay the institutional foundations for a future world space programme, that alone will be one of its most important legacies."[2]

What we understand today about the Moon and Mars has been primarily the result of visits to those bodies by teams of specialists that followed robust initial visits by robots.

The life sciences have advanced as a result of the manned exploration and settlement of our Solar System. The Moon provides a stable and ideal laboratory for the life sciences. "Among the disciplines of interest are exobiology, radiation biology, ecology and human physiology. ... The Moon provides a unique laboratory for radiation biology studies, with built-in radiation sources, covering the whole solar ultraviolet, visible and infrared spectrum as well as solar and galactic protons, α particles and heavier ions, the so-called HZE particles ... [which] are of special importance for radiation biology, because their spectrum is unique and, as a whole, cannot be simulated on Earth."[3] The settlement of the Moon required "the maintenance and propagation of life at an altered g-level and with increased radiation with limited shielding ... a fundamental scientific research project." Additional research areas include bioregenerative life support systems in the lunar, Martian and microgravity environments.

Before we better understood how living organisms adapted to extra-terrestrial extreme environments, many questions were raised. "One point of concern [was] human stay at $1/6\ g$, which [will] trigger a string of adaptational processes, in the field of neurophysiology, the cardiovascular system, oxygen metabolism, calcium turn-over, and in the blood forming system." Humans evolved over millions of years and the functioning of the human body in all of its processes was in full Earth gravity. There was little first-hand knowledge of how the human body would operate at other gravitational intensities. There was only limited knowledge of human physiology in the microgravity of Earth orbit.

One serious concern was that bodily processes could not adjust in a partial way to a change in gravity. That is, if humans are living in a $1/6\ g$ field, their biological processes will not operate at 1/6th effectiveness. It was conjectured that even a slight reduction in gravitational intensities would lead to large variations in the ability of the body to perform its living functions. These *gravity thresholds* needed to be identified.

---

[3]G. Horneck, "Life Sciences on the Moon," *Advances in Space Research*, Vol. 18, No. 11, 1996, pp. 95–101.

Fig. 10.10 H.-H. (Jack) Schmitt exploring the North Massif. "Along with the discovery of the orange volcanic glass, this boulder at the base of the North Massif constitutes one of the most important discoveries of the Apollo 17 mission. It provided a sample of intrusive melt breccia that gave lunar science a probable date for the Serenitatis impact event. That event was one of the last four large basin-forming events on the Moon and helps to define the end of the ancient Hadean, pre-biotic period on Earth." Jack Schmitt, personal communication. (Courtesy NASA)

Curing illness in space or on the Moon relies on the effectiveness of pharmaceuticals in gravities less than 1 g. Knowledge of pharmacokinetics in low gravity environments was limited at the beginning of the 21st century. "Studies under weightlessness have shown increased blood concentrations of some pharmaceuticals."[3]

Planetary geology and geophysics found the lunar surface to be a perfect laboratory to investigate the origin and evolution of the Solar System, a laboratory available nowhere else. Each generation of scientists working on the Moon, with their deeper knowledge and more sophisticated instruments, led to new insights on our origins. Except for lunar planetary studies, large sections of the lunar surface have been placed in protective isolation, for both study by future generations of scientists as well as for preservation.

These unique regions can never be replicated. The Moon contains the oldest rocks in the Solar System. Being able to observe and scientifically test them in their natural environment has been tremendous for our understanding. It has been possible to precisely determine the absolute age of different regions of the Moon.

Knowing the probability distributions of lunar craters and their sizes, and knowing that the rate and directions of impacts were statistically similar throughout the Solar System, allowed us to use the lunar data and dates as benchmarks for the estimation of the ages of planetary surfaces elsewhere.

10.0.2 An historical interview with David Livingston (March 2009)

Can you give us a one or two paragraph bio?

I am the founder and host of The Space Show®, the nation's only talk radio show focusing on increasing space commerce, developing space tourism, and facilitating our move to a space-faring economy and culture. In addition, I am an adjunct professor at the University of North Dakota Graduate School of Space Studies. I earned my B.A. from the University of Arizona, my M.B.A. in International Business Management from Golden Gate University in San Francisco, and my Doctorate in Business Administration also at Golden Gate University. My doctoral dissertation was titled "Outer Space Commerce: Its History and Prospects." I am a business consultant, financial advisor, and strategic planner.

It looks as though the U.S. will be without heavy launch capability beginning in 2010 when the Shuttle is retired. It seems almost unbelievable that we have gotten into this situation. What do you think will happen?

Now if the government does not live up to its promises to keep us all safe and happy, I assume we will go without human spaceflight for 5–7 years and depend on the Russians for a ride. If Arianne 5 gets human rated, we might be able to get rides with it as well but the progress on human rating that rocket seems slow if not at all.

Why is settling the Moon so important for civilization?

I am not sure that it is. I am not so sure I agree with this statement.

What do you see as the major hurdles for our return to the Moon for permanent manned settlements? Are they technical, financial, physiological, psychological?

There is no commercial market for anything returning to the Moon that is real and not decades if longer off into the future. There is no cost effective, affordable transportation in the cis-Lunar environment. Thus, all of what we hear is costly, has no commercial value and is of questionable sustainability because only a government program can afford what is at least currently the plan in the vision for space exploration (VSE). Until a business case can be made that is real regarding the cis-Lunar environment, I remain skeptical and would consider this to be a major hurdle facing this type of a plan. Couple this with our leadership and economic crisis and I suspect that commercial sustainability is a long way into our future.

Fig. 10.11 Lunar surface garage. (Drawing by Pat Rawlings. Courtesy NASA)

Is it possible to summarize your space radio show guests' vision of our path to lunar settlement?

Over half of the guests do not think it's worth returning to the Moon. They claim that the real action is a human to Mars mission and that is where the resources should go as it will inspire, motivate, etc. Among those that want to go to the Moon, some cite getting there before others, most all cite it will be a good training and experience ground for deeper space missions, and most of all they say it's only a few days from Earth so it's safer until we learn how to do things in space better than we currently know how to do it. But let me stress, over half the guests do not care about returning to the Moon for any reason.

In their views, what are the chief stumbling blocks?

No cost effective transportation. No leadership. No national will or support for returning and no real mission for doing so.

We appear to have a number of very motivated competitors for the return to the Moon: China, Japan, Korea, Russia, and the European Space Agency, not to mention dozens of national space programs that

are quite competent at placing objects in orbit. Do you think the U.S. and Americans in general take this seriously?

Some do, some don't. For me, I will take it seriously when I see these nations start putting a significant part of their national wealth and budget into space and Moon programs. Until I see the money trail going to their space and lunar programs, I think its largely rhetoric. I have no doubt that technically most, if not all, can go to the Moon. The question remains when will they decide to pay for such efforts and with how much.

When you envision the future, where do you see us in 50 years? In 100 years? How do you see this evolving from the present?

I suspect that we will be more spacefaring in 50 years and at the end of this century but for those of us alive today, were we alive in a hundred years we would think progress was not fast enough. But I suspect there will be so many new concepts and opportunities that do not exist today and are not even in our consciousness today, that to try to predict the future is probably not that easy. How many distractions and side tracks will we take? What new direction will a new opportunity or discovery open up for us? What we want today may not be what we want 50 years from now. I would just say that we will evolve in many ways, some very rapidly, some so new and different we won't recognize the new reality or the status quo. I think the thing to do is remain flexible, remain open to change and to be willing to accept change no matter what it looks like or where it comes from.

You were one of the first to discuss space ethics, and you wrote a code of ethics for spacefarers. Can you summarize your thoughts?

To ward off opposition to commercial space development when it becomes something that is real and not just rhetoric for the most part, it would behoove the commercial space industry to operate under codes of ethics, similar to how businesses operate here on Earth. Small businesses don't use them but large businesses do. Perhaps an effective trade or biz group will emerge that will operate with a code of ethics for all of its members on the order of the American Bar Association or the National Association of Realtors. There are probably a minimum of 8–12 code statements that could be enacted that would go far in avoiding unnecessary regulation, objection, and fear from the public and the government.

Given the recent financial meltdown worldwide, does the return to the Moon become more tenuous?

Yes. Especially if one cannot identify a commercial reason for going and how the private sector can create wealth from going. If it's going to be a government project, how is wealth creation for the nation going to be created and flow through the economic system as a multiplier to each dollar the government invests? This type of cost benefit analysis, be it for public money or private money, is now more important than ever.

With President Obama in office, how do you see President Bush's vision evolving?

As President Obama is now in power, I see the Bush vision in trouble. It was not an effective vision, it was not funded, and the dots for doing it were never connected within the government let alone with the American people. If a form of the VSE is to survive in the Obama administration, it needs some dramatic changes, it needs a realistic cost benefit assessment for both public and private funding, and bloated programs need to be abandoned or modified. The American people have to understand the value for them in spending this money to go to the Moon and beyond. I hope the Obama Administration does this type of evaluation for a new and revised statement of this important vision.

* * *

Solar power from space and from the lunar surface has been promoted as the reason to be in space and on the Moon. Limitless energy has been promised. All we needed to do was to place large solar collectors in space and on the surface of the Moon, and energy would be plentiful and almost free. Such access to energy by all was deemed to be a catalyst for major political and economical changes on Earth – and it was!

Between the time we had a permanent colony on the Moon in 2029, and the time we put the first nuclear fusion reactor online using Helium-3 as fuel in 2070, solar power was a significant fraction of our sources of energy. Nuclear fission reaction was the bigger fraction.

But during those 41 years, not only did the Sun power many of our systems, it also became a major export to the Earth. While we could have become 100% solar energy reliant, we had a critical need to export in order that we could support ourselves economically. Those with a longer view decided that as our nascent infrastructure matured we would increasingly rely on nuclear fission power while simultaneously building a solar power grid that included panels on both sides of the Moon – we needed to always have a source of sunlight.

Every year we beamed energy to Earth and funded a great deal of imports to the Moon. Even with the onset of Helium-3 fusion energy, we continued to export energy to certain regions of the Earth that were not ready to invest in fusion. Today, all of our energy needs on the Moon – as well as on Mars – are met by that "golden dust," Helium-3.

10.1 Space solar power

We have discussed possible commercial activities on the Moon throughout this volume. Tourism ranks high. Energy from Helium-3 deposits is a tremendous resource. A solar power initiative called *Luna Ring* was proposed in the early 21st century, this one by the Japanese construction company Shimizu.

Fig. 10.12 The "Luna Ring" concept to generate solar power from the lunar surface for transmission to the Earth. (Courtesy Shimizu Corporation)

The concept was to generate solar power from a ring of solar panels on the lunar surface.[4] Electric power generated by the ring of solar cells around the lunar equator would be converted into microwave power and laser power and would be transmitted and beamed to the Earth from the near side of the Moon. That the ring circles the Moon means that there will always be a part of the ring that receives solar energy.

On Earth there will be microwave power receiving antennas – rectennas[5] that convert microwave power to DC electricity. There will also be laser power receiving facilities. The transmitted energy can be converted into hydrogen for use as a fuel. Shimizu estimated that 98% of the 20 GHz microwaves from the Moon would be transmitted through Earth's atmosphere to its receivers on a clear day. Similarly, 98% of lunar laser power of amplitude 1 μm would reach energy conversion facilities on the Earth's surface on a clear day. Transmitted laser power would be concentrated by a Fresnel lens and numerous mirrors to generate power via photoelectric conversion elements that release electrons.

The lunar ring was planned to be constructed to a great extent using lunar *in situ* resources. "Water can be produced by reducing lunar soil with hydrogen that is imported from the Earth. Cementing material can also be extracted from lunar resources. These materials will be mixed with lunar soil and gravel to make concrete. Bricks, glass fibers and other structural materials can also be produced by solar-heat treatments."

The Shimizu timeline was to begin construction of the power ring in 2035, with research and demonstration projects during 2010–2035.

Oxygen, metals, silicon and glass are raw materials that can form the foundation for long-term habitation on the Moon, providing materials for structures and solar

[4] "Clean Energy Innovation – The Luna Ring – Lunar Solar Power Generation," Shimizu Corporation Brochure, Copyright 2009.

[5] A rectenna is a rectifying antenna, a special type of antenna that is used to directly convert microwave energy into DC electricity. Semiconductors and inverters convert microwave power into electrical power at high efficiency.

Fig. 10.13 A closeup of the "Luna Ring" concept for lunar solar power generation. (Courtesy Shimizu Corporation) See Plate 11 in color section.

arrays.[6] Processes had been developed to derive propellants from *in situ* resources.[7] And the South Pole had been suggested as a prime location for a base that would have access to significant *in situ* resources.[8]

Figure 10.14 shows a pilot plant sized to produce 2 metric tons of liquid oxygen per month. In this early concept, the plant would react hydrogen gas with ilmenite to produce water and residual solids. The water would be subsequently electrolytically separated into hydrogen and oxygen. The hydrogen would then be recycled to react with more ilmenite while the oxygen is liquefied and stored. The feed stock for the depicted conceptual design would be basalt rock mined from the bottom of a nearby crater. Oversized and undersized rocks are seen to be rejected at the pit site.

The transported rock would be loaded into a hopper and conveyed into a three-stage crushing and grinding circuit that reduces the rock size. Fine particles are then separated by screens and discarded to a return conveyor. Ilmenite in the sized solid stream would be concentrated prior to the reactor unit in a multi-roll high intensity magnetic separator. The reactor was conceived to be a three-stage fluidized bed reactor. Reactor auxiliaries included low and high pressure feed hoppers, gas-solid

[6]G.A. Landis, "Materials Refining on the Moon," *Acta Astronautica*, Vol. 60, 2007, pp. 906–915.

[7]S.W. Siegfried and J. Santa, "Use of Propellant From the Moon in Human Exploration and Development of Space," IAA Paper IAA-99-IAA.13.2.02, 50th International Astronautical Congress, 4-8 October 1999, Amsterdam.

[8]M.B. Duke, "Lunar Polar Regolith Mining and Materials Production," IAA-99-IAA.13.02.05, 50th International Astronautical Congress, 4–8 October 1999, Amsterdam.

Fig. 10.14 Eagle Engineering artist concept of a NASA-sponsored project called the Lunar Base Systems Study. The NASA study revealed that oxygen propellant derived from lunar raw materials could play a key role in reducing the amount of mass launched into low Earth orbit to support a lunar base program and thus cut costs. A program to manufacture lunar oxygen would involve a series of development stages on Earth and on the Moon to demonstrate process feasibility and to generate engineering data to support optimum one-sixth g design of a full-scale production plant. (This painting was done for NASA by Eagle Engineering artist Mark Dowman. The concept's principal investigator was Eric Christianson. S88-33648, April 1988. Courtesy NASA)

cyclone separators, a solid residual hopper, a gas electric heater, blowers and a hydrogen makeup system. A high-temperature, solid ceramic electrolysis cell was conceived to split reaction water into oxygen and hydrogen. The oxygen would then be liquefied and stored. After product certification, possible uses for the pilot plant oxygen included LOX reactant for fuel cells and to supplement oxygen requirements for life support systems. This production scheme is based on the carbotex process. Telerobotic application is extensive in this conceptual design to allow long-term operation of the plant without on-site human involvement.[9]

Space solar power satellites have been discussed since the dawn of the space age, and have gained in importance as the Earth's energy needs soared during the 21st century. Their feasibility also soared as the component technologies became more efficient, light, and small. Many of the satellites were placed into orbit from manufacturing sites on the Moon using maglev technologies.

Solar power, the mining of Helium-3, the development of fusion reactors, the launch infrastructure, the education and training facilities, the maintenance fa-

[9]This description appeared with the image of Figure 10.14. (Courtesy NASA)

Fig. 10.15 Phobos propellant processing facility in the shadow of Mars. (Courtesy Phillip Richter, Fluor Daniel and Rockwell International) See Plate 12 in color section.

cilities, the creation of a supporting infrastructure – food, housing, medical care, entertainment, waste disposal – all created significant challenges and business opportunities for entrepreneurs as well as the multi-world corporations. Economic growth was exponential.

10.2 Implications for investors and entrepreneurs

The bottom line for investors and entrepreneurs is that they must be as creative as they have always been. Never has the creativity of the investor community been called to a more difficult and challenging task, and never to a greater calling. This was true in the 21st century, and it is true today as we approach 150 years of permanent habitation off-Earth in 2179. Entrepreneurial activity is proceeding today at a breakneck pace. Vast resources are being pumped into off-Earth locations.

The last example of such vision and courage was that of the great explorers of the 13th through 16th centuries and their benefactors. They were in it for the long haul – it took hundreds of years to map and settle the New World. The uncertainties the great explorers faced make ours appear miniscule. They went to places they did not know existed, on seas that some people of the day believed led to waterfalls at the edge of a flat Earth – with no communications, no weather predictions, no health care, hardly enough food and in ships barely the size of a lunar coal carrier, with an unpredictable wind as the only propulsion.

Yet, look at the result of that boldness.

Fig. 10.16 An artist's rendering gives a possible preview of 21st-century lunar base activity. A lunar surface crane removes a newly-arrived habitation module from an expendable lunar lander. The crane operator would place the module on the flat trailer for hauling back to the main base. Other expendable landers (background) tell-tale earlier shipments as a buildup plan gradually progresses. (The artwork was done by Patrick Rawlings of Eagle Engineering and was unveiled during an 29–31 October 1984, conference in Washington, D.C. titled Lunar Bases and Space Activities of the 21st Century. It was originally done for a NASA report titled "Impact of Lunar and Planetary Missions on Space Station." S84-43855, October 1984. Courtesy NASA)

10.3 Quotes

- "Innovation is the specific instrument of entrepreneurship. The act that endows resources with a new capacity to create wealth." Peter F. Drucker
- "Innovation is not the product of logical thought, although the result is tied to logical structure." Albert Einstein
- "Once again, this nation has said there are no dreams too large, no innovation unimaginable and no frontiers beyond our reach." John S. Herrington
- "Perfect freedom is as necessary to the health and vigor of commerce as it is to the health and vigor of citizenship." Patrick Henry

11 Additional visions

> *"Earth is where humanity incubated and Space is where we meet our full potential."*
>
> Yerah Timoshenko

While it was the Americans and the Soviets who opened up space to mankind, and the Americans who landed men on the Moon, all peoples and all nations have yearned for the hopeful future for which space is iconic. Had it been known in the first decade of the 21st century what awaited us in space, on the Moon and on Mars, then the meager financing of NASA would have been multiplied by a factor of 10. Not enough people outside the space community realized the bounty of space. Many did not want to know and had no curiosity.

Some people realized how important it was for humankind to extend its activities beyond Earth during Sputnik and Mercury. Some people – and even some politicians – recognized the bounty that awaited those who invested in the science and technology that could support man outside the Earth's atmosphere on the surface of its Moon.

Ironically, this was common knowledge in the 1960s. Every child in elementary school understood that a vibrant and active society was based on an energized population of engineers and scientists – people who were drawn to challenging professions and who honed their talents on projects that had a deep meaning for them.

And yet, in the 1970s, success was met with disinterest by the people. The day's problems demanded all the attention of the leaders and the followers. Tomorrow did not matter as much. And so mankind – as far as space was concerned – was in a staying pattern with close-to-Earth challenges. While important, these were only a small part of what should have been on our list of things to do in space.

Fortunately, with time, we reinvigorated our interest, and chopped away at the vines and brush that covered the path that we should have never left.

11.1 Russian concepts

Only after the dissolution of the U.S.S.R. did Russia create a civilian organization for space activities. Formed in February 1992, the Russian Space Agency acted as a central focus for the country's space policy and programs. Although it began

Fig. 11.1 Just a few kilometers from the Apollo 17 Taurus Littrow landing site, a lunar mining facility harvests oxygen from the resource-rich volcanic soil of the eastern Mare Serenitatis. Here a marketing executive describes the high iron, aluminum, magnesium, and titanium content in the processed tailings, which could be used as raw material for a lunar metals production plant. (This image was produced for NASA by Pat Rawlings, SAIC. Technical concepts for NASA's Exploration Office, Johnson Space Center, S99-04195, 1995. Courtesy NASA)

as a small organization that dealt with international contacts and the setting of space policies, it quickly took on increasing responsibility for the management of nonmilitary space activities and, as an added charge, aviation efforts. It was later renamed the Russian Aviation and Space Agency and then the Russian Federal Space Agency, commonly known as "Roskosmos", RKA, or RSA.

Russia has been a space powerhouse since the mid-20th century, with many scientific and engineering achievements – many matching and some surpassing those of the United States. Plans for the 21st century also included a return to the Moon, a return to Mars, new launchers, and numerous satellites and science missions.

Russia was a major player during the past almost 200 years of major space initiatives and has outposts as well on the Moon, Mars, and the key planetary bodies. The Russians are a part of all the entrepreneurial teams that have ventured forth in all directions from Earth.

The following historical interview recalls the Russian view of the Apollo era, and includes images of Russian concepts for advanced lunar cities.

11.1.1 An historical interview with Vladislav Shevchenko (June 2008)

Can you give us a one or two paragraph bio?

I have been Head of the Department of Lunar and Planetary Research, Sternberg State Astronomical Institute, Moscow University, Moscow, Russia since 1978. Previously, from 1964 I was at the Sternberg Astronomical Institute after graduating from the Astronomy and Geodesy faculty. A candidate of Physics and Mathematics in 1969 and Doctor of Physics and Mathematics in 1982, I am a Professor at Moscow University, a planetary scientist currently investigating potential locations and designs for lunar and Martian bases. I participated in the creation of lunar and planetary maps and globes, and in astrophysical telescopic and space research of the Moon (the Zond and Lunokhod Soviet programmes, ESA mission SMART-1). I was involved in projects of the Soviet lunar manned base (during the 1970s to 1980s). Now I participate in the NASA project LRO-LEND[1] and in Russian projects PHOBOS-GRUNT[2] and LUNA-GLOB.[3] I have authored 215 papers and several books: *The Modern Selenography* (1980), *Observation of the Moon* (1982), *Lunar Base – Project of the 21st Century* (1989), *Lunar Base* (1991), and coauthored *Far Side Atlas of the Moon* (1967, 1975), *Optical and Thermal Parameters of the Moon* (2001), and *Model of Space* (2007).

What was the mood in the Soviet Union when Gagarin went into space?

Yuri Alekseyevich Gagarin was a Soviet cosmonaut. On 12 April 1961, he became the first person in space and the first to orbit the Earth. Of course, the mood of the Soviet people was enthusiastic. I remember all of the staff members of our organization (I was a student in that time) went to Red Square near the Kremlin in Moscow to celebrate with copies of Gagarin's photo that were cut from the newspaper. Figure 11.2 is the front page of the newspaper *Evening Moscow* from 12 April 1961.

In Figure 11.3 you can see a photo of such celebrations. The next picture (Figure 11.4) is very interesting. It's a photo from the newspaper *Evening Moscow* of 15 April 1961. This photo shows a meeting in the Kremlin. From left to right you can see Nikita Khrushchev, head of the Soviet government, Valentina Gagarina (Gagarin's wife), Yuri Gagarin, Nina Khrushcheva (Khrushchev's wife), Anastas Mikoian (one of the Soviet leaders in that time) and Sergei Pavlovich Korolev (!!!). It's very surprising. I believe it was the first and last example of the publication in the press of Korolev's photo. During all of his life he was a secret person. The quality of the photo is not good, but it's a copy from a real issue of the old newspaper from my collection.

What were the thoughts and feelings after the Kennedy Moon speech?

To my regret I don't remember. It seems to me that the Kennedy Moon speech was not published in the Soviet Union at that time. In any case, I learned about it later along with the Apollo flights.

In addition to Korolev, the Soviet counterpart to von Braun, who were some of the key people of the Soviet lunar effort? In what ways were these people special? What were their special talents?

[1] Lunar Reconnaissance Orbiter – Lunar Exploration Neutron Detector.

[2] Phobos-Grunt (soil) was an unmanned lander sent in 2011 by the Russians to study Phobos and then return a soil sample to Earth.

[3] Luna-Glob (sphere), an unmanned mission to the Moon sent by Russia in 2012 that includes an orbiter with ground penetrating sensors.

СОВЕТСКИЙ ЧЕЛОВЕК—ПОКОРИТЕЛЬ КОСМОСА!

ВЕЧЕРНЯЯ МОСКВА

СООБЩЕНИЯ ТАСС

СООБЩЕНИЕ ТАСС

Fig. 11.2 Photo of Gagarin on front pages of the *Evening Moscow* from 12 April 1961. (Courtesy Vladislav Shevchenko)

Fig. 11.3 Muscovites celebrating Gagarin's flight on 12 April 1961. (Courtesy Vladislav Shevchenko)

Fig. 11.4 Kremlin celebration of 4 April 1961. Photo from the *Evening Moscow* of 15 April 1961 shows a meeting in the Kremlin. From left to right you can see Nikita Khrushchev, head of the Soviet government, Valentina Gagarina (Gagarin's wife), Yuri Gagarin, Nina Khrushcheva (Khrushchev's wife), Anastas Mikoian (one of the Soviet leaders in that time) and Sergei Pavlovich Korolev. (Courtesy Vladislav Shevchenko)

Sergei Pavlovich Korolev (1907–1966) is widely regarded as the founder of the Soviet space program. The key people were the Council of the Chief Designers. In a photo from 1957 at the Baykonur Space Center (Figure 11.5) you can see the main members of the Council.

Of these Chief Designers, I knew Sergei Korolev and Valentin Glushko personally. According to my personal opinion, I believe that Valentin Pertrovich Glushko was the most talented specialist and space scientist. The engines designed by him were used to launch numerous Earth and Moon satellites and also to get aloft automated probes designed to head for the Moon, Venus, Mars, and send up Sputnik, Vostok, Voskhod, Soyuz, Proton, etc. Remember, on May 1961, President J.F. Kennedy said these words of the Soviet-built rocket engines: "We have come to witness that initial space achievements of the Soviet Union have been secured through the availability of high-power rocket engines, which has placed the USSR in the lead." In 1974, Glushko was appointed General Designer of the Energia Association. He was the leader of the Soviet lunar manned base project (1970s to 1980s). During that time he was the head of the group that created the Energia-Buran system.

Numerous previous test flights of the classic Soviet rocket R-7 (Sputnik, Vostok, etc.) were failures. All the test flights of the N-1 (lunar) super rocket were failures too. The very first test flight of Energia LV designed by Glushko was successful! The first test flight of the Energia-Buran space transportation system was successful too! The RD-107 and RD-108 rocket propulsion systems, created over four decades ago, continue to be functional in support of Russian cosmonautics, and they truly

Fig. 11.5 From left to right: Nikolay Alekseevich Pilugin (1908–1982) – Chief Designer of the control systems for rocket and spacecraft complexes; Sergei Pavlovich Korolev (1907–1966) – Chief Designer of the rockets and spacecrafts; Valentin Pertrovich Glushko (1908–1989) – Chief Designer of the rocket engines; Vladimir Pavlovich Barmin (1909–1993) – Chief Designer of the launch complexes (including the Baykonur space center). (Courtesy Vladislav Shevchenko)

can be dubbed "eternal" engines. The RD-108 is now stated to power the upgraded American Atlas LV.

The illustration from Gabon (Figure 11.6) is of Valentin Glushko, Sergei Korolev and Russian rockets created by different Russian designers; but all these rockets have Glushko's motors.

Is settling the Moon so important for civilization? How do we answer critics who say space is too expensive and that there are numerous problems on Earth to take care of first?

A number of ecological investigations and some results of paleoclimatology have found that the permissible level of energy production inside the Earth's environment is about 0.1% of solar energy received by the Earth's surface. This value is about 90 TW. On the other hand, the general prognosis shows that the total energy use (and production, accordingly) in the world will be about 16 TW soon after 2010. This value will increase by a factor of two (about 34 TW) by the year

Fig. 11.6 Gabon stamp depicting Valentin Glushko (top), Sergei Korolev (bottom) and Russian rockets created by different Russian designers. (Courtesy Vladislav Shevchenko)

2050. If this tendency continues, the total energy production in the world will approach 98 TW by the year 2100 (maybe by 2150). It means that the permissible level of energy production within the Earth's environment will be exceeded and the destruction of the Earth's environment will not be reversible.

But it is obvious that the processes destroying the Earth's environment on a global scale will begin before that time – after the middle of the century. It may be that we are now observing some of the signs of these processes as the global change of the Earth's climate and the unusual natural catastrophes in different regions of the Earth. Hence, our efforts to rescue the Earth's environment must see practical results no later than between 2020–2030 in order for the environment to survive.

The only way to resolve this problem consists in the use of extraterrestrial resources. The nearest available body – a source of space resources – is the Moon. The best known space energy resource is lunar Helium-3. Very likely, the lunar environment contains new resource possibilities unknown now. So, lunar research space programs must have priority not only in fundamental planetary science, but in practical purposes too. Now it's needed to consider the new lunar research space programs for practical purposes to rescue mankind in the 21st century.

Of course, space programs (and lunar manned settlements) are too expensive. However, this cost is not larger than the value of mankind!

What do you see as the major hurdles for our return to the Moon for permanent manned settlements?

I believe the hurdle is one: mankind does not understand that the current situation in Earth's environment is very tragic. About 10 years ago I tried to explain these problems to members of a commission of our parliament (State Duma) – in that time I was a member of an expert group on space sciences. The reaction to my speech was: "You try to frighten us to give money for your science." That was 10 years ago! Now we know that global changes of the Earth's climate and the unusual natural catastrophes in different regions of the Earth are realities. But I am not glad about my rightness! Many people think now as we did 10 years ago.

Fig. 11.7 Valery Vladimirovich Polyakov spent 438 continuous days aboard Mir, from 8 January 1994 to 22 March 1995. (Courtesy Vladislav Shevchenko)

Are human physiological and psychological factors being taken as seriously as the engineering factors in the return to the Moon? Given the numerous hours spent by Russians in space, are you confident that we know enough about human physiology in low gravity and under space radiation to assure astronaut safety on the Moon for extended periods of time?

It seems to me that the activity of astronauts and cosmonauts in space during many months provides as a basis for extended stays for man on the Moon. Valery Vladimirovich Polyakov (born in 1942), a medical doctor (Figure 11.7), spent 438 continuous days aboard the Mir orbital station. Additional modules of the station that housed scientific equipment and expanded the living space were attached to Mir in subsequent years.

When you envision human settlement of the Solar System, where do you see us in 50 years? In 100 years?

The following chart shows my estimates for the proposed time-line for lunar industrialization:

Stage	Years
Deadline for mankind to colonize the Moon	2100–2150
Space power and lunar resources utilization	2040–2050
Initiation of space industrialization	2020–2030
Preliminary elaboration and appraisal	2010
Principle decision	yesterday

The table gives the proposed stages for the creation of a space industrial system during the century from the future critical date ("deadline for mankind") to the present. If destruction of the Earth's environment is to be reversed by the end of the century, the first results of the practical actions taken to rescue the environment must be observed no later than during the 2020 to 2030 time frame. It means that general decisions to return to the Moon must be approved now – at the beginning of the century.

* * *

Fig. 11.8 Manned Lunar Base 2050: Energia-Sternberg Project, residential zone in crater, general view. (Courtesy Vladislav Shevchenko) See Plate 13 in color section.

By the present-day – in 2169 – the majority of our facilities are under the lunar surface. It was known very early on that once we had the infrastructure in place we would be excavating and erecting our facilities under regolith – especially the habitats. If you were to be taking a Sunday drive on the lunar surface around our cities, you would only see occasional towers and antennae protruding the surface, perhaps a few low-lying buried structures.

The Russian concepts from the early 21st century presented on these pages are representative of such structural expanses. We are not laid out in exactly the same way, and we were nowhere near such complexes in 2050, as depicted in the images.

A base in a crater is shown in plan view in Figure 11.9 and in side view in Figure 11.10. There are inhabited modules (1), clean room facilities (3), rehabilitation centers (5), medical facilities (6), a control center (7), and laboratories (10).

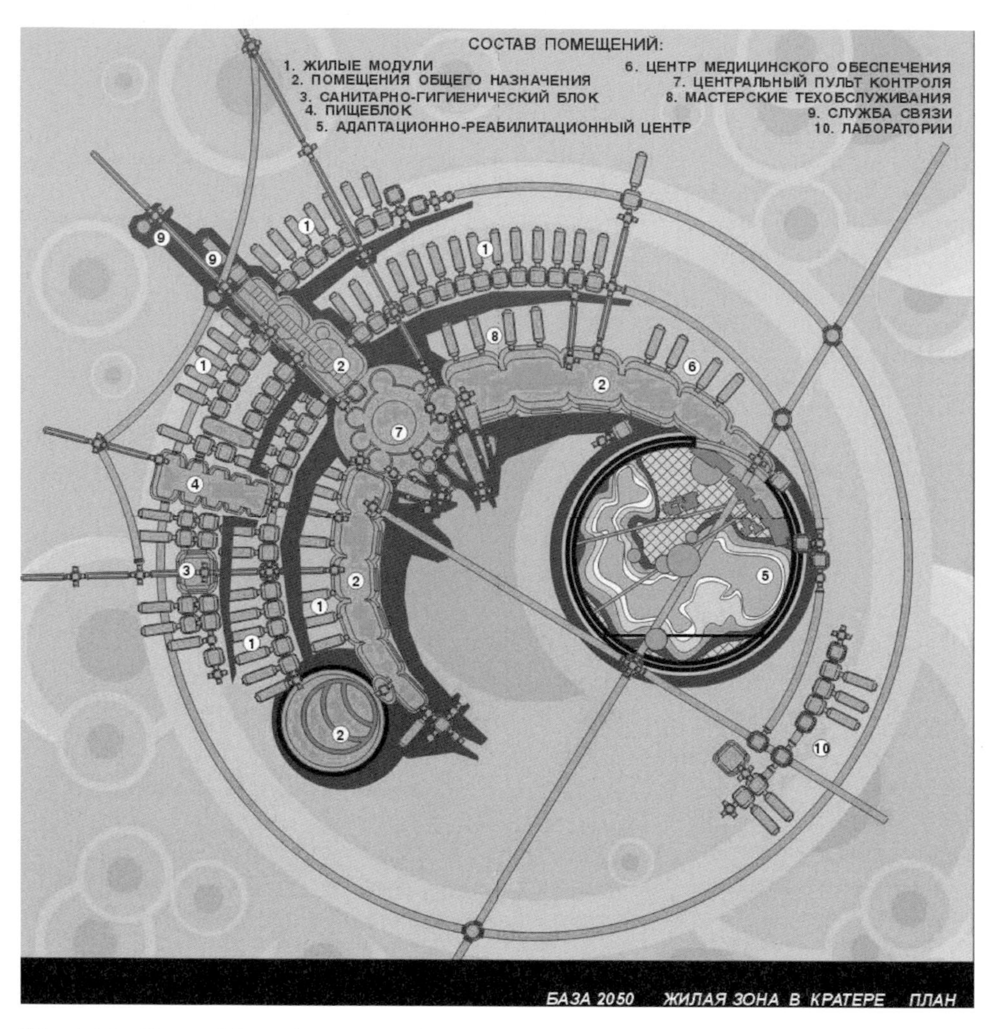

Fig. 11.9 Manned base in crater: plan view. **Legend:** 1. inhabited modules 2. general purposes 3. clean facilities 4. kitchens 5. adaptation and rehabilitation 6. medical facilities 7. control center 8. maintenance 9. communications 10. labs. (Courtesy Vladislav Shevchenko) See Plate 14 in color section.

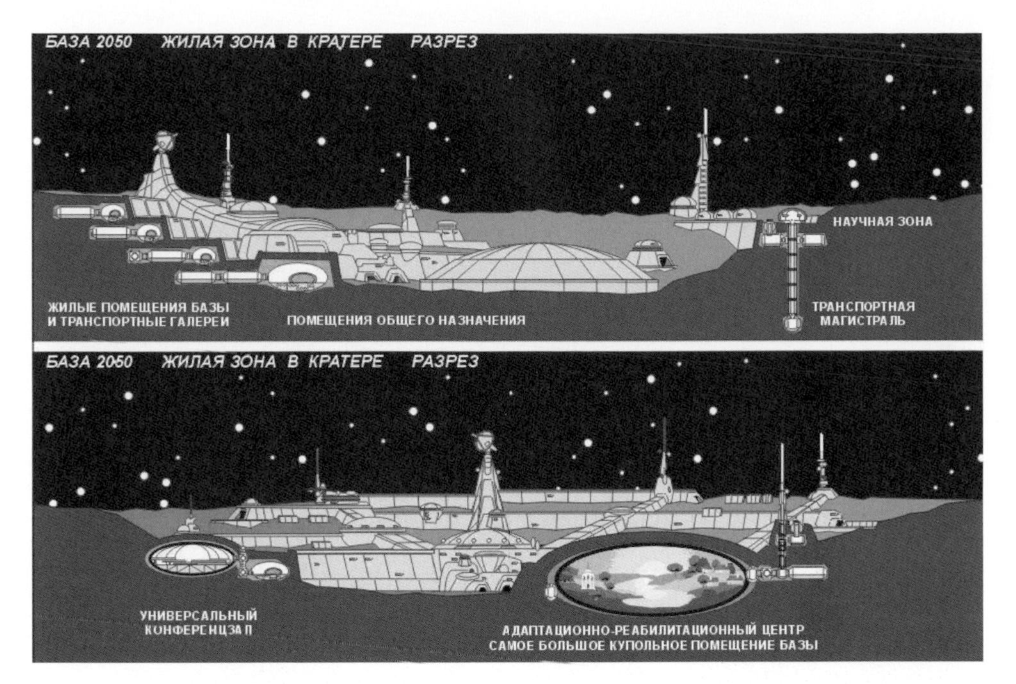

Fig. 11.10 Manned Base in Lava Tube. Energia-Sternberg Project. Side Views. The base is mostly within the lava tubes for shielding. In the top figure: on the left side are general purpose premises; the lower left are residential facilities; on the right side is the scientific zone and a transport line. In the bottom figure: in the lower left is a general conference hall, and in the lower right side is a rehabilitation center and the largest cupola-shaped premises on the base. (Courtesy Vladislav Shevchenko)

The rehabilitation facilities (5) are quite remarkable. In an ellipsoidal volume we have what looks like a small town with trees and a body of water. In this concept drawing, technologies have been assumed to permit the containment of the water. The volume is lit by natural light. This must be where it was conceived that the inhabitants of the lunar settlement go to rest and decompress.

Lava tubes were considered to be prime locations for large lunar settlements, providing a natural defense against radiation, temperature variations and micrometeorites. Figure 11.11 depicts an advanced concept, shown from three perspectives and at different levels of detail. In the present day we have buried facilities, but nothing so extensive.

The Japanese have been a prime mover when it came to technology development for the return to the Moon and for allocating the resources necessary to make it happen. They have worked with the United States and Korea in support of the Bush Vision of 2004. Here on the Moon, in 2169, the Japanese are embedded in the life of the settlements in all ways, from research to commerce. They are also major participants in our solar power generation efforts as well as the nuclear fusion developments.

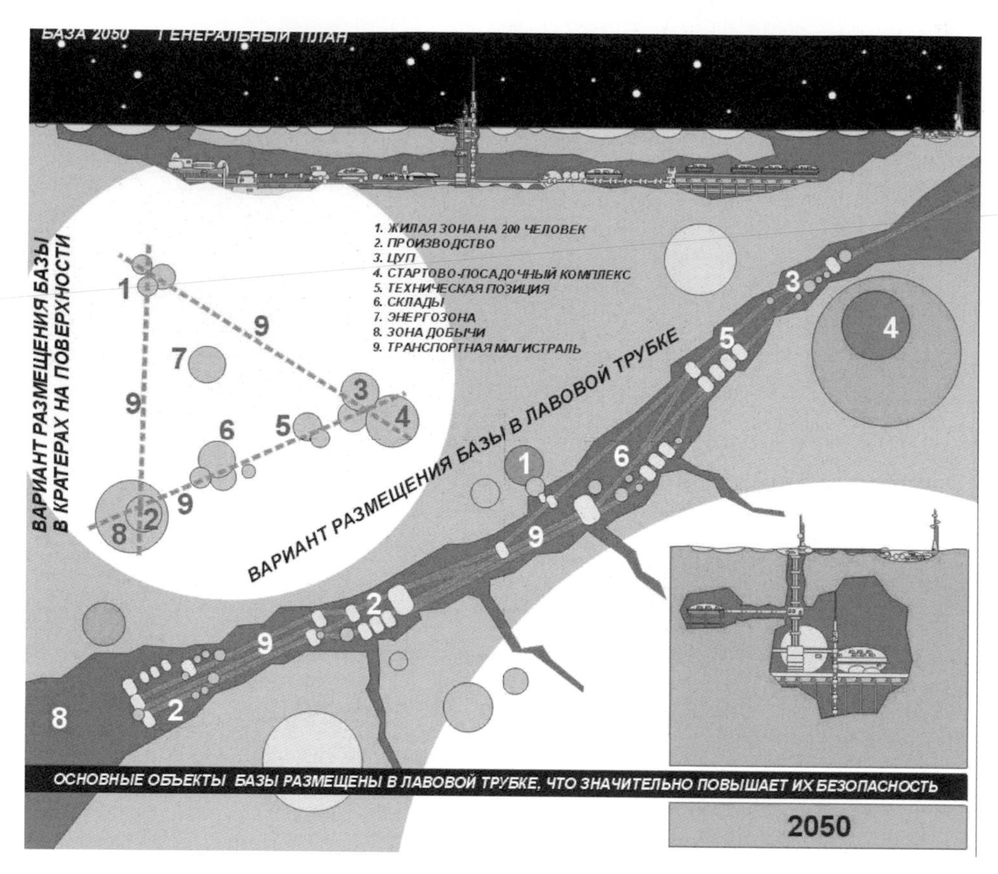

Fig. 11.11 Manned Base in Lava Tube. Energia-Sternberg Project. The base is mostly within the lava tubes for shielding. **Legend:** 1. residential facilities for 200 people 2. manufacturing 3. control 4. launch complex 5. technical 6. warehouses 7. power systems 8. shelters 9. roads. (Courtesy Vladislav Shevchenko)

11.2 Japanese concepts

On 1 October 2003, the Institute of Space and Astronautical Science, the National Aerospace Laboratory of Japan and the National Space Development Agency of Japan (NASDA) were merged into one independent administrative institution: the Japan Aerospace Exploration Agency (JAXA). While space development and utilization, and aviation research and development are the measures to achieve the nation's policy objectives, JAXA's corporate message is "reaching for the skies, exploring space," while contributing to the peace and happiness of humankind.

11.2.1 An historical interview with Hiroshi Kanamori (April 2009)

Can you give us a one or two paragraph bio?

I was born in January 1958 and majored in concrete material engineering at Waseda University. After finishing the post-graduate school of Waseda, I began to study

Fig. 11.12 A preliminary outpost on the Moon as originally proposed by NASDA. Visible are telescopes on the left-rear, a solar farm on the right, and activity by two astronauts in the foreground. (Courtesy JAXA)

various construction methods of concrete structures at the Shimizu Corporation. I have been conducting studies on the utilization of concrete on the Moon, and utilization of resources of the Moon, since I joined the Space Project Office established in Shimizu in 1987. I also developed a lunar soil simulant called 'FJS-1,' which was the first lunar soil simulant produced in Japan.

Shimizu is a very interesting company – a construction company with an Institute and an interest in space and lunar settlement. Can you tell us a bit about the history of this interest?

In the second half of the 1980s, Japanese business was very good, and most of the major construction companies like Shimizu were looking for new business frontiers. Space, polar regions, deep sea, ultra skyscrapers, deep underground, and deserts were the targets of their challenge. In this circumstance, Shimizu began to study the possibilities of successful space businesses for construction, and it established the Space Project Office, which was the first space-related division among construction companies in the world. We first proposed several unique concepts, such as the space hotel and the lunar base, to express the differences between construction companies and space industries. We particularly focused our studies on construction and living on the Moon, and conducted various investigations relating to construction methods, utilization of indigenous resources, life support, etc.

Is the goal of JAXA and private space in Japan the settlement of the Moon?

Many researchers may know that the settlement of Mars will be much more comfortable and meaningful than the settlement of the Moon. However, JAXA and private industries do not have any plan regarding Mars at this point, mainly because Japan does not have human flight technologies so far. Consequently, our first target should be the settlement of the Moon.

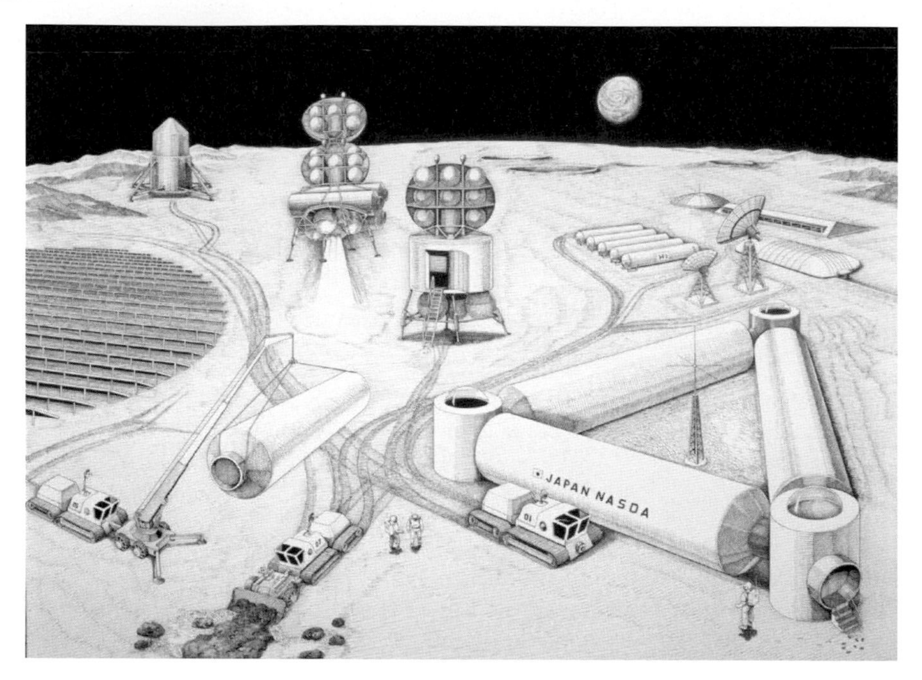

Fig. 11.13 A NASDA graphic depicting a well-developed lunar surface base undergoing construction. A cylindrical module is being placed by a crane on the left. There are solar energy panels on the left. An existing older facility is seen in the upper right of the image. (Courtesy JAXA)

Why is settling the Moon so important for civilization?

The Moon could be a place to visit for vacation, to observe Earth and other stars, to produce useful materials for various space activities, and to be an energy station for the people of the Earth and the Moon. Civilization of the Moon is definitely required to realize these functions.

How do the Japanese people view space exploration by humans? Do they support government spending on space, or do they prefer private expenditure?

Most Japanese believed that Japan should promote robotic missions to space, because Japan has been good at robot technologies. These days, however, some people notice that this may be a limitation, and we also need to develop technologies for human missions. JAXA also announced early this year that Japan would begin new studies for human mission technologies. These studies will definitely require governmental funding, because they will need tremendous amounts of money.

What do you see as the major hurdles for our return to the Moon for permanent manned settlements? Are they technical, financial, physiological, psychological?

Every item listed above might be a major hurdle. I think it will be a question of order. First, we need to develop technologies required for the mission, and then we need funding to execute the mission. When humans will start to settle on the Moon, various investigations regarding physiological and psychological subjects will be performed.

Basically, transportation costs from the Earth to the Moon will be the biggest hurdle, because so many supplies will be needed on the Moon to keep the people on the Moon alive.

How do we answer critics who say space is too expensive and that there are numerous problems on Earth to take care of first?

Of course, it may be true, but Shimizu will soon announce that the Moon will be an ultimate energy station for all of the people on Earth. That will certainly reduce territorial and environmental problems on Earth. Details of this concept will be disclosed soon.

Where do you see the space station evolving?

I think Earth-Moon L1 point will be one of the possible places for the station. Although the stability of L1 is not so high, it will be a good place for an Earth-Moon logistics system. Lunar oxygen can also be refueled to various space vehicles at this L1 gas station. For people to live in a colony, L3 and L4 points will be the best because they are much more stable places.

What are the strongest connections between the Japanese and American space initiatives?

I do not know much about the government connections, but Japan (or JAXA) always tends to rely on the U.S.A. (or NASA) in every matter. However, component technologies of mission devices produced by the Japanese have excellent workmanship and contribute to the space industry of the U.S.A.

When you envision the future, where do you see us in 50 years? In 100 years? How do you see this evolving from the present?

This is a very important and difficult question to answer. Shimizu will be doing some construction work on the Moon 50 years from now. At that time, however, not so many people will be living on the Moon, and various types of construction robots will assist the workers. A lunar hotel will be open, and a small number of rich people will enjoy the "Moon walk."

Fig. 11.14 An advanced lunar settlement which pictures concepts that have been under study for many years, including a payload shooting off the end of a rail-launch system, an astronomical observatory in the upper left and a nuclear power plant in the upper right. (Courtesy JAXA) See Plate 15 in color section.

It will take about 100 years for the lunar industry to contribute to the economy and life on Earth. Hopefully, space elevators, lunar Helium-3 fusion power, lunar colonies and energy stations, Mars habitation, and so on will be realized.

Given the recent financial meltdown worldwide, does the return to the Moon become more tenuous?

I think that this may be true as long as lunar missions do not create any new jobs that can help in the recovery from this recession. We need to make the next lunar mission much more exciting and attractive not only for the scientists but also for the business investors.

* * *

JAXA, the Japanese Aerospace Exploration Agency – literally the Independent Administration on the Exploration and Aviation of Space Study and Development Organization – was created when three organizations were merged: Japan's Institute of Space and Astronautical Science (ISAS), the National Aerospace Laboratory of Japan (NAL), and Japan's National Space Development Agency (NASDA).

Before the merger, ISAS was responsible for space and planetary research while NAL was focused on aviation research. NASDA, which was founded on 1 October

Fig. 11.15 An advanced concept of a lunar base. (Courtesy JAXA)

1969, had developed rockets, satellites, and also built the Japanese Experiment Module for the International Space Station. The old NASDA headquarters was located at the current site of the Tanegashima Space Center, on Tanegashima Island. NASDA also trained the Japanese astronauts that flew with the United States Space Shuttles.

In 2005 JAXA released its long-term vision "JAXA 2025" on how the Japanese planned to engage with efforts to return to the Moon.[4] These two decades of goals focused on the lunar leg of manned space settlement. The Japanese have participated with American efforts and have harnessed their own industries to develop and utilize space technologies.

The Japanese roadmap was to send an orbiter, called Kaguya, to map the Moon topologically and gravitationally. It was launched in 14 September 2007 (JST) from Tanegashima Space Center. The major achievements of the Kaguya mission were the gathering of data on lunar origin and evolution and developing the technology for future lunar exploration. Kaguya consisted of a main orbiting satellite at about

[4]K. Matsumoto, N. Kamimori, Y. Takizawa, M. Kato, M. Oda, S. Wakabayashi, S. Kawamoto, T. Okada, T. Iwata, M. Ohtake, "Japanese Lunar Exploration Long-term Plan," *Acta Astronautica*, Vol. 59, 2006, pp. 68–76.

Fig. 11.16 A NASDA graphic showing a mature early facility. Of particular note is the rail system that is used to launch payloads into space. An electromagnetic launch system can take advantage of the Moon's low gravity to launch payloads into orbit by bringing them to an escape velocity before they reach the end of the rail guides that point into space. (Courtesy JAXA)

Fig. 11.17 Selene-2, the first Japanese Moon lander concept. (Courtesy JAXA)

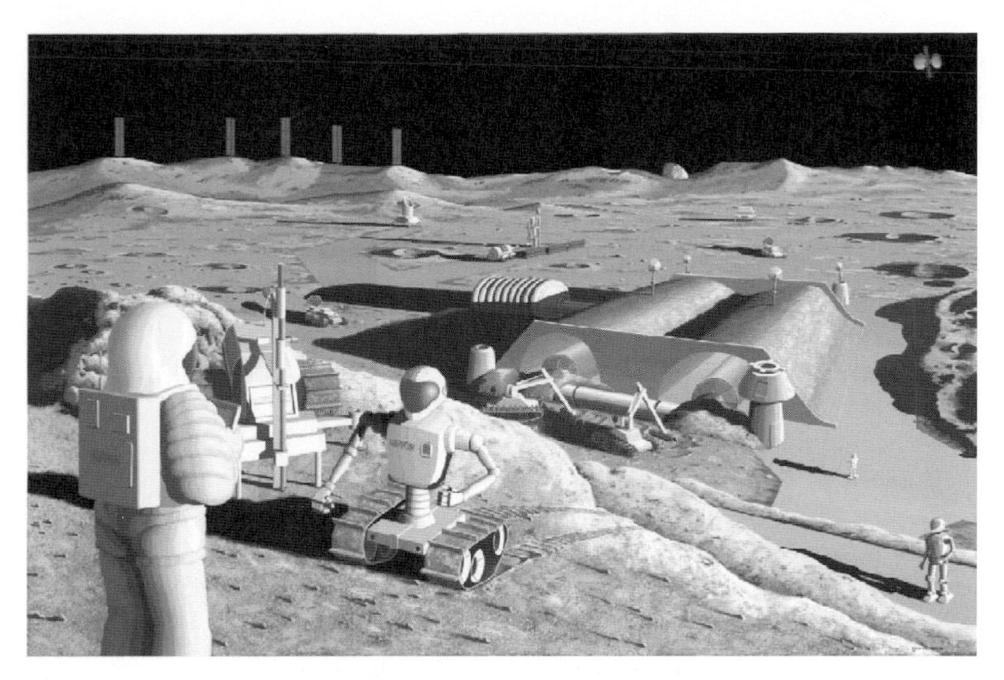

Fig. 11.18 An international lunar base supported by autonomous robots. (Courtesy JAXA)

100 km altitude and two small satellites in polar orbit. The Kaguya mission was concluded with a controlled impact to the south-east of the near side of the Moon on 10 June 2009.

Selene-2 landed on the Moon in 2015 in the southern polar region and explored for available resources, in particular water/ice, with a rover, extending the discovery of water molecules by the Chandrayaan-1 in 2009. Selene-3 landed in 2020 with advanced robotics technologies. The Japanese plan was to participate with other national efforts to land people on the Moon, to provide technical support and to assist with the robotic missions. Additionally, the Japanese were major participants in the American human presence on the Moon beginning with the initial return in 2024, which became permanent in 2029.

That first lunar presence was at the permanently shadowed regions at the pole. Hopes of water/ice in large quantities existed. The possibility of placing a manned facility that would be close to both permanent dark and near-permanent light was exciting from the energy-generation perspective. Difficulties associated with being in the polar regions included a very low sun angle resulting in long shadows and issues with light contrast that affect human and machine.

The Japanese sent teams of robots – small ones the size of a lawn mower – to explore the southern regions. They were able to roam for weeks at a time using battery and solar energy, and send back telemetry on regolith composition and detailed terrain maps.

11.3 Quotes

- “Once you’ve been in space, you appreciate how small and fragile the Earth is.” Valentina Tereshkova
- “Treading the soil of the Moon, palpating its pebbles, tasting the panic and splendor of the event, feeling in the pit of one’s stomach the separation from Terra – these form the most romantic sensation an explorer has ever known ... this is the only thing I can say about the matter. The utilitarian results do not interest me.” Vladimir Nabokov, regarding the first Moon landing
- “Every cubic inch of space is a miracle.” Walt Whitman
- “Anyone who has spent any time in space will love it for the rest of their lives. I achieved my childhood dream of the sky.” Valentina Tereshkova
- “Once the hatch was opened, I turned the lock handle and bright rays of sunlight burst through it. I opened the hatch and dust from the station flew in like little sparklets, looking like tiny snowflakes on a frosty day. Space, like a giant vacuum cleaner, began to suck everything out. Flying out together with the dust were some little washers and nuts that dad got stuck somewhere; a pencil flew by. My first impression when I opened the hatch was of a huge Earth and of the sense of unreality concerning everything that was going on. Space is very beautiful. There was the dark velvet of the sky, the blue halo of the Earth and fast-moving lakes, rivers, fields and clouds clusters. It was dead silence all around, nothing whatever to indicate the velocity of the flight . . . no wind whistling in your ears, no pressure on you. The panorama was very serene and majestic.” Valentin Lebedev

12 Mars 2034–2169

> *"The thing that sets Mars apart is that it is the one planet that is enough like Earth that you can imagine life possibly once having taken hold there."*
>
> Steven Squyres

The first people on Mars landed there in 2034. They were trained on the Moon. Some of the systems used to transport the teams to Mars were manufactured on the Moon. Typical cylindrical habitats were sent to Mars beforehand, as were ISRU systems, to begin the production of water and oxygen. The first team was comprised of six people – three men and three women – from three countries. They planned to stay nine months and then rotate out back to the Moon, being replaced by another team.

Many in the early 21st century lobbied for a direct path to Mars without the stop-over on the Moon. They glossed over the "man-machine" show stoppers. These included the fact that at that time we could not build a rocket to carry astronauts to Mars with high probability (reliability). The main hurdles were the human physiological responses to 1–2 years of radiation and microgravity exposure during the flight. The lethal space radiation environment was not given the weight it deserved – it deserves respect to this day! We still have some serious challenges.

A National Research Council committee report[1] "... finds that [the] lack of knowledge about the biological effects of and responses to space radiation is the single most important factor limiting prediction of radiation risk associated with human space exploration." Radiation was a potential show stopper in the 21st century.

By the launch to Mars in 2033, we had a better understanding of the radiation risks that the astronauts would be subjected to on that trip. One component of our radiation protection systems was the storage of water on the outside of our craft. We were also better able to predict solar activity and to include that factor into the decision of a launch time frame. Regarding microgravity, once the spacecraft was on its trajectory, a spin was initiated to allow a low, Moon-like gravity to develop. Even though the craft was small, it did have enough room to allow for locomotion and exercise.

---

[1]From Reuters: "What's Keeping Us from Mars? Space Rays, Say Experts," online 1 April 2008, 9:36am.

Fig. 12.1 A Mars settlement with much activity. (Courtesy JAXA) See Plate 16 in color section.

The same compelling arguments were made in support of the settlement of Mars as were made in support of the Moon. It was understood that exploration alone could not justify the significant risks and major costs associated with a manned Mars mission. Three factors taken together could justify such a mission: economics, education, and exploration.[2] The argument was made that while the costs of a Mars manned mission were small as compared to the rest of the U.S. Federal budget, the benefits of such an effort would be enormous:

> "The health of a nation's economy and its international competitiveness are in part a measure of the national investment in research and development in science and engineering. NASA has devoted its facilities, labor force, and expertise to generating innovative technologies that overcome the challenges of space and to sharing mission technologies with U.S. industries. These countless technologies have successfully contributed to the growth of the U.S. economy. For example, satellite technology has created an $85 billion industry that improves our daily lives through a myriad of communication, navigation, and weather-forecasting services."

[2]B.L. Ehlmann, J. Chowdhury, T.C. Marzullo, R.E. Collins, J. Litzebberger, S. Ibsen, W.R. Krauser, B. DeKock, M. Hannon, J. Kinnevan, R. Shepard, F.D. Grant, "Humans to Mars: A Feasibility and Cost-Benefit Analysis," *Acta Astronautica,* Vol. 56, 2005, pp. 851–858.

Table 12.1 Dual-use technologies from human Mars missions: Mars need.

Challenge	Mars technology
1. Harmful effects of microgravity and radiation	1. Pharmacological and mechanical preventive treatments
2. Scarce air, water, resources	2. Closed-loop life support systems
3. Scarce energy supplies	3. Advanced energy sources, low energy-use technologies
4. Safety and health risks	4. Automation and robotics
5. Hardware for extreme conditions	5. Ultra-reliable, low-maintenance systems

Table 12.2 Dual-use technologies from human Mars missions: Earth use.

Challenge	Earth application
1. Harmful effects of microgravity and radiation	1. Prevention, detection, and treatments of illnesses
2. Scarce air, water, resources	2. Resource management
3. Scarce energy supplies	3. Energy-conserving, high-efficiency products
4. Safety and health risks	4. Automated technologies
5. Hardware for extreme conditions	5. Ultra-reliable, low-maintenance systems

The idea of dual-use technologies cited earlier in support of human lunar settlements have also been valid for manned Martian settlements. The Mars Cosmic Study[3] summarized some of the technological benefits from a human Mars mission. See Tables 12.1 and 12.2 from this study.

12.1 The Martian environment

Between 2024 and 2033 we were on the Moon building, studying, surviving, and also preparing for the next step, the sojourn to Mars. A successful trip to Mars required a major increase in our understanding of the Martian environment. We accelerated our rate of sending probes to the Martian surface to test and report back on the needed soil properties. Meteorological data was accumulated. There was concern about the dust storms and the damage they could do to equipment and space suits.

Weather/windblown dust

Mars has a tenuous atmosphere (mostly CO_2); surface pressure averages about 0.7% of Earth's sea-level pressure, and greatly varies with the change in seasons. Wind speeds get quite high (of the order 100 m/s), but direct wind-loading by the thin

[3]International Academy of Astronautics, "International Exploration of Mars: A Mission Whose Time has Come," *Acta Astronautica*, Vol. 31, 1993, pp. 1–101.

Table 12.3 Comparison of Earth and Martian physical parameters.

Property	**Earth**	**Mars**
Surface area [km^2]	510.1×10^6	144.1×10^6
Radius [m]	6,371	1,699
Gravity at Equator [m/s^2]	9.81	3.69
Escape Velocity at Equator [km/s]	11.2	5.02
Surface temperature range [$^\circ C$, $^\circ F$]	$\begin{bmatrix} -89 \text{ to } 58 \\ -128 \text{ to } 136 \end{bmatrix}$	$\begin{bmatrix} -87 \text{ to } -5 \\ -125 \text{ to } 23 \end{bmatrix}$
Surface atmospheric pressure [kPa, psi]	101.3, 14.7	0.6, 0.09
Day length [Earth Days]	1	1.03
Communication delay [s]	0	Max 2300

atmosphere is not a dominant problem. However, lofted surface fines can achieve relatively high kinetic energy leading to erosive and penetrative capabilities. Local dust storms, and occasional global, long-lived dust storms are a regular occurrence. They are difficult to predict with accuracy.

Unlike on the airless Moon, contamination avoidance cannot be attained simply by locating components off the ground. The surface of Mars is in constant motion; sand-dune mobility has been observed on the surface. Thus, the simple foundation pads that are possible on the Moon are generally impractical on Mars. We developed a tentacled foundation design that is able to fix our structures to location.

We also regularly collect and sinter dunes that have collected near our structures.

Temperature

Mars' surface temperature varies between about −120 °C and −25 °C daily and rarely exceeds the freezing point of water. The Mars diurnal cycle is 1.03 Earth days long. Mars receives less than half of the specific solar flux that the Moon or Earth do because of its greater distance from the Sun; its elliptical orbit causes a 39% annual insolation variation. Together, the shorter diurnal cycle and moderating atmosphere make Mars a more hospitable place than the Moon for materials systems.

Solar power is not a viable source of energy on Mars. We depend solely on nuclear power – initially nuclear fission, but since 2071 on nuclear fusion reactors.

Gravity level

Mars' gravitational acceleration is roughly 3/8 of Earth's. Dead loading is thus 2.25 times the lunar case for the same equipment. This fact required some redesigning of our lunar habitats and lunar mechanical equipment for use on Mars.

Another consequence of Mars' greater gravity is its ability to retain an atmosphere. Mars landers used aerobraking for entry, which limited the allowable delivered payload dimensions as compared to lunar delivery systems. Maglev systems were quickly deployed to Mars, even though they could not be as efficient as

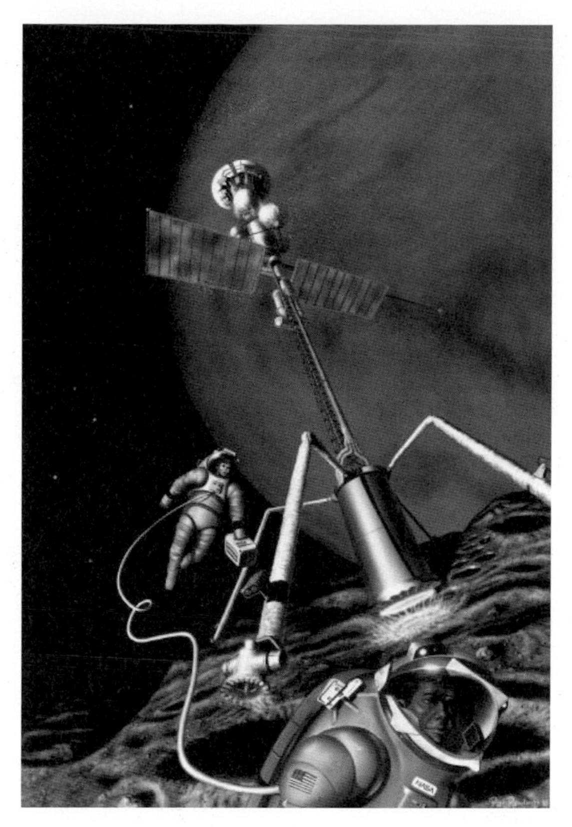

Fig. 12.2 On Phobos, the innermost moon of Mars and likely location for extraterrestrial resources, a mobile propellant-production plant lumbers across the irregular surface. Using a nuclear reactor the large tower melts into the surface, generating steam that is converted into liquid hydrogen and liquid oxygen. (Artwork by Pat Rawlings of Eagle Engineering. S86-25375, 1986. Courtesy NASA) See Plate 17 in color section.

they are on the Moon. Space elevators were also placed in orbit as soon as possible. All of these transport systems have their niche in the Martian economy.

The maglev systems are optimal for lunar gravity and less effective on Mars. But the space elevators are our workhorses – they were easier to deploy and operate on the Moon because of the lack of atmosphere and the lack of a significant magnetic field.[4]

Substrate properties

For a long time not much was known about the seismic characteristics of Mars, its static soil mechanics (except in two places where Vikings landed), or its soil chemical properties. Anomalous reactions detected in one of the Viking landers' three

[4]K. Zala, "Did Mars's Magnetic Field Die With a Whimper or a Bang?" *ScienceNOW Daily News*, 30 April 2009.

experiments seeking signs of life had led some scientists to speculate that chemical reactions in Martian soil would be detrimental to some materials, including polymers.

However, both Viking landers outlived design lives (although they had few vulnerable polymeric materials). Surface frost had been photographed at dawn on Mars. Permafrost occurs through most of the Martian surface (except for the equatorial belt); foundations require special thermal-management designs. The Martian poles are particularly active regions seasonally, with massive alternating between freezing and sublimation of CO_2 ice; it is difficult to place permanent structures on Mars.

Fig. 12.3 The first humans on Mars revisited the landing site of the Viking 2 Lander in order to study the effects that the Martian surface and atmosphere have had on the spacecraft. (These images produced for NASA by Pat Rawlings. Technical concepts from NASA's Planetary Projects Office, Johnson Space Center. S91-52337, 1991. Courtesy NASA)

Forward contamination

It was thought that Mars may harbor native life; Viking results remained controversially inconclusive, and in any case were limited to two exposed, windswept points on the planet. Human field work started soon after our arrival. Concerns were great about the possibility of contaminating a Martian ecology with Earth/lunar organisms when human operations began.

Protocols were developed to assure that contamination was extremely unlikely. This requirement had implications for the structural systems that we could initially build. Special decontamination provisions to preclude any contact whatsoever between man-systems or cabin air and the native Martian environment was an extreme engineering challenge. With time we became confident in our ability to avoid such problems.

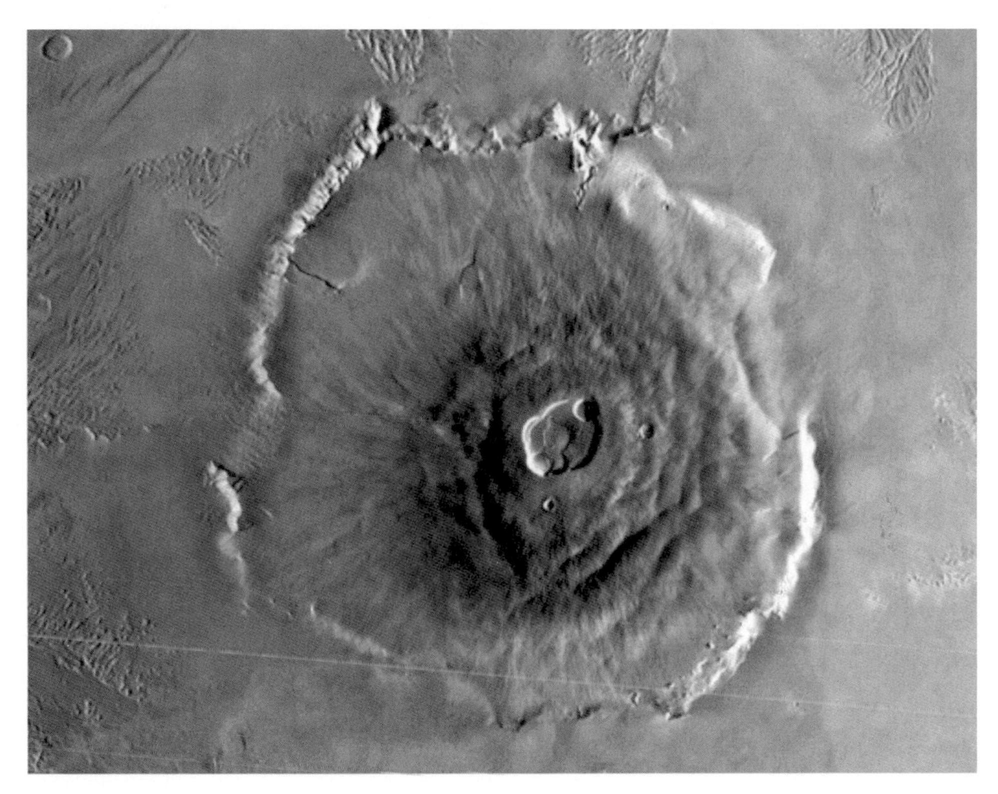

Fig. 12.4 Martian Olympus Mons is the tallest mountain in the Solar System. Its peak is over 26 kilometers (16.2 miles) above its base. Olympus Mons is called a shield volcano because of its shape. The volcano is very tall, but it has a very gentle slope. Olympus Mons is over 20 times wider than it is high. It is the same kind of volcano as the active volcanoes currently making the Hawaiian Islands. The line around Olympus Mons is a "basal cliff" which is as high as 6 km in some places. In others, lava from the formation of Olympus Mons flowed over the cliff, smoothing it out. (Courtesy NASA)

Mars' material combinations

One of the most exciting discoveries of the early exploration of Mars was that of the Phoenix Mars Lander digging into the Martian soil and finding water ice.[5] The

[5]K. Chang, "White Patches Found in Mars Trench are Ice, Scientists Say," *nytimes.com*, 20 June 2008.

raw materials available for fabricating structures on Mars were found to be largely those comprising silicate rocks and regolith, as on the Moon. Mars has iron in the form of FeO_2, and the availability of water made conventional concrete a more promising option than on the Moon.

The planet has an extremely diverse geology, with abundant evidence of past hydrologic and volcanic processes, which has provided us with concentrations of useful compounds. The presence of lighter elements (notably deficient on the Moon) meant that organic polymers and related materials can be manufactured on Mars. The supply of elements is well balanced for human uses.

We can make anything we need on Mars, just as we can on Earth. This includes Martian settlement structures, outfitting equipment, consumables, and biomass. In terms of environment and resources, Mars is the next most hospitable planet to Earth for humans.

Martian soil mechanics

Martian soil mechanics at the dawn of the 21st century was even more problematic than lunar regolith mechanics. At least for the Moon, samples were in-hand from the Apollo missions. These samples were guarded like the crown jewels that they were – a remnant of an era of exploration that demonstrated that humans could still achieve the loftiest of goals if they set their minds to the task.

Data from spacecraft and telescopes were initially used to estimate the composition and particle size of the Martian regolith.[6] Some soil simulants were created, such as the Mojave Mars Simulant developed at the Jet Propulsion Laboratory and made of sand and dust derived from the Mojave Desert in California.

Also discovered by the Phoenix Mars Lander was that "the dirt on the planet's northern arctic plains [is] alkaline, though not strongly alkaline, and full of the mineral nutrients that a plant would need. ... The sort of soil you have there is the type of soil you'd probably have in your backyard. The kind of soil asparagus would like."[7]

Nickel-iron meteorites are a common rock type found on the Martian surface and can be processed into a steel for *in situ* construction.[8] The Martian CO_2 atmosphere provides the elements needed to purify the iron and nickel at temperatures well below the melting point of each, and to refine cobalt and platinum.

It was within this environment that we had to build habitats and roads and powerplants. In many ways the Martian environment was "an average" of the Earth and lunar environments. So even though we learned much on the Moon, some tailoring of the technology and the designs was needed to meet the needs of the Martian surface. However, it is important to note that the lunar experience was extremely valuable to Martian settlement.

---

[6] R. Courtland, "How to Make a Martian Mud Pie," *NewScientist.com*, 19 June 2008.

[7] K. Chang, "Alkaline Soil Sample From Mars Reveals Presence of Nutrients for Plants to Grow," *nytimes.com*, 27 June 2008.

[8] G.A. Landis, "Meteoric Steel as a Construction Resource on Mars," *Acta Astronautica*, Vol. 64, 2009, pp. 183–187.

Fig. 12.5 The Surface Stereo Imager on NASA's Phoenix Mars Lander took this image (original in false color) on 21 October 2008, during the 145th Martian day, or sol, since landing. The white areas seen in these trenches are part of an ice layer beneath the soil. The trench on the upper left, called "Upper Cupboard," is about 60 cm (24 in) long and 3 cm (1 in) deep. The trench in the middle, called "Ice Man," is about 30 cm (12 in) long and 3 cm (1 in) deep. The trench on the right, called "La Mancha," is about 31 cm (12 in) and 5 cm (2 in) deep. (Courtesy NASA) See Plate 18 in color section.

12.2 Habitats

The first Martian structures were identical to the first lunar structures; they were "tin cans" like the ones shown in Figure 12.6. They were our bases from which an infrastructure could be built. The Habitat Design Workshop considered Martian habitats[9] in addition to lunar habitats.[10] The external structure followed the inter-

[9]Project: Mars 2 Studentteam: E. Mac Donald, N. Fischer, J. Lamamy, N. Mair, G. Messina, N. Pattyn, © ESA Habitat Design Workshop 2005 (M. Aguzzi, R. Drummond, S. Häuplik-Meusburger, J. Hendrikse, J. van der Horst, S. Lorenz, E. Laan, K. Özdemir, D. Robinson, G. Sterenborg and P. Messino / ESA.)

[10]N. Fischer, J. Lamamy, E. MacDonald, N. Mair, G. Messina, N. Pattyn, *Space Habitation Design Workshop*, ESTEC 2005.

Fig. 12.6 The crew's ascent vehicle and propellant production facility can be seen 1 km away from the completed outpost. (These images produced for NASA by John Frassanito and Associates. Technical concepts for NASA's Exploration Office, Johnson Space Center. S97-07847, July 1997. Courtesy NASA)

Fig. 12.7 Modular base on Mars. (Courtesy Project Mars 2 Studentteam)

Fig. 12.8 A colony of mated modules. (Courtesy Project Mars 2 Studentteam)

nal requirements, and architectural and engineering requirements were developed concurrently. Each module could stand alone or be combined into an expanded facility.

12.2.1 ISRU Mars base design

The Mars base concept outlined in this section is a summary of an early 21st-century concept based on the ISRU-robotic methodology described in Section 7.2. The premise was that a group of robots could use *in situ* resources to create the base over some length of time before astronauts would arrive. The robots would erect the framework of a base structure, utilizing Martian *in situ* resources as well as additional stock brought to Mars from the Moon. The completed structure is seen in Figure 12.9.

This base consisted of two levels, one above ground and one below ground. In addition, a safe haven against fatal doses of solar radiation was buried well beneath the surface. The images in this section are from this early design.[11] Dimensions of the habitat are based on accepted percentages, with the enclosed space comprised of 40% general storage space, 20% laboratory space, 20% living area and 20% for other general use facilities.

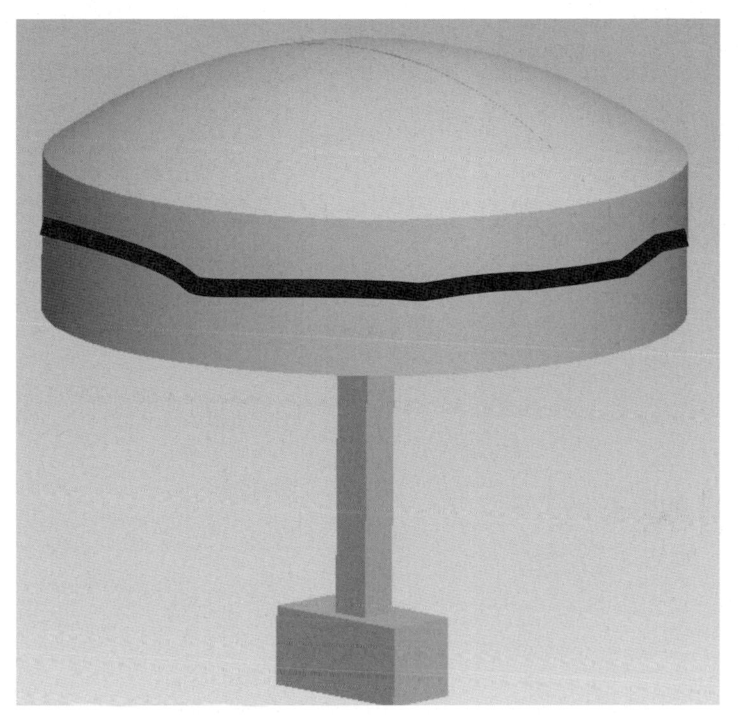

Fig. 12.9 Complete structure showing two levels plus dome, with a vertical shaft leading from the lower level to the deep shelter. The dark line is the demarcation between buried structure and exposed structure.

[11]S. Rajaram, E. Sosnov, A. Suszko, I. Uzicanin, H. Benaroya, "Design of Martian Surface Structure Using In-situ Resource Utilization and Rapid Prototyping Techniques," Rutgers School of Engineering, Department of Mechanical and Aerospace Engineering, 1 May 2009. All the images in this section are courtesy of this group, the Rutgers Design Team.

The study focused on the use of freeform fabrication technologies as the basis for utilizing autonomous mini-robots that can, as a team, build a Martian structure for habitation using *in situ* resources and primarily nuclear power over a 6–12 month period prior to the arrival of astronauts.

Figure 12.10 was an artistic conceptualization of how such a construction site on Mars would look with several robots capable of processing Martian soil from beneath and then precision-ejecting the *in situ* created construction material to the appropriate location on the evolving structure.

The 150 ft diameter dome would be created using cast basalt. The outer wall would be 6 ft thick for additional radiation shielding. The dome was designed to be dual-layered cast basalt with regolith inside, while the inner part of the wall would be coated with aluminum for additional support and aesthetic purposes.

The lower level is the primary storage area that the astronauts will use, and which houses the entrances and exits to the structure, not seen in these drawings. Access is available to stored materials and parked vehicles. The lower level also has a diameter of 150 ft and will have the same 6 ft thickness as the dome.

There are two main entrances by foot into the structure. Upon entering the structure the astronauts will be in a room that will be pressurized. The astronauts will be able to enter the rest of the structure without the need of a spacesuit. All the internal walls will be one foot thick and created using cast basalt.

Fig. 12.10 Artistic conceptualization of robotic ISRU construction of Mars habitats. These robots use ballistic particle manufacturing procedures. (Courtesy Ana Benaroya) See Plate 19 in color section.

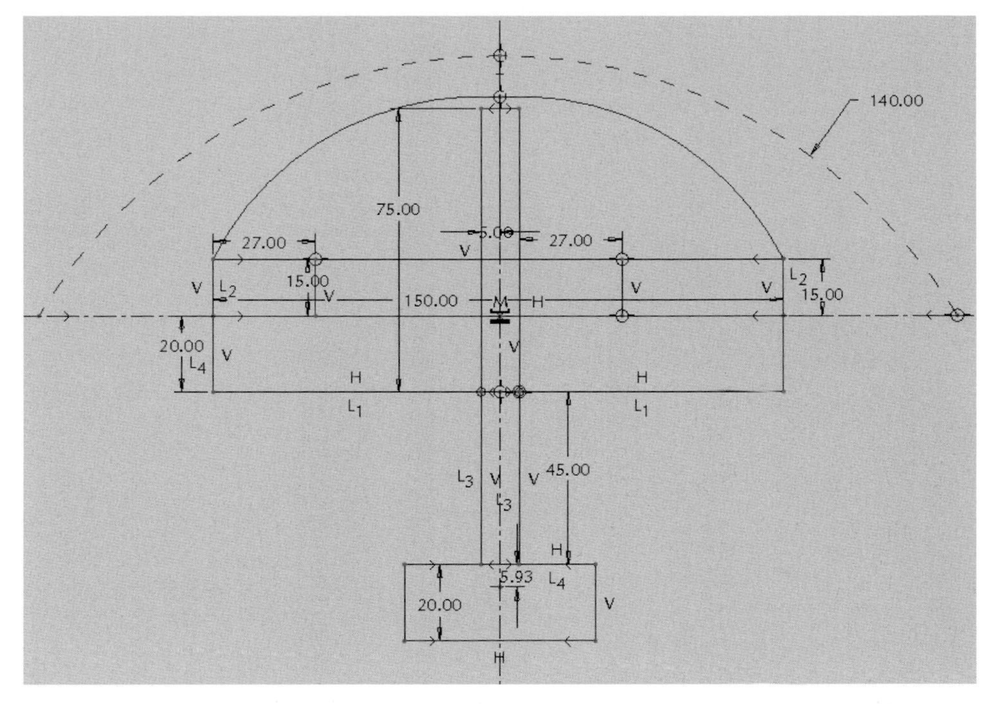

Fig. 12.11 Schematic of side view of structure. Dimensions are in feet.

Level 1 is the main living area for the astronauts. At one end there is a room for two labs and a sickbay. The other side of Level 1 contains the ward room and entertainment room, a bathroom, three living areas for the astronauts, in addition to a dining room and gym. A central space surrounded by the stairwell that spirals around the elevator shaft provides space for growing plants.

Sketches included here show the dimensions of the structure in profile, Figure 12.11, and in plan, Figure 12.12.

A team of freeform manufacturing mini-robots were envisioned to work as a team, with the group comprising the skill set of capabilities needed to construct all the parts of the structure. As an example, ballistic particle manufacturing robots, of the kind shown in Figure 12.10, can be placed on a ridge to project particles to the higher locations of the lunar structure.

Figure 12.13 shows a section of the structure. Two levels are shown – stairs between the levels are evident – as is the vertical shaft down to an emergency shelter, which is also connected to a horizontal tunnel that can be used by occupants to leave the general area of the facility in the case of impending and potentially catastrophic events.

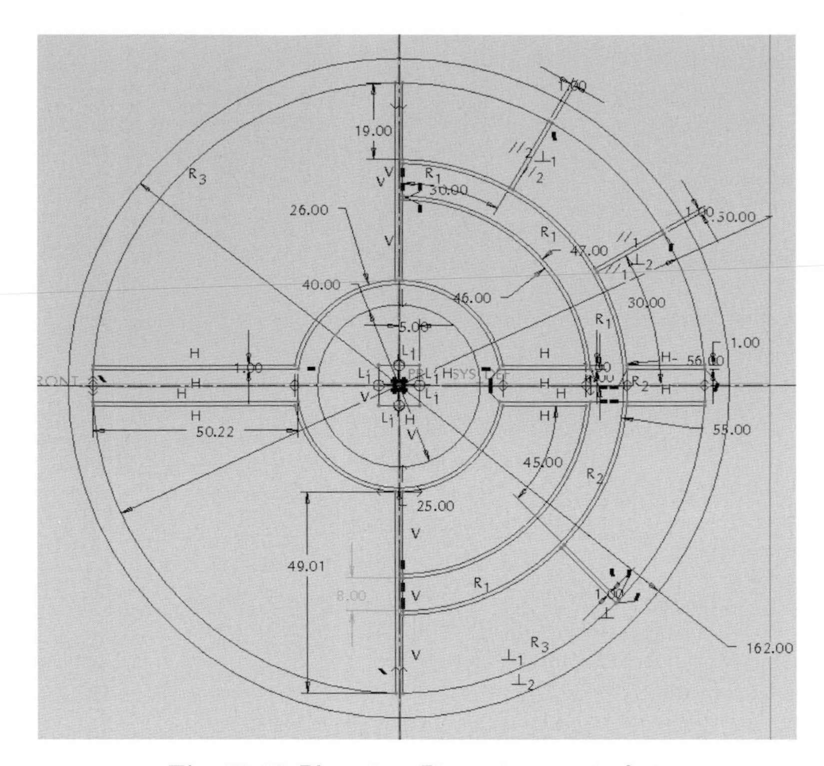

Fig. 12.12 Plan view. Dimensions are in feet.

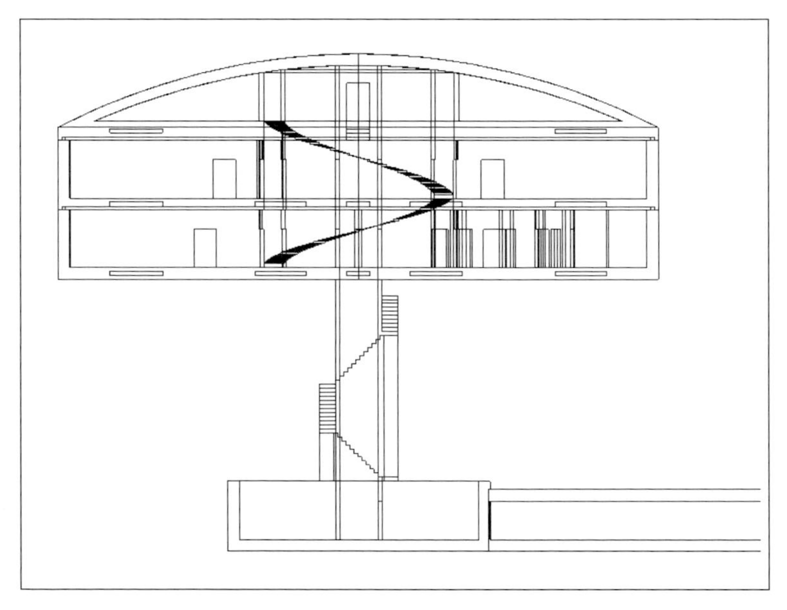

Fig. 12.13 Side view of the structure, the radiation bunker and the escape tunnel.

Table 12.4 Composition of Martian basalt.

Basalt components	**Chem. symbol**	**% in basalt**	**% in Martian soil**
Silica	S_1O_2	45	49.6
Alumina	Al_2O_3	12	7.1
Calcium oxide	CaO	11	10.9
Magnesium oxide	MgO	10	8.8
Iron oxide	Fe_2O_3	13	–
Sodium oxide	NaO	3	2.3
Titanium oxide	TO	3	1.0
Potassium peroxide	K_2O_2	1.5	–
phosphorus	P_2O	1.5	–
Manganese oxide	MnO_5	0.5	–

Table 12.5 Elementary properties of cast basalt.

Property	**Value**
Specific gravity	2.9–3.0 g/cm^3
Compressive strength	450 MPa
Bending strength	40 MPa
Tensile strength	10 MPa
Thermal conductivity	0.7 KCal/m^2/Hg °C
Specific heat	0.2 BTU/lb °C
Thermal expansion	$\begin{bmatrix} 20\text{ - }100\ ^\circ\text{C}(77 \times 10^{-7})\text{deg} - 1 \\ 20\text{ - }400\ ^\circ\text{C}(86 \times 10^{-7})\text{deg} - 1 \end{bmatrix}$
Abrasion resistance	3–3.5cm^3/50cm^2 - DIN 52 108
Hardness	8.5 mohs scale

Figure 12.14 provides the floor plan showing several large rooms for testing and manufacturing, as well as smaller rooms. The stairs are shown in the central core region.

Figure 12.15 shows a computational analysis of a single floor to the environmental loads. The deformation of the structure is shown in an exaggerated scale.

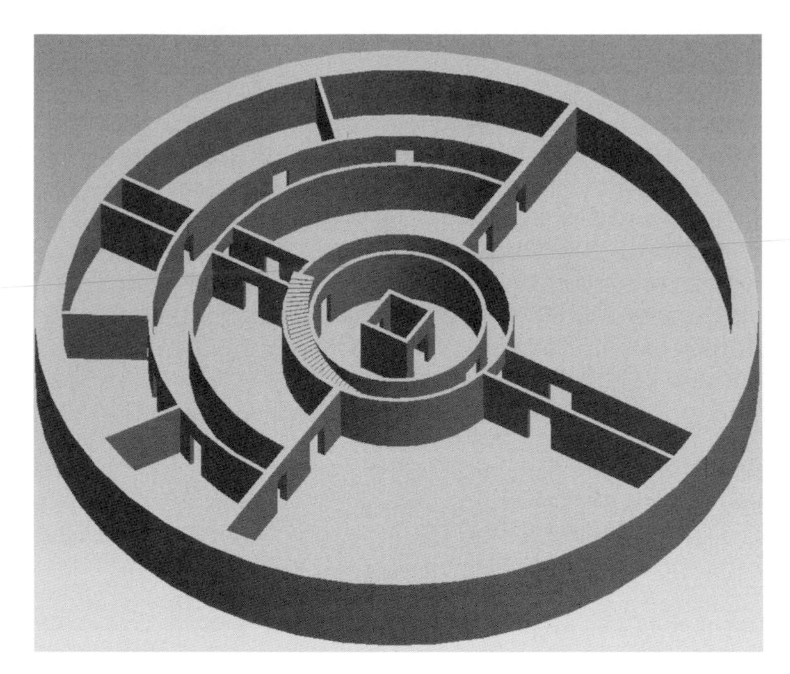

Fig. 12.14 Floor plan showing a single level of the facility. Shown in this cutaway drawing are the central core, the spiraling stairs, and the various rooms.

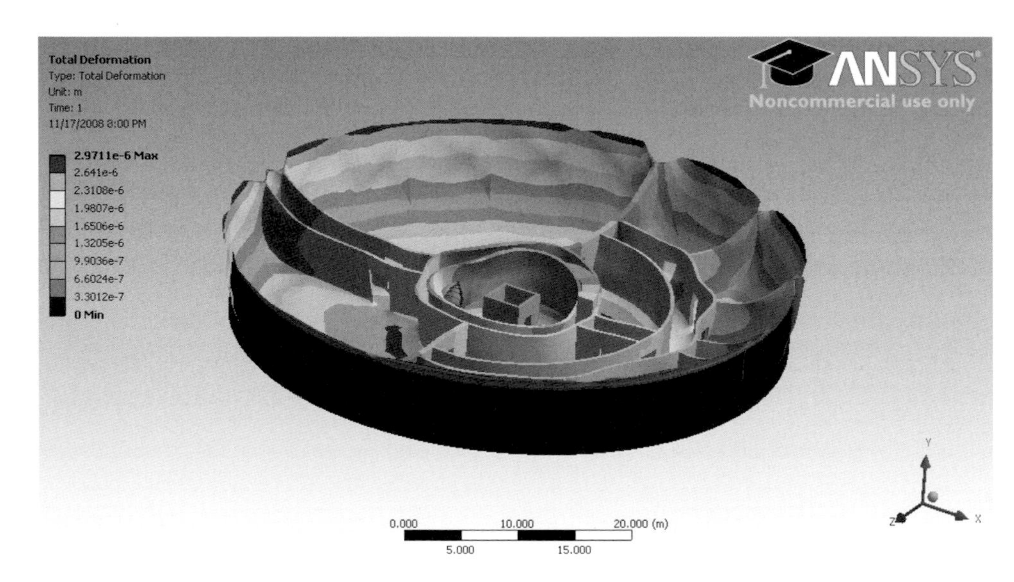

Fig. 12.15 ANSYS deformation analysis. Deformations shown in exaggerated scale. (Image courtesy ANSYS, Inc.)

12.3 Quotes

- "Ladies and gentlemen, I have a grave announcement to make. Incredible as it may seem, strange beings who landed in New Jersey tonight are the vanguard of an invading army from Mars." Orson Welles
- "It might be helpful to realize, that very probably the parents of the first native born Martians are alive today." Harrison 'Jack' Schmitt

13 Issues for the next generation

> "*We are successful today because of our parents who made their vision a reality. Our children's success tomorrow depends on the efforts of their parents to build upon that earlier vision.*"
>
> Yerah Timoshenko

Since humans returned to the Moon, we went from being an outpost on a barren rock orbiting Earth to being a nascent civilization on the Moon and Mars, with outposts on dozens of asteroids, outer planets and their moons. We went from a population of under a dozen to one approaching 300,000 extraterrestials.

But our impact is larger than what one may expect of that many people. A city on Earth of that population is considered small, perhaps peripheral to the main avenues of power and influence. The 300,000 people who are today distributed throughout large sectors of the Solar System are all prime movers. Each of us has significant responsibilities. We oversee an infrastructure that is vast and very wealthy. We supply 80% of the energy needs of Earth and an increasing percentage of its raw material needs.

Earth still overshadows us, of course. But there is talk of greater freedom, more autonomy. The further one is from Earth, the louder is the talk of independence. Space has that effect on people. We view ourselves as self-reliant. Many feel that Earth needs us and we don't need them. I don't fully agree – but greater autonomy is desirable.

13.1 Environmental Issues on the Moon and Mars

That early lunar mining operation that erased the topology of the surface as it exhumed valuable ore became an unacceptable form of mining to the population that now calls the Moon home. It became unacceptable to many on Earth as well, who were enthusiastic in their support of human evolution beyond the confines of one g, and eager to protect the stark beauty of the new worlds – the new worlds many of us also call home. Mankind has affection for the Moon.

We need to examine these issues from the perspectives of those who are settling our Solar System. While we, in the 22nd century, have had reasonable success in balancing the settling and development of the Moon and more recently Mars, these issues were only beginning to be discussed in the 21st century. The following is a fusion of thoughts on this delicate balance during the period before the second return to the Moon, as well as current thinking.

When, at the beginning of the 21st century, human exploration and colonization of the Moon and the planets appeared far off, that time was viewed by some as an appropriate time to discuss issues concerning the safeguarding of the integrity of these planetary bodies. It was much easier to do this in advance of the economic development that was to be explosive once it began. The question was asked: how do we, the human race, ensure that the Moon in particular, and the planets in general, survive with integrity after colonies and industrial facilities begin to be planted there during the next hundred years? The Moon tested our abilities to learn from the mistakes of all kinds made by civilizations on Earth. Even as a lifeless body the Moon held up contrasts between options that faced us.

On our second return to the Moon, with a patchwork lunar legal system, with different nations of various views on the rights of the individual, questions arose as to how the non-biological environment would be preserved. This included the Apollo landing sites.[1] There was a fear and a risk that two camps would evolve – both supporting space exploration and settlement but one supporting the development of extraterrestrial resources as a way to reduce such exploitation of Earth, and the other seeking ways to develop but in a more environmentally conscious way so that lessons learned on Earth could be applied in space.

What deserves saving on the Moon? Most agreed that the Apollo sites and surrounding areas were worth preserving. A radius of no contact was established where no person and no vehicle could approach, not even overhead since rockets can destroy the footprints and locations of items on the surface.[1]

"The place where the two astronauts landed, resided and worked – a roughly 60 m^2 area named 'Tranquility Base' – is a unique Solar System physical location. Tranquility Base, and what was left behind there when the astronauts departed, should be preserved and protected for all, for all time."[2] It was also suggested that the UN declare the site a World Heritage Site.[3] But there was little public support for the UN in the U.S. – the United States viewed Tranquility Base as historic for the world but still felt it to be a distinctive American accomplishment before the era of multinational space efforts.

> "The Apollo programme truly widened the horizons of humanity at large. The image of an Earthrise over the lunar surface, taken by Apollo 8, brought home in a very visual way the relative position of Earth and demonstrated the fragility of our planet, 'space ship Earth.' It soon became a symbol for the environmental movement. The 'blue marble' photograph of a complete and only slightly cloud-covered Earth taken during Apollo 17 has become the icon of Earth in space. People were fascinated by these missions ... about

---

[1] E.C. Hargrove, "The Preservation of Non-Biological Environments in the Solar System," 2162.pdf, *NLSI Lunar Science Conference*, 2008.

[2] A. Chaikin, *A Man on the Moon*, p. 200, Penguin Books, 1998.

[3] T.F. Rogers, "Safeguarding Tranquility Base: Why the Earth's Moon Base Should Become a World Heritage Site," *Space Policy*, Vol. 20, 2004, pp. 5–6.

Fig. 13.1 *Earthrise* is the name given to a photograph of the Earth taken by astronaut William Anders in 1968 during the Apollo 8 mission. (Courtesy NASA) See Plate 20 in color section.

a quarter of the human population at the time watched the live television broadcast as Neil Armstrong stepped onto the lunar surface."[4]

To this day we support the value of the Apollo sites and have placed hemispherical glass domes over all of them. We have also protected the landing sites of the early probes on the Moon and on Mars.

There are sites on the Moon, Mars and some of the other bodies that have been viewed as worth preserving for their own sake – such objects, like works of art, have intrinsic value. The intrinsic value may be for scientific reasons or for aesthetic reasons. On the Moon, the harsh environment makes it difficult to erase mistakes, and there are no minor mistakes.

Some views on the quality of the environment, that were learned the hard way on Earth have been taken to heart on the Moon, are summarized:

[4]D.H.R. Spennemann, "The Ethics of Treading on Neil Armstrong's Footprints," *Space Policy*, Vol. 20, 2004, pp. 279–290.

Fig. 13.2 The Arabian Peninsula can be seen at the northeastern edge of Africa. The large island off the coast of Africa is Madagascar. The Asian mainland is on the horizon toward the northeast. This photograph is known as the *Blue Marble* and was taken on 7 December 1972 at a distance of about 29,000 km (18,000 miles) as the Apollo craft was heading to the Moon. NASA officially credits this photo to all three astronauts, Eugene Cernan, Ronald Evans and Harrison Schmitt. Some credit Harrison Schmitt as the photographer. See ehartwell.com. (Courtesy NASA)

Pollution: Our survival as individuals and as a species is directly linked to the quality of our environment. We breathe the air, drink the water, and eat food that at some point grew in the earth. It is clear that the quality of the air, water, and the earth directly affects our health. In this connection many mistakes have been made by ignoring the damage done to the environment, even after it had been well established that a degraded environment leads to less healthy humans. On the Moon, we are painfully aware how our existence rests on the quality of our artificial environment.

Exploitation of natural resources: There is general agreement that economically and socially, natural resources are to be used by a local population. An essential aspect of this must be that resources need to be managed so that both current and future generations have access. A general desire exists amongst people to ensure that successive generations are not deprived or burdened for the purposes of the present. Thus, there is a general opposition to wholesale depletion of forests, farmlands, and mines. This understanding is crucial for the Moon as well.

Quality of the environment: Given a choice, humans prefer attractive and comfortable surroundings. However, if we look around at the surroundings in which many live on Earth, one might be led to the opposite conclusion. As population demands increase, the incentives are to fit many more people in smaller areas and to disregard their visual and psychological needs. Due to the stark nature of the Moon and its hostile environment, extra effort have been expended to make human habitation comfortable and enjoyable in addition to viable.

Fig. 13.3 A 16 m diameter inflatable habitat is depicted and could accommodate the needs of a dozen astronauts living and working on the surface of the Moon. Depicted are astronauts exercising, a base operations center, a pressurized lunar rover, a small clean room, a fully equipped life sciences lab, a lunar lander, selenological work, hydroponic gardens, a wardroom, private crew quarters, dust-removing devices for lunar surface work and an airlock. (This artist concept reflects the evaluation and study at Johnson Space Center by the Man Systems Division and Johnson Engineering personnel. S89-20084, July 1989. Courtesy NASA) See Plate 21 in color section.

13.2 Principles of a spacefaring civilization

The essential principles of any new civilization must evolve from the basic democratic principles many of us cherish. On Earth we have seen the need to safeguard our environment and to respect the integrity of the planet that sustains us with air, water, whose atmosphere protects us from the harsh environment of the Solar System, and whose soil provides us sustenance. Some of these essential principles were transferred to the colonization and industrialization of the Moon and the Solar System. Here is a summary of the founding principles at the time of the second return to the Moon in the early 21st century:

- Democratic principles form the basis for interaction between people, and in the commercial aspects of the new societies. Some objected to the West imposing its values on the new settlements on the Moon and the planets. But what are Western values? The indispensable achievement of the West was the concept of individual rights. It is the idea that individuals have certain inalienable rights and individuals do not exist to serve government, but governments exist to protect these inalienable rights. It took until the 17th century for that idea to arrive on the scene, mostly through the works of English philosophers such as John Locke and David Hume. One need not be a Westerner to hold dear Western values. It is no accident that Western values of reason and individual rights have produced unprecedented health, life expectancy, wealth and comfort for the ordinary person. There is an indisputable positive relationship between liberty and standard of living.
- The integrity of the "land" must be reasonably maintained; that is, the terrain should not be destroyed in order to mine and extract material. Early in the development of the Moon, techniques must be in place for resource recovery without the devastation inherent in strip mining, even if costs rise. If society values the lunar landscape, then it must be willing to share in the economic burden of its maintenance. It is unreasonable to expect businesses to shoulder the complete burden.
- Because the environment is of an extreme nature, care must be taken to avoid its degradation. Any environmental repairs would be extremely difficult and costly, if they are at all possible. The Moon must be viewed as a second home for humanity, not only a big rock for mining. This is also true for Mars.
- Ownership of and on the Moon requires the input of the legal profession. Initially, we would expect that those who finance the colonization and development of the Moon have certain rights. One question is whether there are national rights. For example, if the U.S. creates a colony on the Moon, what will be its rights on the Moon? One can expect that two sets of rights will evolve, one for the individual (and corporation) and the other for national bodies such as governments. Eventually, however, lunar settlements will become autonomous regions, just as did the American colonies. Hopefully, revolutionary wars can be avoided if a sufficiently long-term framework can be set up that safeguards property rights but recognizes that autonomy and eventually independence are unavoidable, as they were in the New World.

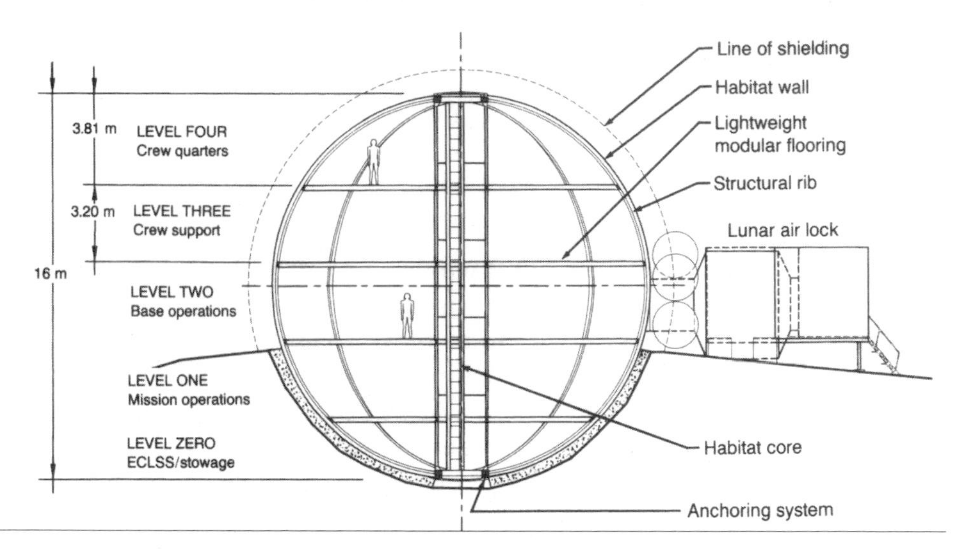

Fig. 13.4 A schematic of the inflatable habitat of Figure 13.3, detailing the levels. (Courtesy NASA)

* * *

All of these issues warn us that we need to understand the constraints under which we continue to colonize and industrialize the Moon in 2169. While some critics decry the added costs associated with maintaining the environmental integrity of the Moon, we only need to look at the added costs on Earth of pollution – air and water pollution on Earth added enormous economic costs due to added health care, lowering the quality of our food, water, air, and thus our quality and length of life. One can easily justify a careful environmental policy purely on the economic and health benefits that result, not to mention quality of life. Similarly, if one takes a long view, all will benefit by taking care of the planets upon which we are placing settlements and proceeding to extract resources.

In essence, it is necessary to reaffirm that only rarely do the ends justify the means. The rapid development of the Moon does not justify its physical destruction. In 2169, we believe that we have brought mankind's best qualities into space. These will be difficult issues to adjudicate.

Much thought was given to these issues. We are fortunate today that those discussions began a long time before there were vested interests to lobby in opposition. Similarly, a paper from the SETI Institute[5] initiated some of the discussions that continue to the present-day, two centuries later:

> "Some important questions must be addressed in considering future human exploration of space, questions that spacefaring nations have given

[5] L. Billings, "How Shall We Live in Space? Culture, Law and Ethics in Spacefaring Society," *Space Policy*, Vol. 22, 2006, pp. 249–255.

insufficient attention. How will extending the human presence into the Solar System affect society and culture on Earth? What legal, ethical and other value systems should govern human settlements and other activities in space? Do humans have rights to exploit extraterrestrial resources and alter extraterrestrial environments? Do spacefaring nations have an obligation to share the benefits of access to space with those nations that do not have access? Do those nations with early access to space have a right to impose their social and cultural norms on space-based civilization?"

There were discussions about property rights, on how pristine to keep the planetary bodies that had not yet been settled, and how to share, if at all, the "spoils" of the early ventures to the Moon and the asteroids. Scientists were worried that rampant economic development would preclude scientific exploration. The frontier model was used to describe the extension of human civilization to space – space as a frontier leads to homesteading and unlimited opportunities for those who are able to take advantage of the new possibilities.

Some suggested a world space organization so that cultural differences between the spacefaring nations could be part of the dialogue as space law, ethics and economic development issues were addressed. There was a push to share in the resources recovered evenly while the risks and costs were borne by a few. But we re-learned the lessons of the past millennium – initiative matters, work matters and rewards are proportional to risk. Economically, we live at the edge: we all must contribute in order to survive. There is little room for non-productive activity.

The lessons of the past millennium are that the Western view of life, while of course with much imperfection, has evolved into an egalitarian system. There were and continue to be enormous Western contributions to science, medicine and engineering. Also, the most basic Western concepts of legal process and civil liberties have inspired reformers all around the world. Western freedoms – having been slowly adopted everywhere – seem to be a reasonable basis for the evolution of humans beyond Earth. These values have been fundamental to our evolution into a spacefaring species.

13.3 Ethical Codes and Treaties

The question that dominated discussions about the settlement of the Moon – after wondering how to do it and how to pay for it – was the definition of the legal framework within which such settlement would occur.

The Agreement Governing the Activities of States on the Moon and Other Celestial Bodies, better known as the Moon Treaty, was an effort to use the United Nations as a vehicle to codify what activity was allowed – really what was not allowed – to nations and others on the lunar surface. It failed as such a mechanism because the major spacefaring nations of the time did not agree with its provisions and did not ratify the treaty. It was viewed as a way to grab the benefits of lunar activity from those who created them.

Fig. 13.5 A major concern of planners is the fine dust which covers the lunar surface and collects easily on astronauts' garments, as evidenced by six crews of Apollo Moon explorers. This special annex to the 16 m diameter inflatable habitat provided a possible solution to the dust problems, according to teams studying possible lunar expeditions. As much dust as possible must be removed before re-entering the habitat, shown in full in Figure 13.3. The astronauts might pass through wickets (far left) which remove much of the dust. A perforated metal porch would allow dust to fall through. Once inside the dust lock (center) the astronauts would remove their white coveralls. This outer garment would provide an extra layer of dust control and protection for the precision moving joints of the space suit from gritty dust. An air shower could remove remaining dust with strong jets of air. An astronaut at right, after having removed as much dust as possible, would be able then to move into the airlock to doff his suit. The airlock could accommodate up to four astronauts at one time. Suits could be stored there when not in use. (This artist concept reflects the evaluation and study at the Johnson Space Center by the Man Systems Division and Johnson Engineering personnel. S89-20088, July 1989. Courtesy NASA)

Various codes of ethics were created at the time. One focused on commerce,[6] and a European Union code of conduct was more elaborate and legalistic.[7] The commercial code was an extension of how it was expected that the business codes

[6]D.M. Livingston, "A Code of Ethics for Off-Earth Commerce," *Space 2002*, American Society of Civil Engineers 2002.

[7]European Union, "Draft Code of Conduct for Outer Space Activities," *Space Policy*, Vol. 25, 2009, pp. 144–146.

of Earth would be extended to the space venue. It supported a free market economy, that business decisions take into consideration future generations and that "the highest level of integrity, honesty, fairness, and ethics" be met. It supported appropriate environmental protection. Many of today's space businesses adhere to this strict but reasonable code, one that improves business rather than being a burden to it.

Some of the general codes on the fair and responsible use of space[8] proposed that supranational organizations such as the United Nations or some all-encompassing Port Authority would be able to set rules and tariffs on nations and companies. A difficulty with such utopian visions was that they were to be imposed by such an authority – after all, to this day we still have nations with their differing views on economics and politics; even though all of us are linked we don't think alike. Another problematic element in the codes of conduct, when they started looking like charters of world organizations, was that they were generally written from the perspectives of those who have not participated much in the space enterprise.

Some of these codes had the philosophy that any entrepreneur or nation could spend their money and use all their brainpower to go to the Moon – risk life and limb and treasure – but if successful, any knowledge and recovered resources must be shared with everyone on Earth. This philosophy did not build the great economies that brought the world out of threat of disease and famine. By the beginning of the 21st century, any such disease and famine were the results of politics and wars, not the ability of engineering and science to ameliorate these problems.

"If Homo sapiens is the first space faring species to have evolved on Earth, space settlement would not involve acting 'outside nature,' but legitimately 'within our nature.' "[9] This was the shot across the bow of those who viewed humanity as external to nature rather than as being integral to nature. "... [If] spacefaring is a legitimate activity for microbes, why should it not be so for humans?"

The various ethical theories created a philosophical framework that defined how people should behave with respect to other people and in nature. Thus, on the lifeless Moon the interaction was with the natural environment as well as the Apollo man-made environment. On a Mars with presumed life – at the beginning of the 21st century there were only speculations – extra dimensions of ethical considerations were needed.

We can loosely define the spectrum of ethical frameworks as those that believe that "everything that humans do is fair and valid" on one end, to "everything that humans do is wrong and unacceptable" at the other end. Somewhere in that range of possibilities is where we, in the present, have gravitated. Our nascent terraforming activities on Mars are viewed by some as vicious. But we view it as an improvement allowing life to evolve.

Large endeavors, such as Apollo or the Return to the Moon, required a major infusion of financial resources – both corporate and national – and thus needed a justification. In a few instances, extremely wealthy people pushed their vision

[8]N.-L. Remuβ, "The Fair and Responsible Use of Space: An International Perspective," *Space Policy*, Vol. 25, 2009, pp. 63–64.

[9]M. Fogg, "The Ethical Dimensions of Space Settlement," IAA-99-IAA.7.1.07, 50th International Astronautical Congress, 4–8 October 1999, Amsterdam.

of space without being accountable to anyone. Generally though, accountability meant that significant decisions had to demonstrate that the benefits outweighed the costs. Such cost-benefit analyses were quantitative methods for listing and comparing the financial and other costs of a venture with its financial and other benefits.

Since all costs and benefits do not start and end with money, efforts were made to generalize the cost-benefit analyses tools to include other measures. "The cost-benefit analysis paradigm itself is flexible regarding how costs and benefits are defined. Indeed, costs and benefits can be and often are defined in many ways: there can be monetary costs and benefits, human costs and benefits, environmental costs and benefits, and so on."[10] Instead of relating costs and benefits to market valuation, non-market valuation attempts to quantify seemingly qualitative values, such as clean air and water and even preservation of regions in their natural state. "Another non-market benefit of space exploration is reduction in the risk of the extinction of humanity and other Earth-originating life."

13.4 Independence for the colonies

The inhabitants of my lunar settlement have mixed loyalties. Since the first lunar birth in 2099, many have been born here and we have a majority of lunarians. Many, while retaining a kind view about the planet of their parents, have a decidedly lunar physiology and psychology. Their hearts and minds naturally have different perspectives on how their mother planet, the Moon, as well as Mars and the outer Solar System, should be developed.

What may have once been viewed as the property of Earth-based interests, the lunar settlements have become entities in their own right, much like children who grow up to become individuals, unique and independent. It hasn't happened yet but it is inevitable that with time the residents on the extraterrestrial bodies will decide that they are descendents of Earth, but not owned by Earth. In the same way that the Americas became independent of their mother countries, and where properties developed by trading companies evolved into autonomous entities, cities on the Moon and Mars will demand independence from Earth. At best, a confederation may become the evolution of the link between Earth and its former colonies.

While we have much autonomy on the Moon, we are still viewed as extensions of Earth-based nations and companies. A reasonable approach still needs to be created for governments and private investors to take as they prepare the groundwork for the evolution of the settlements.

A paradigm shift is in the works and we need to be cognizant of its existence.

13.5 Final thoughts: Yerah Timoshenko

I hope this volume has provided you with an overview of how we got here, how in the two hundred years since men landed on the Moon mankind became a spacefaring

[10]S.D. Baum, "Cost-benefit Analysis of Space Exploration: Some Ethical Considerations," *Space Policy*, Vol. 25, 2009, pp. 75–80.

civilization. Of course, many details were left out. My goal has been to provide a mix of specifics and generalities so that you will go away with the feeling of how it was and how it is.

If you knew nothing about our lives in space except what you read here, you would start to have an understanding.

I would like to end this commemorative volume with the J.F.K. Rice speech of 1962.

13.5.1 J.F.K. at Rice University

(Address at Rice University on the Nation's Space Effort by President John F. Kennedy, Houston, Texas, 12 September 1962.)

President Pitzer, Mr. Vice President, Governor, Congressman Thomas, Senator Wiley, and Congressman Miller, Mr. Webb, Mr. Bell, scientists, distinguished guests, and ladies and gentlemen:

I appreciate your president having made me an honorary visiting professor, and I will assure you that my first lecture will be very brief.

I am delighted to be here and I'm particularly delighted to be here on this occasion.

We meet at a college noted for knowledge, in a city noted for progress, in a State noted for strength, and we stand in need of all three, for we meet in an hour of change and challenge, in a decade of hope and fear, in an age of both knowledge and ignorance. The greater our knowledge increases, the greater our ignorance unfolds.

Despite the striking fact that most of the scientists that the world has ever known are alive and working today, despite the fact that this Nation's own scientific manpower is doubling every 12 years in a rate of growth more than three times that of our population as a whole, despite that, the vast stretches of the unknown and the unanswered and the unfinished still far outstrip our collective comprehension.

No man can fully grasp how far and how fast we have come, but condense, if you will, the 50,000 years of man's recorded history in a time span of but a half a century. Stated in these terms, we know very little about the first 40 years, except at the end of them advanced man had learned to use the skins of animals to cover them. Then about 10 years ago, under this standard, man emerged from his caves to construct other kinds of shelter. Only five years ago man learned to write and use a cart with wheels. Christianity began less than two years ago. The printing press came this year, and then less than two months ago, during this whole 50-year span of human history, the steam engine provided a new source of power.

Newton explored the meaning of gravity. Last month electric lights and telephones and automobiles and airplanes became available. Only last week did we develop penicillin and television and nuclear power, and now if America's new spacecraft succeeds in reaching Venus, we will have literally reached the stars before midnight tonight.

This is a breathtaking pace, and such a pace cannot help but create new ills as it dispels old, new ignorance, new problems, new dangers. Surely the opening vistas of space promise high costs and hardships, as well as high reward.

Fig. 13.6 Attorney General Kennedy, McGeorge Bundy, Vice President Johnson, Arthur Schlesinger, Admiral Arleigh Burke, President Kennedy, Mrs. Kennedy watching the 15 minute historic flight of Astronaut Shepard on television, 5 May 1961 becoming the first American in space. (Cecil Stoughton, photographer. Courtesy John Fitzgerald Kennedy Library, Boston, MA)

So it is not surprising that some would have us stay where we are a little longer to rest, to wait. But this city of Houston, this State of Texas, this country of the United States was not built by those who waited and rested and wished to look behind them. This country was conquered by those who moved forward – and so will space.

William Bradford, speaking in 1630 of the founding of the Plymouth Bay Colony, said that all great and honorable actions are accompanied with great difficulties, and both must be enterprised and overcome with answerable courage.

If this capsule history of our progress teaches us anything, it is that man, in his quest for knowledge and progress, is determined and cannot be deterred. The exploration of space will go ahead, whether we join in it or not, and it is one of the great adventures of all time, and no nation which expects to be the leader of other nations can expect to stay behind in the race for space.

Those who came before us made certain that this country rode the first waves of the industrial revolutions, the first waves of modern invention, and the first wave of nuclear power, and this generation does not intend to founder in the backwash of the coming age of space. We mean to be a part of it – we mean to lead it. For the eyes of the world now look into space, to the moon and to the planets beyond, and we have vowed that we shall not see it governed by a hostile flag of conquest, but by a banner of freedom and peace. We have vowed that we shall not see space filled with weapons of mass destruction, but with instruments of knowledge and understanding.

Yet the vows of this Nation can only be fulfilled if we in this Nation are first, and, therefore, we intend to be first. In short, our leadership in science and in industry, our hopes for peace and security, our obligations to ourselves as well as others, all require us to make this effort, to solve these mysteries, to solve them for the good of all men, and to become the world's leading space-faring nation.

We set sail on this new sea because there is new knowledge to be gained, and new rights to be won, and they must be won and used for the progress of all people. For space science, like nuclear science and all technology, has no conscience of its own. Whether it will become a force for good or ill depends on man, and only if the United States occupies a position of pre-eminence can we help decide whether this new ocean will be a sea of peace or a new terrifying theater of war. I do not say the we should or will go unprotected against the hostile misuse of space any more than we go unprotected against the hostile use of land or sea, but I do say that space can be explored and mastered without feeding the fires of war, without repeating the mistakes that man has made in extending his writ around this globe of ours.

There is no strife, no prejudice, no national conflict in outer space as yet. Its hazards are hostile to us all. Its conquest deserves the best of all mankind, and its opportunity for peaceful cooperation many never come again. But why, some say, the moon? Why choose this as our goal? And they may well ask why climb the highest mountain? Why, 35 years ago, fly the Atlantic? Why does Rice play Texas?

We choose to go to the moon. We choose to go to the moon in this decade and do the other things, not because they are easy, but because they are hard, because that goal will serve to organize and measure the best of our energies and skills, because that challenge is one that we are willing to accept, one we are unwilling to postpone, and one which we intend to win, and the others, too.

It is for these reasons that I regard the decision last year to shift our efforts in space from low to high gear as among the most important decisions that will be made during my incumbency in the office of the Presidency.

In the last 24 hours we have seen facilities now being created for the greatest and most complex exploration in man's history. We have felt the ground shake and the air shattered by the testing of a Saturn C-1 booster rocket, many times as powerful as the Atlas which launched John Glenn, generating power equivalent to 10,000 automobiles with their accelerators on the floor. We have seen the site where five F-1 rocket engines, each one as powerful as all eight engines of the Saturn combined, will be clustered together to make the advanced Saturn missile, assembled in a new building to be built at Cape Canaveral as tall as a 48 story structure, as wide as a city block, and as long as two lengths of this field.

Within these last 19 months at least 45 satellites have circled the earth. Some 40 of them were "made in the United States of America" and they were far more sophisticated and supplied far more knowledge to the people of the world than those of the Soviet Union.

The Mariner spacecraft now on its way to Venus is the most intricate instrument in the history of space science. The accuracy of that shot is comparable to firing a missile from Cape Canaveral and dropping it in this stadium between the 40-yard lines.

Transit satellites are helping our ships at sea to steer a safer course. Tiros satellites have given us unprecedented warnings of hurricanes and storms, and will do the same for forest fires and icebergs.

We have had our failures, but so have others, even if they do not admit them. And they may be less public.

To be sure, we are behind, and will be behind for some time in manned flight. But we do not intend to stay behind, and in this decade, we shall make up and move ahead.

The growth of our science and education will be enriched by new knowledge of our universe and environment, by new techniques of learning and mapping and observation, by new tools and computers for industry, medicine, the home as well as the school. Technical institutions, such as Rice, will reap the harvest of these gains.

And finally, the space effort itself, while still in its infancy, has already created a great number of new companies, and tens of thousands of new jobs. Space and related industries are generating new demands in investment and skilled personnel, and this city and this State, and this region, will share greatly in this growth. What was once the furthest outpost on the old frontier of the West will be the furthest outpost on the new frontier of science and space. Houston, your City of Houston, with its Manned Spacecraft Center, will become the heart of a large scientific and engineering community. During the next 5 years the National Aeronautics and Space Administration expects to double the number of scientists and engineers in this area, to increase its outlays for salaries and expenses to $60 million a year; to invest some $200 million in plant and laboratory facilities; and to direct or contract for new space efforts over $1 billion from this Center in this City.

To be sure, all this costs us all a good deal of money. This year's space budget is three times what it was in January 1961, and it is greater than the space budget of the previous eight years combined. That budget now stands at $5,400 million a year – a staggering sum, though somewhat less than we pay for cigarettes and cigars every year. Space expenditures will soon rise some more, from 40 cents per person per week to more than 50 cents a week for every man, woman and child in the United Stated, for we have given this program a high national priority – even though I realize that this is in some measure an act of faith and vision, for we do not now know what benefits await us. But if I were to say, my fellow citizens, that we shall send to the moon, 240,000 miles away from the control station in Houston, a giant rocket more than 300 feet tall, the length of this football field, made of new metal alloys, some of which have not yet been invented, capable of standing heat and stresses several times more than have ever been experienced, fitted together with a precision better than the finest watch, carrying all the equipment needed for propulsion, guidance, control, communications, food and survival, on an untried mission, to an unknown celestial body, and then return it safely to earth, re-entering the atmosphere at speeds of over 25,000 miles per hour, causing heat about half that of the temperature of the sun–almost as hot as it is here today – and do all this, and do it right, and do it first before this decade is out – then we must be bold.

I'm the one who is doing all the work, so we just want you to stay cool for a minute. [laughter]

However, I think we're going to do it, and I think that we must pay what needs to be paid. I don't think we ought to waste any money, but I think we ought to do the job. And this will be done in the decade of the sixties. It may be done while some of you are still here at school at this college and university. It will be done during the term of office of some of the people who sit here on this platform. But it will be done. And it will be done before the end of this decade.

I am delighted that this university is playing a part in putting a man on the moon as part of a great national effort of the United States of America.

Many years ago the great British explorer George Mallory, who was to die on Mount Everest, was asked why did he want to climb it. He said, "Because it is there."

Well, space is there, and we're going to climb it, and the moon and the planets are there, and new hopes for knowledge and peace are there. And, therefore, as we set sail we ask God's blessing on the most hazardous and dangerous and greatest adventure on which man has ever embarked.

Thank you.

* * *

13.6 Chronological summary

1969 First men on the Moon
2009 Chandarayaan-1/Moon Mineralogy Mapper reveals H_2O molecules
2014 Space tourism reaches $1B threshold
2024 Humans return to Moon
2029 Permanent colony
2034 Humans land on Mars
2041 Permanent Mars colony
2046 Space elevator prototype construction begins over Earth
2049 Lunar space elevator construction begins
2059 Yerah Timoshenko's (YT) great-grandparents go to Moon
2060–61 YT's grandparents born on Earth (conceived on Moon)
2070 Fusion reactors go online on Moon, a year later on Mars
2084 First families on Moon
2089–90 YT's parents born on Earth
2094 First lunar Olympics
2099 First human birth on Moon
2115 YT's parents move to Moon
2119 YT born on Moon
2142–43 YT's boy and girl born on Moon

2169 The present – 200th anniversary of the first men on the Moon
Now Terraforming of Mars
2179 150th anniversary of off-Earth permanent habitation

* * *

Fig. 13.7 *From the Earth to the Moon* (French: *De la Terre à la Lune*, 1865) is a humorous science fantasy novel by Jules Verne and is one of the earliest entries in that genre. It tells the story of a Frenchman and two well-to-do members of a post-American Civil War gun club who build an enormous sky-facing cannon, the Columbiad, and launch themselves in a projectile/spaceship from it to a Moon landing.

Index

Printing: Mercedes-Druck, Berlin
Binding: Stein+Lehmann, Berlin